COLLECTED PAPERS OF

JOHN MILNOR

COLLECTED PAPERS OF

JOHN MILNOR

III

Differential Topology

American Mathematical Society
Providence, Rhode Island

2000 *Mathematics Subject Classification.* Primary 55Q45, 57R55, 57R65, 57R75, 57R77.

Library of Congress Cataloging-in-Publication Data

Milnor, John Willard, 1931–
III. Differential topology / edited by the author.
p. cm. — (Collected papers of John Milnor)
Includes bibliographical references and index.
ISBN-13 978-0-8218-4230-0 (alk. paper)
ISBN-10 0-8218-4230-7 (alk. paper)
1. Differential topology. I. Title.

QA613.6.M549 2007
514′.72—dc22

2006048014

∞ The paper used in this book is acid-free and falls within the guidelines
established to ensure permanence and durability.
Visit the AMS home page at `http://www.ams.org/`

10 9 8 7 6 5 4 3 2   16 15 14 13 12 11

CONTENTS

† Not previously published.

[†] Not previously published.

Preface

This is the third volume of my collected papers to appear.[1] Most of the material which follows was written during a ten year period in the late 1950s and early 1960s. It was my good luck to begin working on the topology of manifolds, and especially on the topology of smooth manifolds, at a time when the field was ripe for development. The study of cohomology and homotopy theory and the study of fiber bundles were in full flower, and I was particularly fortunate to learn about these subjects from Norman Steenrod. Three important developments during the 1950s played a key role. The first was Serre's thesis, which opened up the study of homotopy groups. The second was Thom's work on cobordism, which provided powerful new tools towards understanding smooth manifolds, and in particular led to a proof of Hirzebruch's signature formula. The third was Bott's work on the homotopy groups of classical groups, which again played an essential role. I hope that the present collection of papers will help to keep this circle of ideas alive for the next generation of students.

In general the previously published papers have been copied without change, although a few minor errors have been corrected and a few contemporary references filled in. In the case of unpublished papers, there has been some editing for readability. In both cases, where some significant modification of the original text seemed needed, it has been clearly indicated.

I want to thank Rob Kirby and Larry Siebenmann for their help in describing further developments, and Araceli Bonifant for substantial help in preparing this manuscript.

Stony Brook, NY, October 2006
John Milnor

[1] "Collected Papers: Volume 1, Geometry," and "Volume 2, The Fundamental Group," were published by Publish or Perish in 1994 and 1995 respectively, and are now available through the American Mathematical Society.

Dedicated to
Georges de Rham and Hassler Whitney,
and to their beloved mountains.

Georges de Rham and Hassler Whitney in the 1980s.

Georges de Rham (1903-1990) and Hassler Whitney (1907-1989) were two of the founders of the field of differential topology; however, their first love was for the mountains. There is a third name which is closely linked with theirs: James Alexander (1888-1971) was another pioneer of topology whose heart was in the mountains.

Let me begin by describing an encounter which took place in September 1926 at the Refuge de la Charpoua, a climbing hut high above the Mer de Glace near Chamonix. Two parties of two were both planning to traverse the Aiguilles des Drus, but in opposite directions. They agreed to leave their ice axes at the edge of the glacier where their respective rock climbs began, so that each party could avoid the awkwardness of carrying axes on the rock, but would have them waiting for the descent. (There was probably a great deal more ice in 1926.) Everything went as planned: the two parties crossed near the summit, picked up the waiting ice axes on the way down, and recovered their own axes when they returned to the refuge that evening, without ever introducing themselves. A few years later one of the climbers, Georges de Rham, happened to run into a member of the other party, a well known musician, and asked who his companion had been. He discovered that it was James Alexander, whose mathematical work he knew well. De Rham met Alexander again in 1932, at the International Congress in Zürich. Then in 1933, near the Weisshorn in Valais, de Rham encountered Alexander who was climbing with Hassler Whitney. This meeting was the beginning of a long friendship between Whitney and de Rham, who visited or climbed together frequently for more than forty years.

Unfortunately I never met James Alexander (1888-1971), although he lived in Princeton during all of my days as a student and faculty member at Princeton University. In fact he was very reclusive, and refused to have anything to do with mathematics after his retirement, although quite likely he would have been happy to talk about climbing. Among American climbers, he is famous as the first person to climb the east face of Long's Peak, in a solo ascent (circa 1922) on a route now known as Alexander's chimney.

I first got to know Hassler Whitney in 1956, at an International Symposium on Algebraic Topology, which was held in new and beautiful buildings of the University of Mexico. During the conference, several of us planned a climb of Ixtaccihuatl, one of the two great volcanos which dominate the skyline of Mexico City on the (nowadays all too rare) days when the air is clear. At that time, I was about 25 years old and Whitney was about 50. Perhaps the first hint that I was in trouble came when I noticed that Hass always took the stairs to the conference on the 12-th floor in order to keep in shape, while I was perfectly happy with the elevator. When the time came for our expedition, we drove for some miles and then hiked up to a tiny climbing hut which resembled a tin can lying on its side. It was high and cold, the air was very thin, and it was crowded with far too many people. I slept very little. Before dawn the next morning we started the actual climb, and it soon became clear that I was completely out of shape. After hours of hiking up beautiful snowfields, we reached the saddle. The climbing conditions were perfect, but I couldn't go any further. We agreed that I should wait at the saddle, while Hass and the others completed the climb. Thus I was left sitting on my ice axe in brilliant tropical sunshine on a high snowfield. Five minutes after the others went on, I realized that I didn't have any sun cream. The descent was uneventful, but I returned to the conference with a magnificent sunburn.

I first went rock climbing with Georges de Rham in the 1960s, on the Salève near Geneva, I was rather apprehensive about climbing with such an old person. He walked very deliberately, keeping careful track of his pulse as we went up the trail. However, once we started actually climbing, he was flying up the rock and I worried only about myself. I have particularly fond memories of climbing with Georges on the Argentine, a mountain with lovely silvery walls in his home Canton of Vaud, and I can remember a fine day there with Georges and with Hass, followed by a delicious raclette in the little village of Solalex. I also have happy memories of a topology conference which was held some years later in the nearby village of Les Plans sur Bex, high in the mountains of Vaud. This area was very special to Georges; and I treasure a copy of the guidebook to the Argentine which he wrote.

The Stockhorn from the south

But my most vivid memory is of a climb of the south ridge of the Baltschieder Stockhorn in northern Valais in 1966, with Georges together with Oscar Burlet and Jacques Boéchat. This was long and steep, a series of five stepped towers, with excellent rock, and almost too exciting for me. (I probably learned more French that day than I have in the rest of my life: There is nothing that focuses the mind so much as a little terror.) Starting out at dawn, I assumed that we would return by daylight, and brought only dark glasses. But by the time we got to the foot of the formidable fifth tower, it was already well after noon. I was overwhelmed, and ready to give up and turn back; but Georges explained that it was too late—it would be much safer to go ahead and return by the relatively easy east ridge than to try to retrace our steps. We got over the crux of the climb, an overhang with what seemed like miles of air under our feet, with the aid of a small rope ladder. It was pitch dark when I finally stumbled down the trail at the end of the day, still wearing my sun glasses, but very happy. In Georges' words:

"Car l'alpinisme n'est pas seulement un exercice physique, c'est aussi un travail de l'esprit, et il permet de réaliser une merveilleuse harmonie entre la nature, l'âme, et le corps"

Here are some references for further information.

James Alexander

Cohen L., *James Waddell Alexander, 1888-1971,* Bull. Amer. Math. Soc. **79** (1973) 900–903.

James I. M., *Portrait of Alexander (1888–1971),* Bull. Amer. Math. Soc. **38** (2001) no. 2, 123–129.

Jones C., "Climbing in North America," Univ. Calif. Press 1976.

Lefschetz S., *James Waddell Alexander (1888-1971),* Biogr. Mem. Amer. Phil. Soc. **79** (1973), Philadelphia (1974) 663–680.

Georges de Rham

Bach D., Burlet O. and de la Harpe P. (editors), "Georges de Rham, 1903-1990," Imprimerie Dupuis S.A., Le Brassus, Switzerland 1995.

Bott R., *Georges de Rham,* Notices Amer. Math. Soc. **38** (1991) 114–115.

Cartan H., *La vie et l'œuvre de Georges de Rham,* Comptes Rend. Acad. Sci. Paris, Série Gén. **9** (1992) 453–455.

Eckmann B., *Georges de Rham, 1903-1990,* Elemente Math. **47** (1992) 118–122.

Enseignement editors, *Georges de Rham, 1903-1990,* Enseign. Math. **36** (1990) 207–214.

de Rham G., "L'Argentine, Description de Vingt Itinéraires d'Escalade," Librairie Roth, Lausanne 1944.

de Rham G., "Œuvres Mathématiques," Enseign. Math., Genève 1981.

Hassler Whitney

Eells J. and Toledo D. (editors), "The Collected Papers of Hassler Whitney" (two volumes), Birkhäuser 1992.

Lax A., *Hassler Whitney, 1907-1989 — Some Recollections, 1979-1989,* 301–309 of "A Century of Mathematical Meetings" (B.A. Case, ed.), Amer. Math. Soc., 1996.

Thom R., *La vie et l'œuvre de Hassler Whitney,* Comptes Rend. Acad. Sci. Paris, Série Gén. **7** (1990) 473–476.

Whitney H., *Letting research come naturally,* Mathematical Chronical **14** (1985) 1–19.

Zund J. D., *Hassler Whitney,* American National Biography **23** (Oxford, 1999) 303–304.

Photo Credits

Page

ix. Georges de Rham: photo courtesy of Oscar Burlet.

ix. Hasler Whitney: Herman Landshoff photographer. Courtesy of the Archives of the Institute for Advanced Study, Princeton, New Jersey, USA.

xi. Stockhorn: photo courtesy of Oscar Burlet.

3. Friedrich Hirzebruch: Courtesy of the Archives of the Mathematisches Forschungsinstitut Oberwolfach.

5. René Thom: photograph courtesy of the Institut des Hautes Études Scientifiques.

6. John Milnor: Herman Landshoff photographer. Courtesy of the Archives of the Institute for Advanced Study, Princeton, New Jersey, USA.

223. Raoul Bott: photo courtesy of Dr. Wolfgang Vollrath.

251. Relaxing at the 1959 Arbeitstagung: photo courtesy of F. Hirzebruch.

Acknowledgments

The American Mathematical Society gratefully acknowledges the kindness of these institutions in granting the following permissions:

Association Nicolas Bourbaki

By René Thom, *Travaux de Milnor sur le cobordisme*, Séminaire Bourbaki 1958/59, no. **180**, February 1959, pp. 169–177, © Association Nicolas Bourbaki.

Cambridge University Press

with M. Kervaire, *Bernoulli numbers, homotopy groups, and a theorem of Rohlin*, Proc. Intl. Congr. Math. 1958, Cambridge University Press, 1960, pp. 454–458.

L'Enseignement Mathématique. Revue Internationale

A survey of cobordism theory [erratum added], L'Enseignement Mathématique. Revue Internationale. IIe Série, **8** (1962), pp. 16–23.

Spin structures on manifolds, L'Enseignement Mathématique. Revue Internationale. IIe Série, **9** (1963), pp. 198–203.

Elsevier

On the Stiefel-Whitney numbers of complex manifolds and of spin manifolds, Topology, Vol. 3, 1965, pp. 223–230. © 1965, with permission from Elsevier.

James Munkres

By J. Munkres, *Lectures on differential topology*, unpublished lecture notes, 1958.

John Wiley & Sons, Inc.

Differential topology, Lectures on Modern Mathematics, Vol. II, edited by T. L. Saaty, 1964, pp. 165–183. © 1964 John Wiley & Sons, Inc. Reprinted with permission of John Wiley & Sons, Inc.

Princeton University Press

On manifolds homeomorphic to the 7-sphere, Annals of Mathematics, Vol. 64, No. 2, 1956, pp. 399–405, © Princeton University Press.

Groups of homotopy spheres: I, Annals of Mathematics, Vol. 77, No. 3, 1963, pp. 504–537, © Princeton University Press.

Some consequences of a theorem of Bott, Annals of Mathematics, Vol. 68, No. 2, 1958, © Princeton University Press.

Remarks concerning spin manifolds, Differential and Combinatorial Topology, A Symposium in Honor of Marston Morse, 1965, pp. 55–62. © Princeton University Press.

Société Mathématique de France

Sommes de variétés différentiables et structures différentiables des sphères, Bull. Soc. Math. France **87** (1959), pp. 439–444, © Société Mathématique de France.

The Johns Hopkins University Press

On the cobordism ring Ω^* and a complex analogue, Part I, American Journal of Mathematics **82:6** (1960), pp. 505–521. © The Johns Hopkins University Press. Reproduced with permission of The Johns Hopkins University Press.

Differential structures and spheres, American Journal of Mathematics **81:4** (1959), pp. 962–972. © The Johns Hopkins University Press. Reproduced with permission of The Johns Hopkins University Press.

Universitad Nacional Autonoma de Mexico

Smooth manifolds with boundary: A translation and extension of "Variedades lisas con frontera", Lecture notes by H. Cardenas, E. Lluis, F. Recillas, and R. Vázquez, as translated by I. E. Colon-Roldan and S. Simanca, from Lectures at the University of Mexico in the summer of 1960, An. Inst. Mat. Univ. Nac. Auton. Mexico **1** (1961), Universitad Nacional Autonoma de Mexico, pp. 82–116.

PART 1. EXOTIC SPHERES

This section will consist of the following eight papers:

On manifolds homeomorphic to the 7-sphere, Annals of Mathematics **64** (1956) 399–405.

On the relationship between differentiable manifolds and combinatorial manifolds. Unpublished notes, Princeton University 1956.

Sommes de variétés différentiables et structures différentiables des sphères, Bulletin de la Société Mathematique de France **87** (1959) 439–444.

Differentiable structures on spheres, American Journal of Mathematics **81** (1959) 962–972.

A procedure for killing homotopy groups of differentiable manifolds, in "Differential Geometry," Proceedings Symposia in Pure Mathematics **III**, American Mathematical Society (1961) 39–55.

Differentiable manifolds which are homotopy spheres. Unpublished notes, Princeton University 1959.

Groups of homotopy spheres: I (with Michel Kervaire), Annals of Mathematics **77** (1963) 504–537.

Differential topology, in "Lectures on Modern Mathematics **II**," edited by T. L. Saaty; Wiley, New York (1964) 165–183.

Introduction: How these papers came to be written

During the 1950's, I worked on an ongoing project of trying to understand one particularly simple class of manifolds, namely $2n$-dimensional manifolds which are $(n-1)$-connected. Although my intended paper on this subject was never finished, the project none the less led to the eight papers which follow (as well as the paper "On Simply Connected 4-Manifolds," Milnor [1958][1]).

[1]Names in small caps refer to the bibliography at the end of this volume.

To understand how this came about, let me first describe the original plan. The homotopy theory of a closed manifold M^{2n} which is $(n-1)$-connected is relatively easy to describe. If the middle Betti number is β, then the manifold can be obtained (up to homotopy type) by attaching a $2n$-cell to a bouquet

$$W^n = S^n \vee \cdots \vee S^n$$

of β copies of the n-sphere. The attaching map can be described as an element of a homotopy group $\pi_{2n-1}(W^n)$ which is reasonably well understood. More explicitly, this group splits as the direct sum of β copies of the group $\pi_{2n-1}(S^n)$ together with $\beta(\beta-1)/2$ free cyclic groups which correspond to Whitehead products of distinct generators of $\pi_n(W^n)$.

Better still, this space can be described in terms of cohomology theory. The cohomology groups $H^k = H^k(M^{2n};\mathbb{Z})$ are zero with the exception of H^0 and H^{2n} which are free cyclic and $H^n \cong H^n(W^n;\mathbb{Z})$ which is free abelian of rank β. The bilinear pairing

$$H^n \otimes H^n \;\to\; H^{2n}$$

is either symmetric or skew according as n is even or odd, and has determinant ± 1 by Poincaré duality. Finally the "stable" attaching map can be described by a cohomology operation

$$\psi \;:\; H^n \;\to\; H^{2n}(M^{2n};\Pi_{n-1}) \;\cong\; \Pi_{n-1}\;,$$

where Π_{n-1} is the stable homotopy group $\pi_{k+n-1}(S^k)$ for $k > n$. This operation can be described as follows. Any element $\eta \in H^n \cong H^n(W^n;\mathbb{Z})$ corresponds to a homotopy class of maps $W^n \to S^n$. Composing with the attaching map in $\pi_{2n-1}(W^n)$, we obtain an element of $\pi_{2n-1}(S^n)$ which stabilizes to the required $\psi(\eta) \in \Pi_{n-1}$.

The problem arises when one tries to flesh out this homotopy picture by constructing actual manifolds realizing specified homotopy invariants. In the simplest case $\beta = 1$, we must attach a $2n$-cell to a single n-sphere in such a way as to obtain a manifold. It seems that the best chance of carrying out this construction is to first "thicken" the n-sphere, replacing it by a tubular neighborhood in the hypothetical M^{2n}. In other words, we must form an n-disk bundle

$$D^n \;\hookrightarrow\; E^{2n} \;\twoheadrightarrow\; S^n\;.$$

In the hoped for situation, the boundary sphere bundle $\partial E^{2n} = \Sigma^{2n-1}$ will be homeomorphic to the $(2n-1)$-sphere. Hence we will be able to glue on a $2n$-disk by a boundary homeomorphism h so as to obtain a closed manifold $M^{2n} = E^{2n} \cup_h D^{2n}$ with the required homotopy type. Thus we are led to the following problem:

For which sphere bundles $S^{n-1} \hookrightarrow \Sigma^{2n-1} \twoheadrightarrow S^n$ *is the total space* Σ^{2n-1} *homeomorphic to the sphere* S^{2n-1}*?*

Three basic examples had been discovered by Heinz Hopf, namely the fibrations

$$S^1 \hookrightarrow S^3 \twoheadrightarrow S^2\,, \quad S^3 \hookrightarrow S^7 \twoheadrightarrow S^4\,, \quad \text{and} \quad S^7 \hookrightarrow S^{15} \twoheadrightarrow S^8\;.$$

(In fact it is now known that such bundles can exist only in these particular dimensions. Compare "Some Consequences of a Theorem of Bott," pages 233–238

FIG. 1. *Hirzebruch at Erlangen in 1976.*

below.) The corresponding $(n-1)$-connected $2n$-manifolds $E^{2n} \cup_h D^{2n}$ were respectively the complex projective plane, the quaternion projective plane, and the Cayley projective plane. For circle bundles over the 2-sphere, classified by elements of the homotopy group $\pi_1(\mathrm{SO}(2)) \cong \mathbb{Z}$, the Hopf fibration was the only possibility, up to sign. However, for 3-sphere bundles over S^4, classified by elements of $\pi_3(\mathrm{SO}(4)) \cong \mathbb{Z} \oplus \mathbb{Z}$, there are infinitely many bundles which at least have the homotopy type of the 7-sphere. More precisely, such a bundle is classified by two elements of $H^4(S^4; \mathbb{Z}) \cong \mathbb{Z}$, namely the ***Pontrjagin class*** p_1 and the the ***Euler class*** e (denoted by $\bar{c}$ in the first paper), subject only to the relation $p_1 \equiv 2\mathrm{e} \pmod{4H^4}$. It is not difficult to check that the total space Σ^7 has the homotopy type of a 7-sphere if and only if the Euler class generates $H^4(S^4; \mathbb{Z})$. Thus we potentially have infinitely many distinct 3-connected 8-manifolds (twisted versions of the quaternion projective plane), which are distinguished by their Pontrjagin classes, and in some cases by homotopy invariants as well.

It was natural to take a closer look at the structure of these hypothetical manifolds, using the signature theorem (or index theorem) which had recently been developed by René Thom and Fritz Hirzebruch. For any closed oriented $4m$-dimensional manifold M^{4m} we can form the signature (or "index") of the symmetric bilinear form

$$H^{2m} \otimes H^{2m} \to H^{4m} \cong \mathbb{Z}\,.$$

In the differentiable case, as a consequence of his cobordism theory, Thom had shown that this signature could be expressed uniquely as a rational linear combination of the Pontrjagin numbers $p_{i_1} \cdots p_{i_k}[M^{4m}]$ (where $i_1 \le \cdots \le i_k$ with

$i_1 + \cdots + i_k = m$). He worked out the precise formula only in dimensions 4 and 8; but around the same time Hirzebruch had conjectured such a formula and worked out the precise form which it would have to take in all dimensions. (Compare HIRZEBRUCH [1956 or 1966, 1971].)

For an 8-dimensional manifold, the formula reads

$$\text{signature} = (7p_2 - p_1^2)[M^8]/45 \ .$$

In our case, the signature is ± 1, and we can choose the orientation so that it is $+1$. Since $p_2[M^8]$ is an integer, the first Pontrjagin class must satisfy the congruence

$$p_1^2[M^8] + 45 = 7\,p_2[M^8] \ \equiv \ 0 \qquad (\mathrm{mod}\, 7) \ .$$

Thus, for any allowable choice of p_1 which does not satisfy this congruence, we have constructed a homotopy 7-sphere Σ^7 which cannot be diffeomorphic to the standard 7-sphere.

At this point, I believed that I had constructed a counterexample to the Poincaré conjecture in dimension 7. In other words, I assumed that Σ^7 could not even be continuously homeomorphic[2] to the standard S^7. Fortunately however, I did some experimentation, and discovered that this Σ^7 actually is homeomorphic to S^7. In fact, it can be obtained by pasting together the boundaries of two standard 7-disks under a boundary diffeomorphism. (I call such a manifold, a *twisted sphere*.) As an extra bonus, the proof also showed that there exist non-standard diffeomorphisms of the 6-dimensional sphere. These results are described in the paper **On Manifolds Homeomorphic to the 7-Sphere**, on pages 11–17 below. (For an analogous construction in dimension 15, see SHIMADA [1957].)

The next paper **On the Relationship between Differentiable Manifolds and Combinatorial Manifolds**, written in the same year but never published, carries the discussion further by tying it in with Henry Whitehead's theory of C^1-triangulation, and by describing Thom's proof, based on his theory of combinatorial Pontrjagin classes,[3] that the combinatorial manifold $E^8 \cup_h D^8$ described above has no compatible differentiable structure. (A few years later, Michel KERVAIRE [1960] constructed a topological 10-manifold which cannot be given any differentiable structure at all, and Steve SMALE [1961] constructed an analogous example in dimension twelve. Still later, with the proof by Sergei NOVIKOV [1965] that rational Pontrjagin classes are actually topological invariants, it followed that the manifold $E^8 \cup_h D^8$ above has no differentiable structure at all.)

The expository paper **Sommes de Variétés Différentiables et Structures Différentiables des Sphères**, presented (in English) at a conference in Lille a

[2]Actually, there is some question as to what Poincaré meant when he formulated his conjecture for 3-manifolds. The word "homeomorphism" did not have a universally understood meaning at the time, and he may well have intended that homeomorphisms were to be differentiable.

[3]THOM [1958]. For C^1-triangulations, see WHITEHEAD [1940], MUNKRES [1963]. For combinatorial Pontrjagin classes, compare the presentation in MILNOR AND STASHEFF [1974].

FIG. 2. *René Thom at the Institut des Hautes Études Scientifiques in the early 1970's.*

few years later, introduces this group Γ_n of oriented diffeomorphism classes of twisted spheres. Almost by definition, there is a homomorphism from the group $\pi_0 \operatorname{Diff}^+(S^{n-1})$ of smooth isotopy classes of orientation preserving diffeomorphisms onto Γ_n. (In fact this homomorphism extends to an exact sequence

$$\cdots \to \pi_1 \operatorname{Diff}^+(S^{n-1}) \to$$
$$\pi_0 \operatorname{Diff}^+(D^n \operatorname{rel} S^{n-1}) \to \pi_0 \operatorname{Diff}^+(D^n) \to \pi_0 \operatorname{Diff}^+(S^{n-1}) \to \Gamma_n \to 1$$

of abelian groups. Compare the paper "Differential Topology" on pages 123–141 below.) A different characterization, due to THOM [1959] is that a smooth manifold is a twisted sphere if and only if it is combinatorially equivalent to the standard sphere.

This paper also introduced the group Θ_n of smooth oriented manifolds having the homotopy type of S^n, up to the relation which was then called "J-equivalence" but is now known as "h-cobordism". Evidently there is a natural homomorphism $\Gamma_n \to \Theta_n$. In fact we will see later (modulo the Poincaré Conjecture for the case $n = 3$) that Γ_n maps isomorphically onto Θ_n for all n.

The paper **Differentiable Structures on Spheres**, carries out a similar argument based on hypothetical manifolds of the form $(S^p \vee S^q) \cup D^{p+q}$, where the two sub-spheres intersect transversally, with normal bundles described by elements of $\pi_{p-1}(\mathrm{SO}_q)$ and $\pi_{q-1}(\mathrm{SO}_p)$ respectively. This construction is much more robust,

FIG. 3. *The author at the Institute for Advanced Study in 1981.*

not being limited to dimensions 7 and 15. (There is a closely related bilinear pairing

$$\pi_k \mathrm{SO}_\ell \otimes \pi_\ell \mathrm{SO}_\mathrm{k} \rightarrow \pi_0 \mathrm{Diff}^+(\mathrm{S}^{\mathrm{k}+\ell}) \rightarrow \Gamma_{\mathrm{k}+\ell+1} .$$

Compare the introduction to Part 2 on page 143.)

Although I talked with René Thom only a few times during the years when these papers were written, his influence was quite important. For example, his intuitive feeling for the structure of cobordism rings went far beyond his published work. During one particularly decisive conversation, he constructed an interesting example by a technique which I called *surgery*. (The same construction was introduced independently by Andrew WALLACE [1960], who called it *spherical modification*. Both terms have been frequently used in the literature.) I developed this idea in the paper **A Procedure for Killing Homotopy Groups of Differentiable Manifolds**. A fairly easy Morse theory argument shows that one manifold can be obtained from another by a sequence of spherical modifications if and only if the two belong to the same cobordism class. These ideas played a key role in further work on groups of homotopy spheres.[4]

The manuscript **Differentiable Manifolds which are Homotopy Spheres**, written in 1959, was never published since most of its results were absorbed into a larger paper **Groups of Homotopy Spheres I**, written in collaboration with

[4]For further developments in surgery theory, see for example, WALL [1970], BROWDER [1972], MADSEN AND MILGRAM [1979], and CAPPELL ET AL. [2000, 2001].

Michel Kervaire. These papers provided a much more systematic analysis of all possible homotopy spheres of any dimension (other than 3). In particular, they showed that Θ_n is finite for $n \neq 3$. The analysis is based on an exact sequence

$$0 \to bP_{n+1} \to \Theta_n \to \Pi_n/J\pi_n(\mathrm{SO}) .$$

Here bP_{n+1} (denoted by $\Theta_n(\partial\pi)$ in the 1959 manuscript) is the subgroup consisting of those homotopy n-spheres which bound parallelizable manifolds. Using the Thom-Pontrjagin theory of cobordism for manifolds with framed normal bundle, the obstruction to bounding an n-manifold with framed normal bundle is measured by an element of the stable homotopy group Π_n. However, a change in framing will change this obstruction by an element in the image of the stable J-homomorphism $J : \pi_n(\mathrm{SO}) \to \Pi_n$. (Compare the introduction to Part 3 on pages 223–227.) The subgroup bP_{n+1} is trivial for n even and has at most two elements for $n \equiv 1 \pmod 4$. However, this group is quite large when $n \equiv 3 \pmod 4$. In fact bP_{4m} is cyclic of order

$$a_m\, 2^{2m-2}(2^{2m-1} - 1)\ \mathrm{numerator}(B_m/m)\,,$$

where a_m is one or two according as m is even or odd, and B_m is the m-th Bernoulli number. These numbers grow rapidly as $m \to \infty$, as one sees from the identity

$$B_m = 2\left(1 + 2^{-2m} + 3^{-2m} + \cdots\right)(2m)!/(2\pi)^{2m} .$$

(Compare Milnor and Stasheff [1974, Appendix B].) An explicit generator can be constructed as follows.[5] Start with a $2m$-skeleton consisting of eight copies of the $2m$-sphere intersecting in seven points, indicated schematically as follows.

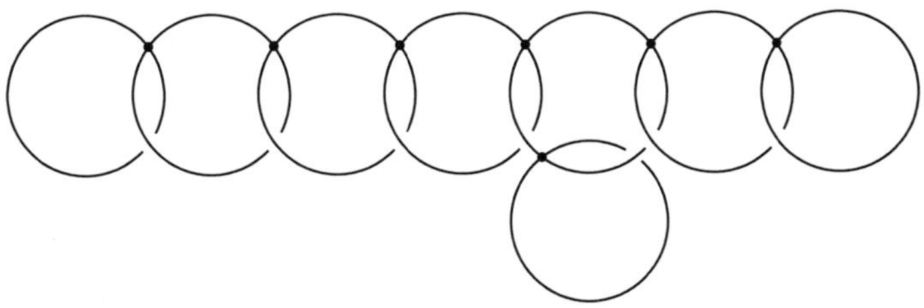

Now thicken each of these spheres by taking a $2m$-disk bundle isomorphic to its tangent disk bundle, with Euler number $+2$. The spheres are to cross each other transversally at each intersection point. (Here the normal bundles should all be non-trivial, but that is not shown in the very schematic picture on the next page.) The result will be a parallelizable $4m$-dimensional manifold W^{4m} of signature $+8$, having a homotopy sphere as boundary, provided that $m > 1$. It is shown that this boundary ∂W^{4m} is a generator for the finite cyclic group consisting of all homotopy spheres of dimension $4m - 1$ which bound parallelizable manifolds.

[5] This form of the construction, based on the E_8-lattice, was suggested by Hirzebruch. Compare the discussion in Hirzebruch [1987, pp. 673, 801]. My original construction was more complicated.

Clearly $\Theta_1 = \Theta_2 = 0$. Using these techniques, the first few groups Θ_n with $n \geq 4$ were described as follows. (Here, for example, the notation (8) stands for some abelian group of order 8.)

$n =$	4	5	6	7	8	9	10	11	12	13	14	15
$\Theta_n \cong$	0	0	0	$\mathbb{Z}/28$	$\mathbb{Z}/2$	(8)	$\mathbb{Z}/6$	$\mathbb{Z}/992$	0	$\mathbb{Z}/3$	$\mathbb{Z}/2$	(16256)

The projected "Groups of Homotopy Spheres: II" was never completed, although a very small part of it found its way into the expository paper **Differential Topology** (pages 123–141 below), which was published a few years later.

Meanwhile, the original project of publishing a paper on $2n$-manifolds which are $(n-1)$-connected, and a closely related project of studying "Spaces with a gap in cohomology" got lost in the shuffle. They were finally abandoned when Terry Wall published a beautiful exposition of the subject which made my attempts unnecessary. (See WALL [1962a], and compare WALL [1962b, 1964].)

Further Developments.

For the differential geometry of exotic spheres, see for example GROMOLL [1966], GROMOLL AND MEYER [1974], WRAITH [1997], GROVE AND ZILLER [2000], BOYER ET AL. [2005]. For exotic spheres via algebraic geometry, see BRIESKORN [1966], HIRZEBRUCH [1966/67], HIRZEBRUCH AND MAYER [1968], and compare MILNOR [1968].

The importance of the groups Γ_n of twisted spheres was brought out by Jim MUNKRES [1960a, 1964] and Moe HIRSCH [1963] who showed that they appear as the coefficient groups for obstructions to the existence and uniqueness of a compatible differentiable structure on a combinatorial manifold.

The more precise results of my paper with Kervaire seemed to be achieved at the cost of replacing these coefficient groups Γ_n by the cruder groups Θ_n. In other words, *twisted spheres* were replaced by *homotopy spheres*, and the relation of *diffeomorphism* by the apparently weaker relation of *h-cobordism*. However, Steve SMALE, in [1961, 1962], had proved that every homotopy sphere of dimension $n \geq 5$ is actually a twisted sphere.[6] In [1962] he proved that simply-connected h-cobordant manifolds of dimension at least five are actually diffeomorphic. (Compare

[6]More precisely, Smale proved this in [1961] for even n, while the general case with $n \geq 6$ follows by applying his h-cobordism theorem [1962] to the n-manifold with two open disks deleted. The 5-dimensional case depends on the h-cobordism theorem together with the additional information that $\Theta_5 = 0$.

the presentation in MILNOR, SIEBENMANN AND SONDOW [1965].) It followed that $\Gamma_n \xrightarrow{\cong} \Theta_n$ for $n \geq 5$.

The groups Γ_1 and Γ_2 are clearly trivial. SMALE [1959b] and MUNKRES [1960b] had shown that $\mathrm{Diff}^+(S^2)$ is connected which implies that $\Gamma_3 = 0$. Jean CERF showed in the [1962/63] Cartan Seminar that $\Gamma_4 = 0$, and indeed that $\mathrm{Diff}^+(S^3)$ is connected. (Compare CERF [1968]; and see ELIASHBERG [1992] for a different proof that $\Gamma_4 = 0$.) Combining these results with the statements above, it follows that Γ_n is trivial for $n < 7$. Thus the Munkres-Hirsch obstruction theory yields the following:

> *Every combinatorial manifold of dimension ≤ 7 possesses a compatible differentiable structure. Furthermore, in dimensions strictly less than 7 this structure is unique up to diffeomorphism.*

An analogous obstruction theory for passing from topological manifolds to combinatorial manifolds was constructed by Rob KIRBY and Larry SIEBENMANN [1969]. In this case, there is only one obstruction: in $H^4(M;\,\mathbb{Z}/2)$ for existence, and in $H^3(M;\,\mathbb{Z}/2)$ for uniqueness. Here the dimension of M must be at least 6 (or at least 5 if M has no boundary).

If we accept the announced proof of the Poincaré Conjecture by PERELMAN [2002, 2003a, 2003b], then it follows that $\Theta_3 = 0$, so that $\Gamma_n \xrightarrow{\cong} \Theta_n$ for all n. However, this simple statement conceals the fact that dimension 4 is a world by itself, different from all other dimensions. Topological 4-manifolds can be wildly non-differentiable, so it was necessary to introduce wildly non-differentiable methods in order to understand them. One key step was the introduction of "flexible handles" by Andrew Casson. (For a description of early steps see SIEBENMANN [1980] and MANDELBAUM [1980], as well as GUILLOU AND MARIN [1986].) The decisive breakthrough came with the work of Mike FREEDMAN [1982], who used Casson handles not only to prove the 4-dimensional topological Poincaré Hypothesis, but also to completely classify closed simply-connected topological 4-manifolds.

Completely different tools were needed to get a grip on the differentiable theory. If we consider only simply-connected 4-manifolds which are differentiable (or combinatorial) then Simon DONALDSON [1983, 1987] showed by gauge theory that there are very strong restrictions on the cohomology. It quickly became apparent to specialists that the contrast between Freedman's results and those of Donaldson led to very strange consequences. (Compare KIRBY [1989], GOMPF [1993].) For example Freedman showed that there exists a differentiable manifold homeomorphic to $\mathbb{R}^4$ which cannot be smoothly embedded in $\mathbb{R}^4$. Then Clifford TAUBES [1987] showed that in fact there are uncountably many such manifolds; while DEMICHELIS AND FREEDMAN [1992] showed that there are also uncountably many distinct examples which *can* be smoothly embedded in $\mathbb{R}^4$. The situation in other dimensions is quite different, since no exotic $\mathbb{R}^n$ can exist for $n \neq 4$. (See MOISE [1952] for $n < 4$ and STALLINGS [1962] for $n > 4$.)

The question of exotic structures on the 4-sphere seems particularly difficult. In fact one can ask whether there exists a non-standard differentiable structure on

S^4 which reduces to the standard structure on $S^4 \setminus$ (point), and one can also ask whether there exists one which does not. As far as I know, both questions remain open.

Since I cannot describe all the work in differential topology which has been done since these papers were written, let me simply list the titles of some books which help to fill in further history. (For complete citations, see the bibliography, starting on page 319.)

MUNKRES, "Elementary Differential Topology," 1963;
MILNOR, SPIVAK AND WELLS, "Morse Theory," 1963;
MILNOR, "Topology from the Differentiable Viewpoint," 1965;
MILNOR, SIEBENMANN AND SONDOW, "Lectures on the h-Cobordism Theorem," 1965;
STONG, "Notes on Cobordism Theory," 1968;
WALLACE, "Differential Topology, First Steps," 1968;
GUILLEMIN AND POLLACK, "Differential Topology," 1974;
MILNOR AND STASHEFF, "Characteristic Classes," 1974;
HIRSCH, "Differential Topology," 1976;
CHILLINGWORTH, "Differential Topology with a View to Applications," 1976;
KIRBY AND SIEBENMANN, "Foundational Essays on Topological Manifolds, Smoothings, and Triangulations," 1977;
BRÖCKER AND JÄNICH, "Introduction to Differential Topology," 1982;
GAULD, "Differential Topology, an Introduction," 1982;
BING, "The Geometric Topology of 3-Manifolds," 1983;
GUILLOU AND MARIN (editors), "A la Recherche de la Topologie Perdue," 1986;
DIEUDONNÉ, "A History of Algebraic and Differential Topology, 1900-1960," 1989;
KIRBY, "The Topology of 4-Manifolds," 1989;
DONALDSON AND KRONHEIMER, "The Geometry of Four-Manifolds," 1990;
FREEDMAN AND QUINN, "Topology of 4-Manifolds," 1990;
AKBULUT AND MCCARTHY, "Casson's invariant for oriented homology 3-spheres. An exposition," 1990;
NASH, "Differential Topology and Quantum Field Theory," 1991;
KOSINSKI, "Differential Manifolds," 1992;
FRIEDMAN AND MORGAN, "Smooth Four-Manifolds and Complex Surfaces," 1994;
MCDUFF AND SALAMON, "Introduction to Symplectic Topology," 1995.
FRIEDMAN AND MORGAN, "Gauge Theory and the Topology of Four-Manifolds," 1998;
GOMPF AND STIPSICZ, "4-Manifolds and Kirby Calculus," 1999.

ANNALS OF MATHEMATICS
Vol. 64, No. 2, September, 1956
Printed in U.S.A.

ON MANIFOLDS HOMEOMORPHIC TO THE 7-SPHERE

BY JOHN MILNOR[1]

(Received June 14, 1956)

The object of this note will be to show that the 7-sphere possesses several distinct differentiable structures.

In §1 an invariant λ is constructed for oriented, differentiable 7-manifolds M^7 satisfying the hypothesis (*) $H^3(M^7) = H^4(M^7) = 0$. (Integer coefficients are to be understood.) In §2 a general criterion is given for proving that an n-manifold is homeomorphic to the sphere S^n. Some examples of 7-manifolds are studied in §3 (namely 3-sphere bundles over the 4-sphere). The results of the preceding two sections are used to show that certain of these manifolds are topological 7-spheres, but not differentiable 7-spheres. Several related problems are studied in §4.

All manifolds considered, with or without boundary, are to be differentiable, orientable and compact. The word *differentiable* will mean differentiable of class C^∞. A closed manifold M^n is *oriented* if one generator $\mu \in H_n(M^n)$ is distinguished.

§1. The invariant $\lambda(M^7)$

For every closed, oriented 7-manifold satisfying (*) we will define a residue class $\lambda(M^7)$ modulo 7. According to Thom [5] every closed 7-manifold M^7 is the boundary of an 8-manifold B^8. The invariant $\lambda(M^7)$ will be defined as a function of the index τ and the Pontrjagin class p_1 of B^8.

An orientation $\nu \in H_8(B^8, M^7)$ is determined by the relation $\partial\nu = \mu$. Define a quadratic form over the group $H^4(B^8, M^7)/(\text{torsion})$ by the formula $\alpha \to \langle \nu, \alpha^2 \rangle$. Let $\tau(B^8)$ be the index of this form (the number of positive terms minus the number of negative terms, when the form is diagonalized over the real numbers).

Let $p_1 \in H^4(B^8)$ be the first Pontrjagin class of the tangent bundle of B^8. (For the definition of Pontrjagin classes see [2] or [6].) The hypothesis (*) implies that the inclusion homomorphism

$$i: H^4(B^8, M^7) \to H^4(B^8)$$

is an isomorphism. Therefore we can define a "Pontrjagin number"

$$q(B^8) = \langle \nu, (i^{-1}p_1)^2 \rangle.$$

THEOREM 1. *The residue class of* $2q(B^8) - \tau(B^8)$ *modulo* 7 *does not depend on the choice of the manifold* B^8.

Define $\lambda(M^7)$ as this residue class.[2] As an immediate consequence we have:

COROLLARY 1. *If* $\lambda(M^7) \neq 0$ *then* M^7 *is not the boundary of any* 8-*manifold having fourth Betti number zero.*

---

[1] The author holds an Alfred P. Sloan fellowship.

Let B_1^8, B_2^8 be two manifolds with boundary M^7. (We may assume they are disjoint.) Then $C^8 = B_1^8 \cup B_2^8$ is a closed 8-manifold which possesses a differentiable structure compatible with that of B_1^8 and B_2^8. Choose that orientation ν for C^8 which is consistent with the orientation ν_1 of B_1^8 (and therefore consistent with $-\nu_2$). Let $q(C^8)$ denote the Pontrjagin number $\langle \nu, p_1^2(C^8)\rangle$.

According to Thom [5] or Hirzebruch [2] we have

$$\tau(C^8) = \langle \nu, \tfrac{1}{45}(7p_2(C^8) - p_1^2(C^8))\rangle;$$

and therefore

$$45\tau(C^8) + q(C^8) = 7\langle \nu, p_2(C^8)\rangle \equiv 0 \pmod 7.$$

This implies

$$2q(C^8) - \tau(C^8) \equiv 0 \pmod 7. \tag{1}$$

LEMMA 1. *Under the above conditions we have*

$$\tau(C^8) = \tau(B_1^8) - \tau(B_2^8) \tag{2}$$

and

$$q(C^8) = q(B_1^8) - q(B_2^8). \tag{3}$$

Formulas 1, 2, 3 clearly imply that

$$2q(B_1^8) - \tau(B_1^8) \equiv 2q(B_2^8) - \tau(B_2^8) \pmod 7;$$

which is just the assertion of Theorem 1.

PROOF OF LEMMA 1. Consider the diagram

$$\begin{array}{ccc} H^n(B_1, M) \oplus H^n(B_2, M) & \xleftarrow[\approx]{h} & H^n(C, M) \\ \big\downarrow i_1 \oplus i_2 & & \big\downarrow j \\ H^n(B_1) \oplus H^n(B_2) & \xleftarrow{k} & H^n(C) \end{array}$$

Note that for $n = 4$, these homomorphisms are all isomorphisms. If $\alpha = jh^{-1}(\alpha_1 \oplus \alpha_2) \in H^4(C)$, then

$$\langle \nu, \alpha^2\rangle = \langle \nu, jh^{-1}(\alpha_1^2 \oplus \alpha_2^2)\rangle = \langle \nu_1 \oplus (-\nu_2), \alpha_1^2 \oplus \alpha_2^2\rangle = \langle \nu_1, \alpha_1^2\rangle - \langle \nu_2, \alpha_2^2\rangle. \tag{4}$$

Thus the quadratic form of C^8 is the "direct sum" of the quadratic form of B_1^8 and the negative of the quadratic form of B_2^8. This clearly implies formula (2).

Define $\alpha_1 = i_1^{-1}p_1(B_1)$ and $\alpha_2 = i_2^{-1}p_1(B_2)$. Then the relation

$$k(p_1(C)) = p_1(B_1) \oplus p_1(B_2)$$

implies that

---

[2] Similarly for $n = 4k - 1$ a residue class $\lambda(M^n)$ modulo $s_k\mu(L_k)$ could be defined. (See [2] page 14.) For $k = 1, 2, 3, 4$ we have $s_k\mu(L_k) = 1, 7, 62, 381$ respectively.

$$jh^{-1}(\alpha_1 \oplus \alpha_2) = p_1(C).$$

The computation (4) now shows that

$$\langle \nu, p_1^2(C) \rangle = \langle \nu_1, \alpha_1^2 \rangle - \langle \nu_2, \alpha_2^2 \rangle,$$

which is just formula (3). This completes the proof of Theorem 1.

The following property of the invariant λ is clear.

LEMMA 2. *If the orientation of M^7 is reversed then $\lambda(M^7)$ is multiplied by* -1.

As a consequence we have

COROLLARY 2. *If $\lambda(M^7) \neq 0$ then M^7 possesses no orientation reversing diffeomorphism*[3] *onto itself.*

§2. A partial characterization of the n-sphere

Consider the following hypothesis concerning a closed manifold M^n (where R denotes the real numbers).

(H) *There exists a differentiable function $f: M^n \rightarrow R$ having only two critical points x_0, x_1. Furthermore these critical points are non-degenerate.*

(That is if $u_1, \cdots, u_n$ are local coordinates in a neighborhood of x_0 (or x_1) then the matrix $(\partial^2 f / \partial u_i \partial u_j)$ is non-singular at x_0 (or x_1).)

THEOREM 2. *If M^n satisfies the hypothesis* (H) *then these exists a homeomorphism of M^n onto S^n which is a diffeomorphism except possibly at a single point.*

Added in proof. This result is essentially due to Reeb [7].

The proof will be based on the orthogonal trajectories of the manifolds f = constant.

Normalize the function f so that $f(x_0) = 0$, $f(x_1) = 1$. According to Morse ([3] Lemma 4) there exist local coordinates $v_1, \cdots, v_n$ in a neighborhood V of x_0 so that $f(x) = v_1^2 + \cdots + v_n^2$ for $x \in V$. (Morse assumes that f is of class C^3, and constructs coordinates of class C^1; but the same proof works in the C^∞ case.) The expression $ds^2 = dv_1^2 + \cdots + dv_n^2$ defines a Riemannian metric in the neighborhood V. Choose a differentiable Riemannian metric for M^n which coincides with this one in some neighborhood[4] V' of x_0. Now the gradien: of f can be considered as a contravariant vector field.

Following Morse we consider the differential equation

$$\frac{dx}{dt} = \text{grad } f / \| \text{grad } f \|^2.$$

In the neighborhood V' this equation has solutions

$$(v_1(t), \cdots, v_n(t)) = (a_1(t)^{\frac{1}{2}}, \cdots, a_n(t)^{\frac{1}{2}})$$

for $0 \leqq t < \varepsilon$, where $a = (a_1, \cdots, a_n)$ is any n-tuple with $\sum a_i^2 = 1$. These can be extended uniquely to solutions $x_a(t)$ for $0 \leqq t \leqq 1$. Note that these solutions satisfy the identity

---

[3] A *diffeomorphism* f is a homeomorphism onto, such that both f and f^{-1} are differentiable.

[4] This is possible by [4] 6.7 and 12.2.

$$f(x_a(t)) = t.$$

Map the interior of the unit sphere of R^n into M^n by the map

$$(a_1(t)^{\frac{1}{2}}, \cdots, a_n(t)^{\frac{1}{2}}) \to x_a(t).$$

It is easily verified that this defines a diffeomorphism of the open n-cell onto $M^n - (x_1)$. The assertion of Theorem 2 now follows.

Given any diffeomorphism $g: S^{n-1} \to S^{n-1}$, an n-manifold can be obtained as follows.

CONSTRUCTION (C). *Let $M^n(g)$ be the manifold obtained from two copies of R^n by matching the subsets $R^n - (0)$ under the diffeomorphism*

$$u \to v = \frac{1}{\| u \|} g\left(\frac{u}{\| u \|}\right).$$

(Such a manifold is clearly homeomorphic to S^n. If g is the identity map then $M^n(g)$ is diffeomorphic to S^n.)

COROLLARY 3. *A manifold M^n can be obtained by the construction* (C) *if and only if it satisfies the hypothesis* (H).

PROOF. If $M^n(g)$ is obtained by the construction (C) then the function

$$f(x) = \| u \|^2/(1 + \| u \|^2) = 1/(1 + \| v \|^2)$$

will satisfy the hypothesis (H). The converse can be established by a slight modification of the proof of Theorem 2.

§3. Examples of 7-manifolds

Consider 3-sphere bundles over the 4-sphere with the rotation group SO(4) as structural group. The equivalence classes of such bundles are in one-one correspondence[5] with elements of the group $\pi_3(\mathrm{SO}(4)) \approx Z + Z$. A specific isomorphism between these groups is obtained as follows. For each $(h, j) \in Z + Z$ let $f_{hj}: S^3 \to \mathrm{SO}(4)$ be defined by $f_{hj}(u) \cdot v = u^h v u^j$, for $v \in R^4$. Quaternion multiplication is understood on the right.

Let ι be the standard generator for $H^4(S^4)$. Let ξ_{hj} denote the sphere bundle corresponding to $(f_{hj}) \in \pi_3(\mathrm{SO}(4))$.

LEMMA 3. *The Pontrjagin class $p_1(\xi_{hj})$ equals $\pm 2(h - j)\iota$.*

(The proof will be given later. One can show that the characteristic class $\bar{c}(\xi_{hj})$ (see [4]) is equal to $(h + j)\iota$.)

For each odd integer k let M_k^7 be the total space of the bundle ξ_{hj} where h and j are determined by the equations $h + j = 1$, $h - j = k$. This manifold M_k^7 has a natural differentiable structure and orientation, which will be described later.

LEMMA 4. *The invariant $\lambda(M_k^7)$ is the residue class modulo 7 of $k^2 - 1$.*

LEMMA 5. *The manifold M_k^7 satisfies the hypothesis* (H).

Combining these we have:

---

[5] See [4] §18.

THEOREM 3. *For $k^2 \not\equiv 1 \bmod 7$ the manifold M_k^7 is homeomorphic to S^7 but not diffeomorphic to S^7.*

(For $k = \pm 1$ the manifold M_k^7 is diffeomorphic to S^7; but it is not known whether this is true for any other k.)

Clearly any differentiable structure on S^7 can be extended through $R^8 - (0)$. However:

COROLLARY 4. *There exists a differentiable structure on S^7 which cannot be extended throughout R^8.*

This follows immediately from the preceding assertions, together with Corollary 1.

PROOF OF LEMMA 3. It is clear that the Pontrjagin class $p_1(\xi_{hj})$ is a linear function of h and j. Furthermore it is known that it is independent of the orientation of the fibre. But if the orientation of S^3 is reversed, then ξ_{hj} is replaced by ξ_{-j-h}. This shows that $p_1(\xi_{hj})$ is given by an expression of the form $c(h - j)\iota$. Here c is a constant which will be evaluated later.

PROOF OF LEMMA 4. Associated with each 3-sphere bundle $M_k^7 \to S^4$ there is a 4-cell bundle $\rho_k: B_k^8 \to S^4$. The total space B_k^8 of this bundle is a differentiable manifold with boundary M_k^7. The cohomology group $H^4(B_k^8)$ is generated by the element $\alpha = \rho_k^*(\iota)$. Choose orientations μ, ν for M_k^7 and B_k^8 so that

$$\langle \nu, (\bar{\imath}^{-1}\alpha)^2 \rangle = +1.$$

Then the index $\tau(B_k^8)$ will be $+1$.

The tangent bundle of B_k^8 is the "Whitney sum" of (1) the bundle of vectors tangent to the fibre, and (2) the bundle of vectors normal to the fibre. The first bundle (1) is induced (under ρ_k) from the bundle ξ_{hj}, and therefore has Pontrjagin class $p_1 = \rho_k^*(c(h - j)\iota) = ck\alpha$. The second is induced from the tangent bundle of S^4, and therefore has first Pontrjagin class zero. Now by the Whitney product theorem ([2] or [6])

$$p_1(B_k^8) = ck\alpha + 0.$$

For the special case $k = 1$ it is easily verified that B_1^8 is the quaternion projective plane $P_2(K)$ with an 8-cell removed. But the Pontrjagin class $p_1(P_2(K))$ is known to be twice a generator of $H^4(P_2(K))$. (See Hirzebruch [1].) Therefore the constant c must be ± 2, which completes the proof of Lemma 3.

Now $q(B_k^8) = \langle \nu, (\bar{\imath}^{-1}(\pm 2k\alpha))^2 \rangle = 4k^2$; and $2q - \tau = 8k^2 - 1 \equiv k^2 - 1$ (mod 7). This completes the proof of Lemma 4.

PROOF OF LEMMA 5. As coordinate neighborhoods in the base space S^4 take the complement of the north pole, and the complement of the south pole. These can be identified with euclidean space R^4 under stereographic projection. Then a point which corresponds to $u \in R^4$ under one projection will correspond to $u' = u/\| u \|^2$ under the other.

The total space M_k^7 can now be obtained as follows.[5] Take two copies of $R^4 \times S^3$ and identify the subsets $(R^4 - (0)) \times S^3$ under the diffeomorphism

$$(u, v) \to (u', v') = (u/\| u \|^2, u^h v u^j/\| u \|)$$

(using quaternion multiplication). This makes the differentiable structure of M_k^7 precise.

Replace the coordinates (u', v') by (u'', v') where $u'' = u'(v')^{-1}$. Consider the function $f: M_k^7 \to R$ defined by

$$f(x) = \Re(v)/(1 + \| u \|^2)^{\frac{1}{2}} = \Re(u'')/(1 + \| u'' \|^2)^{\frac{1}{2}};$$

where $\Re(v)$ denotes the real part of the quaternion v. It is easily verified that f has only two critical points (namely $(u, v) = (0, \pm 1)$) and that these are non-degenerate. This completes the proof.

§4. Miscellaneous results

THEOREM 4. *Either* (a) *there exists a closed topological* 8-*manifold which does not possess any differentiable structure*; *or* (b) *the Pontrjagin class* p_1 *of an open* 8-*manifold is not a topological invariant.*

(The author has no idea which alternative holds.)

PROOF. Let X_k^8 be the topological 8-manifold obtained from B_k^8 by collapsing its boundary (a topological 7-sphere) to a point x_0. Let $\bar{\alpha} \in H^4(X_k^8)$ correspond to the generator $\alpha \in H^4(B_k^8)$. Suppose that X_k^8, possesses a differentiable structure, and that $p_1(X_k^8 - (x_0))$ is a topological invariant. Then $p_1(X_k^8)$ must equal $\pm 2k\bar{\alpha}$, hence

$$2q(X_k^8) - \tau(X_k^8) = 8k^2 - 1 \equiv k^2 - 1 \qquad (\text{mod } 7).$$

But for $k^2 \not\equiv 1$ (mod 7) this is impossible.

Two diffeomorphisms $f, g: M_1^n \to M_2^n$ will be called *differentiably isotopic* if there exists a diffeomorphism $M_1^n \times R \to M_2^n \times R$ of the form $(x, t) \to (h(x, t), t)$ such that

$$h(x, t) = \begin{cases} f(x) & (t \leqq 0) \\ g(x) & (t \geqq 1). \end{cases}$$

LEMMA 6. *If the diffeomorphisms* $f, g: S^{n-1} \to S^{n-1}$ *are differentiably isotopic, then the manifolds* $M^n(f)$, $M^n(g)$ *obtained by the construction* (C) *are diffeomorphic.*

The proof is straightforward.

THEOREM 5. *There exists a diffeomorphism* $f: S^6 \to S^6$ *of degree* $+1$ *which is not differentiably isotopic to the identity.*

Proof. By Lemma 5 and Corollary 3 the manifold M_3^7 is diffeomorphic to $M^7(f)$ for some f. If f were differentiably isotopic to the identity then Lemma 6 would imply that M_3^7 was diffeomorphic to S^7. But this is false by Lemma 4.

PRINCETON UNIVERSITY

REFERENCES

1. F. HIRZEBRUCH, *Ueber die quaternionalen projektiven Räume*, S.-Ber. math.- naturw. Kl. Bayer. Akad. Wiss. München (1953), pp. 301–312.
2. ———, Neue topologische Methoden in der algebraischen Geometrie, Berlin, 1956.

3. M. Morse, *Relations between the numbers of critical points of a real function of n independent variables*, Trans. Amer. Math. Soc., 27 (1925), pp. 345–396.
4. N. Steenrod, The topology of fibre bundles, Princeton, 1951.
5. R. Thom, *Quelques propriétés globale des variétés différentiables*, Comment. Math. Helv., 28 (1954), pp. 17–86.
6. Wu Wen-Tsun, *Sur les classes caractéristiques des structures fibrées sphériques*, Actual. sci. industr. 1183, Paris, 1952, pp. 5–89.
7. G. Reeb, *Sur certain propriétés topologiques des variétés feuilletées*, Actual. sci. industr. 1183, Paris, 1952, pp. 91–154.

On the Relationship Between Differentiable Manifolds and Combinatorial Manifolds.[1]

John Milnor

Princeton University, November 27, 1956

1. Introduction.

J.H.C. Whitehead has shown in [**11**] that every compact differentiable manifold can be assigned a combinatorial structure which is essentially unique. (Precise statements of Whitehead's results are given in §3.) We will see, on the other hand, that a given combinatorial manifold may either (a) have no differentiable structure at all which is compatible with its given combinatorial structure; or (b) have several, essentially distinct, differentiable structures which are compatible with it. (See Theorems 4.2 and 4.1 as well as §5.) Assertion (a) is due to R. Thom [**10**] and is based on his construction of Pontrjagin classes for triangulated manifolds.

An appendix to this paper attempts to set a foundation for the study of differentiable manifolds with boundary. A second appendix is concerned with Whitehead's C^r-triangulations.

2. Definitions.

We remark first that the work "differentiable" will be used to mean differentiable of class C^∞; and that the word "manifold" will allow manifolds with boundary. All manifolds are assumed to be ***separable*** (i.e., with a countable basis for the topology); but are not necessarily connected.

The underlying space of a simplicial complex K is denoted by $|K|$. A ***triangulation*** (K, f) of a space X consists of a simplicial complex K, together with a homeomorphism f of $|K|$ onto X. A second triangulation (K_1, f_1) of X is called a ***subdivision*** of (K, f) if the composition $f^{-1} \circ f_1 : |K_1| \to |K|$ maps simplexes linearly into simplexes. We will also say that the ***triangulated space*** (K_1, f_1, X) is a ***subdivision*** of the triangulated space (K, f, X).

[1] This paper is a continuation of the author's paper [**6**]. By a combinatorial manifold we mean a manifold in the sense of Newman [**7**] and Alexander [**1**]. Definitions will be given in §2. (Note: Numbered footnotes are from the original text; while unnumbered footnotes have been added recently.)

Two triangulated spaces (K_1, f_1, X_1) and (K_2, f_2, X_2) are *isomorphic* if K_1 is isomorphic to K_2. They are *combinatorially equivalent* if they have isomorphic subdivisions. In particular we will say that a triangulated space is a *combinatorial n-cell* (or *$(n-1)$-sphere*) if it is combinatorially equivalent to the n-simplex (or its boundary).

A triangulated space is a *combinatorial n-manifold* if the star neighborhood of every vertex is a combinatorial n-cell. (An equivalent condition would be that the link of every j-simplex, $0 \le j < n$, should be either a combinatorial $(n-j-1)$-sphere or a combinatorial $(n-j-1)$-cell.)

The concept of differentiable manifold (that is C^∞-manifold with boundary) will be defined in Appendix A. A triangulation (K, f) of a differentiable n-manifold is a *C^∞-triangulation* if, for each n-simplex Δ of $|K|$, the map $f|\Delta$ is differentiable (see Appendix A) and has Jacobian of rank n at all points. (This definition is equivalent to that given by Whitehead, although not identical to it. See Appendix B.) Alternatively, we will say that the differentiable structure for M^n is *compatible* with the triangulation (K, f), if (K, f) is a C^∞-triangulation.

3. The Theorems of Whitehead.

(These theorems have been restated in the author's terminology.)

Theorem $\mathbf{W_1}$. *Every compact[2] differentiable manifold has a C^∞-triangulation. In fact every C^∞-triangulation of the boundary can be extended to a C^∞-triangulation of the whole manifold.*

Theorem $\mathbf{W_2}$. *If (K, f) is a C^∞-triangulation of M^n, then the triangulated space (K, f, M^n) is a combinatorial manifold.*

Theorem $\mathbf{W_3}$. *If (K_1, f_1) and (K_2, f_2) are C^∞-triangulations of the same compact[2] differentiable manifold M^n, then the triangulated spaces (K_1, f_1, M^n) and (K_2, f_2, M^n) are combinatorially equivalent.*

Without the hypothesis of differentiability, these assertions would reduce to extremely difficult unsolved problems. (Namely: the triangulability of manifolds; the question of whether every triangulated manifold is a combinatorial manifold; and the Hauptvermutung.*)

---

[2]Whitehead remarks [**11**, p. 823] that the hypothesis of compactness is not necessary.

*In fact the answer to each of these questions is negative. The existence of a compact non-triangulable 4-manifold follows from the work of Andrew Casson and Michael Freedman. See for example S. Akbulut and J.D. McCarthy, "Casson's invariant for oriented homology 3-spheres, An exposition," Mathematical Notes **36**, Princeton U. Press 1990. A counterexample to the Hauptvermutung for simplicial complexes was given by Milnor (Annals of Math. **74** (1961), 575–590 —reprinted in "Collected Papers 2"); while a counterexample for combinatorial manifolds was given by Kirby and Siebenmann (Bull. Amer. Math. Soc. **75** (1969), 742–749). An example of a triangulated manifold which is not a combinatorial manifold was given by R.D. Edwards. Compare F. Latour, Séminaire Bourbaki 30e année (1977/78), **515**, pp. 169–186, Lecture Notes in Math. **710**, Springer 1979.

For the proof of $\mathbf{W_1}$ see [**11**, Theorems 3, 7, 10], together with the remarks at the bottom of [**11**, p. 822]. For $\mathbf{W_3}$ see his Theorem 8, together with these same remarks. In the case of a manifold which is unbounded (that is, without boundary), $\mathbf{W_2}$ follows from his Theorem 5. The extension to a manifold with boundary is straightforward. □

These theorems suggest the following definition. Two compact differentiable manifolds M_1^n and M_2^n are *combinatorially equivalent* if, for some (and therefore for all) C^∞-triangulations (K_1, f_1) and (K_2, f_2) of M_1^n and M_2^n respectively the triangulated spaces (K_1, f_1, M_1^n) and (K_2, f_2, M_2^n) are combinatorially equivalent.

4. The Main Theorems.

THEOREM 4.1. *There exists a differentiable 7-manifold which is combinatorially equivalent to the standard sphere S^7, but not diffeomorphic to S^7.*

THEOREM 4.2. (R. Thom). *There exists a closed combinatorial manifold (K, f, M^8) such that no differentiable structure for M^8 is compatible with the given triangulation (K, f), or with any subdivision of (K, f).*

In order to prove Theorem 4.1 it will be sufficient (comparing [**6**]) to prove the following.

LEMMA 4.3. *Let M^n be a closed differentiable manifold. If there exists a differentiable real valued function g on M^n with only two critical points, both being non-degenerate, then M^n is combinatorially equivalent to S^n.*

Proof. We may suppose that g has maximum $+1$ and minimum -1. Consider the submanifolds M_+^n, M_-^n defined by the inequalities $g \geq 0$ and $g \leq 0$ respectively. The argument given in [**6**] shows that each of these manifolds is diffeomorphic to the solid n-ball. By $\mathbf{W_1}$ we can choose a C^∞-triangulation (K_+, f_+) of M_+^n and extend it to a C^∞-triangulation (K, f) of M^n. Let (K_-, f_-) be the resulting triangulation of M_-^n. By $\mathbf{W_3}$ the triangulated spaces $(K_\pm, f_\pm, M_\pm^n)$ are combinatorial n-cells. But a triangulated space obtained by matching the boundaries of two combinatorial n-cells is clearly a combinatorial n-sphere. This proves Lemma 4.3 and Theorem 4.1. □

Proof of Theorem 4.2. (following [10]). Consider the manifold B_k^8 with boundary M_k^7 as defined in [**6**]. Form a space X_k^8 from B_k^8 by adjoining a cone over M_k^7. Choose a C^∞-triangulation of B_k^8, and extend it to a triangulation (K, f) of X_k^8 by using the standard triangulation for a cone. Then Lemma 4.3, together with $\mathbf{W_2}$, implies that (K, f, X_k^8) is a combinatorial manifold.

Let $p_i \in H^{4i}(X_k^8, \mathbb{Q})$ denote the Pontrjagin classes as defined by Thom [**10**]; and let $\nu \in H_8(X_k^8, \mathbb{Z})$ be a suitably chosen generator. As in [**6**] we obtain $\langle p_1^2, \nu\rangle = 4k^2$; and therefore

$$\langle p_2, \nu\rangle = (45 + 4k^2)/7 .$$

But for $k \not\equiv \pm 1 \pmod 7$, this expression is not an integer; which implies that p_2 is not the Pontrjagin class corresponding to any differentiable structure for X_k^8. This proves Theorem 4.2. □

5. Miscellaneous remarks.

It is interesting to ask whether the phenomena described in Theorems 4.1 and 4.2 can occur in lower dimensions. Cairns has proved in [**5**] that for $n \leq 4$ every unbounded combinatorial n-manifold possesses a differentiable structure. Cairns' proof actually shows that *every unbounded combinatorial n-manifold, $n \leq 4$, has a subdivision which possesses a compatible differentiable structure.* The corresponding question for $n = 5, 6, 7$ remains open.*

It is definitely necessary to allow subdivision, since Cairns has given in [**4**] an example of a combinatorial 4-manifold such that the star neighborhood of a certain vertex can not be embedded in Euclidean 4-space by any linear homeomorphism. Lemma B.1 of Appendix B implies that this combinatorial manifold cannot possess any differentiable structure which is compatible with the specified triangulation.

J.R. Munkres has proved (unpublished) that, if two closed differentiable 2-manifolds are homeomorphic, then they are necessarily diffeomorphic. The corresponding question in dimensions $3 \leq n \leq 6$ remains open.†

Up to this point we have restricted attention to C^∞-manifolds and C^∞-maps. The discussion could equally well have been carried out using C^r-manifolds with $1 \leq r \leq \infty$. For example Whitehead's theorems $\mathbf{W_1}$, $\mathbf{W_2}$, $\mathbf{W_3}$ hold with C^∞ replaced by C^r; and in fact Whitehead considers mainly the C^1-case. Our excuse for sticking to the C^∞ case is provided by the following theorems of Whitney [**13**]: *Every unbounded C^r-manifold ($r \geq 1$) can be given a compatible C^∞-structure. Every C^r-isomorphism between two unbounded C^∞-manifolds can be approximated by a C^∞-isomorphism (or "diffeomorphism").* Thus for the problems which we are considering, it is sufficient to consider the C^∞ case.

The situation is less clear for real analytic manifolds. However suppose we call a real analytic manifold ***embeddable*** if it possesses an analytic embedding in some Euclidean space. (Compare Bochner [**3**]. The existence of non-embeddable analytic manifolds is an unsolved problem.‡) Then (see Whitney [**13**]) *every C^∞-manifold without boundary can be given a compatible real analytic structure which is embeddable; and any C^∞-isomorphism between embeddable analytic manifolds can be approximated by an analytic-isomorphism.* So again it is sufficient to consider the C^∞ case.

---

*In fact this statement remains true for all $n \leq 7$. This follows from the obstruction theory of Munkres and Hirsch, together with Cerf's proof that $\Gamma_4 = 0$.

†This statement is true in dimensions 1, 2, 3, 5, 6, but is definitely false in dimension 4. See GOMPF AND STIPSICZ [1999] for the 4-dimensional case, and see the Introduction to Part 1 on pages 1–9 for the remaining cases.

‡Morrey and Grauert showed two years later that every real analytic manifold is embeddable. See C. B. Morrey Jr., *The analytic embedding of abstract real-analytic manifolds*, Ann. of Math. **68** (1958) 159–201; and H. Grauert, *On Levi's problem and the embedding of real-analytic manifolds*, ibid. pp. 159-201.

Appendix A. Differentiable Manifolds.

Let H^n be a closed half-space in Euclidean space $\mathbb{R}^n$. By an *n-dimensional coordinate system* on a space X we will mean a collection $\{(U_\alpha,\ f_\alpha)\}$ of pairs, where the U_α are open sets covering X, and where each f_α is a homeomorphism of U_α onto either H^n or $\mathbb{R}^n$. An n-manifold is a Hausdorff space which possesses such a coordinate system. We assume that all manifolds considered are separable (i.e., have a countable basis), and hence are paracompact.

The set of points in X which correspond to boundary points of H^n forms the boundary ∂X. A manifold is *unbounded* if its boundary is vacuous. (Note that the boundary of any n-manifold is an unbounded $(n-1)$-manifold.) A manifold is *closed* if it is compact and unbounded.

Let A be any subset of Euclidean space $\mathbb{R}^m$. A map $f : A \to \mathbb{R}^n$ is *differentiable* of class C^r (or briefly, is a *C^r-map*), $1 \leq r \leq \infty$, if for every point $a \in A$ there exists a neighborhood U of a in $\mathbb{R}^m$ and a map $g : U \to \mathbb{R}^n$, such at g is continuous, with continuous j-th partial derivatives for all finite integers $1 \leq j \leq r$, and such that f(x) = g(x) for $x \in A \cap U$. (Compare Whitney [**12**].)

A coordinate system $\{(U_\alpha,\ f_\alpha)\}$ for X is a *C^r-coordinate system* if, for each α_1 and α_2 the composition

$$f_{\alpha_2} \circ f_{\alpha_1}^{-1} \ :\ f_{\alpha_1}(U_{\alpha_1} \cap U_{\alpha_2}) \ \to\ f_{\alpha_2}(U_{\alpha_1} \cap U_{\alpha_2}) \ \subset\ \mathbb{R}^n$$

is a C^r-map. Two C^r-coordinate systems $\{(U_\alpha,\ f_\alpha)\}$ and $\{(V_\beta,\ g_\beta)\}$ are *C^r-equivalent* if the composite system $\{(U_\alpha,\ f_\alpha)\} \cup \{(V_\beta,\ g_\beta)\}$ is also a C^r-coordinate system.

By a *differentiable manifold of class C^r* (or a *C^r-manifold*) we mean a manifold together with a C^r-equivalence class of C^r-coordinate systems. (This equivalence class is called a *C^r-structure*.) Note that the boundary of any C^r-manifold can itself be considered as a C^r-manifold. Also any open subset of a C^r-manifold is a C^r-manifold.

Let M^m and N^n be C^r-manifolds with representative coordinate systems $\{(U_\alpha,\ f_\alpha)\}$, $\{(V_\beta,\ g_\beta)\}$, and let A be any subset of M^m. A map $h : A \to N^n$ is a *C^r-map* if for each α and β the map

$$g_\beta \circ h \circ f_\alpha^{-1} \ :\ f_\alpha\Big(A \cap U_\alpha \cap h^{-1}(V_\beta)\Big) \ \to\ \mathbb{R}^n$$

is differentiable of class C^r (as defined above).

A homeomorphism h of M^m into N^n (or onto N^n) is called a *C^r-embedding* (or a *C^r-isomorphism*) if h is a C^r-map and has Jacobian of rank m at all points. A C^∞-isomorphism is also called a *diffeomorphism*.

Some basic properties of C^r-manifolds follow.

LEMMA A.1. (C^r-extension). *Let A be a subset of the C^r-manifold M^m and let N^n be an unbounded C^r-manifold. Then every C^r-map $h : A \to N^n$ can be extended to a C^r-map of a neighborhood of A into N^n.*

(Compare Whitney [**12**])

LEMMA A.2. (Collar neighborhoods). *If M^n is a C^r-manifold with boundary ∂M^n, then there exists a neighborhood U of ∂M^n and a C^r-isomorphism*

$$g : \partial M^n \times [0,1) \to U$$

satisfying $g(y,0) = y$.

(The corresponding statement for topological manifolds seems very difficult.[§])

LEMMA A.3. (Gluing boundaries). *Let M^n be the unbounded manifold obtained by piecing together two bounded C^r-manifolds M_1^n and M_2^n by means of a C^r-isomorphism $h : \partial M_1^n \to \partial M_2^n$. Then M^n can be given a C^r-structure compatible with that of M_1^n and M_2^n.*

(By "compatible" we mean that the natural homeomorphisms $j_i : M_i^n \to M^n$, $i = 1, 2$, are C^r-embeddings.) This lemma was used without proof in [**6**] and [**9**].

LEMMA A.4. (Extending past the boundary). *Any C^r-manifold M_1^n can be embedded in an unbounded C^r-manifold M^n of the same dimension. If M_1^n is compact then M^n can be chosen as a closed manifold.*

The basic tool used in proving these lemmas is the partition of unity. By a *C^r-partition of unity* on a C^r-manifold M^n is meant a collection $\{p_\beta\}$ of C^r-maps $p_\beta : M^n \to \mathbb{R}$ satisfying

$$(1)\ p_\beta(x) \geq 0\,, \qquad (2)\ \sum_\beta p_\beta(x) = 1\,, \qquad \text{and}$$

$$(3)\ \text{the collection of supports } \{\text{supp}(p_\beta)\} \text{ is locally finite.}^\dagger$$

Here $\text{supp}(p_\beta)$ denotes the closure of the set of x in M^n with $p_\beta(x) > 0$. It will be convenient to say that $\{p_\beta\}$ is *associated* with an open covering $\{V_\beta\}$ of M^n if $\text{supp}(p_\beta) \subset V_\beta$ for each β.

LEMMA A.5. (Partitions of unity). *Given any open covering $\{V_\beta\}$ of a C^r-manifold M^n, there exists an associated C^r-partition of unity $\{p_\beta\}$.*

Proof of Lemma A.5. Since our manifolds are assumed to have a countable basis, it is well known that there exists a C^r-partition of unity $\{q_\gamma\}$ such that each set $\text{supp}(q_\gamma)$ is contained in some set V_β of the given covering. (For the case of an unbounded manifold, see de Rham [**8**, p. 4]. The extension to manifolds with boundary is straightforward.) Now for each index γ choose $\beta(\gamma)$ so that $\text{supp}(q_\gamma) \subset V_{\beta(\gamma)}$, and define p_{β_0} to be the sum of all q_γ such that $\beta(\gamma) = \beta_0$. Then $\{p_\beta\}$ is the required partition of unity. □

[§]The existence of collar neighborhoods for topological manifolds-with-boundary was proved by Morton Brown, *Locally flat imbeddings of topological manifolds*, Ann. of Math. **75** (1962) 331–341. (Compare Essay 1, Appendix A of R. Kirby and L. Siebenmann, "Foundational essays on topological manifolds, smoothings, and triangulations", Annals of Math. Studies **88**, Princeton U. Press 1977.)

[†]These definitions have been rewritten using more standard terminology. Note that at most countably many of the p_β can be non-zero, even when the index set $\{\beta\}$ is uncountable.

Proof of Lemma A.1. First suppose that the target space is the Euclidean space $\mathbb{R}^n$. From the definition of the C^r-map it follows that for each point $a \in A$ we can choose an open neighborhood V_a and a C^r-map $h_a : V_a \to \mathbb{R}^n$ which agrees with h on $V_a \cap A$. Let V be the union of these sets V_a and let $\{p_a\}$ be a C^r-partition of unity on V associated with the covering $\{V_a\}$. Then the required extension $H : V \to \mathbb{R}^n$ is given by $H(x) = \sum_a p_a(x)h_a(x)$ (It is to be understood that $p_a(x)h_a(x) = 0$ for $x \notin V_a$.)

Now let the target space N^n be any unbounded C^r-manifold. According to Whitney [**13**, Lemma 19], there exists a C^r-embedding $i : N^n \to \mathbb{R}^{2n+1}$. By Lemma 23 of [**13**], there exists a neighborhood U of $i(N^n)$ in $\mathbb{R}^{2n+1}$, and a C^r-retraction $\rho : U \to i(N^n)$. By the previous remarks, we can extend $i \circ h : A \to \mathbb{R}^{2n+1}$ to a C^r-map $H : V \to \mathbb{R}^{2n+1}$ where V is some open neighborhood of A. Now $i^{-1} \circ \rho \circ H : H^{-1}(U) \to N^n$ is the required extension of h. □

Proof of Lemma A.2. From Lemma A.1 we see that, for some neighborhood V_1 of ∂M^n, there exists a C^r-retraction $\rho : V_1 \to \partial M^n$. (This follows by substituting M^n, ∂M^n, ∂M^n, and the identity map, respectively, for M^m, N^n, A and h in Lemma A.1.)

Let $\lambda : M^n \to \mathbb{R}$ be a C^r-map which vanishes on ∂M^n but is positive on $M^n \smallsetminus \partial M^n$ and has non-vanishing gradient on ∂M^n. (It is clearly possible to choose a C^r-map $\lambda_\alpha : U_\alpha \to \mathbb{R}$ satisfying this condition for each coordinate neighborhood U_α. If $\{p_\alpha\}$ is a C^r-partition of unity associated with $\{U_\alpha\}$, then

$$\lambda(x) = \sum p_\alpha(x)\lambda_\alpha(x)$$

will be the required function on M^n.)

Define the C^r-map $\mu : V_1 \to \partial M^n \times \mathbb{R}$ by $\mu(x) = (\rho(x), \lambda(x))$. Let V_2 denote the set of points x in V_1 such that the Jacobian of μ at x has the maximal rank n. It is clear that V_2 is a neighborhood of ∂M^n. If μ is not one-to-one on V_2 then we will construct a smaller neighborhood V_3 such that $\mu|_{V_3}$ is one-to-one.

Let D be the subset of $V_2 \times V_2$ consisting of all pairs (y_1, y_2) with $\mu(y_1) = \mu(y_2)$ but $y_1 \neq y_2$. Since $\mu|_{V_2}$ is a local homeomorphism, it follows that D is a closed set. Choose any metric $d(x, y)$ for the space V_2, and define the distance between points (x_1, x_2) and (y_1, y_2) of $V_2 \times V_2$ as $\mathrm{Max}(d(x_1, y_2), d(x_2, y_2))$. Define $\delta(x) > 0$ as the distance of the point (x, x) from the set D. Let V_3 be the set of all points x of V_2 with $d(x, \rho(x)) < \delta(\rho(x))$.

Clearly V_3 is a neighborhood of ∂M^n. If $\mu(x_1) = \mu(x_2)$ with $x_1, x_2 \in V_3$, then $\rho(x_1) = \rho(x_2)$. Since x_1 and x_2 have distance less than $\delta(\rho(x))$ from $\rho(x_1)$, it follows that $(x_1, x_2) \notin D$; and therefore $x_1 = x_2$. Thus $\mu|_{V_3}$ is one-to-one.

Choose a C^r-map $\epsilon : \partial M^n \to \mathbb{R}$ so that $\epsilon(y) > 0$ for all y, and so that the region $W = \{(y, t) \mid 0 \leq t < \epsilon(y)\}$ is contained in $\mu(V_3)$. (This can be constructed using a C^r-partition of unity on ∂M^n.) Let $V_4 = \mu^{-1}(W) \cap V_3$ and define $g : \partial M^n \times [0, 1) \to V_4$ by

$$g(y, t) = \mu^{-1}\Big(y, \epsilon(y)t\Big).$$

Then g is the required C^r-isomorphism. This proves Lemma A.2. □

Proof of Lemma A.3. Choose C^r-isomorphisms $g_i : \partial M_i^n \times [0,1) \to U_i$ as in Lemma A.2 ($i = 1, 2$). Let U be the open subset $j_1(U_1) \cup j_2(U_2)$ of M^n. A homeomorphism $g : \partial M_1^n \times (-1, 1) \to U$ is defined by

$$g(y,t) = \begin{cases} j_1 \circ g_1(y, t), & \text{if} \quad t \geq 0 \\ j_2 \circ g_2(h(y), -t), & \text{if} \quad t \leq 0. \end{cases}$$

In order to define a C^r-structure on a manifold, it is sufficient to define compatible C^r-structures on open sets covering the manifold. But M^n is covered by the sets

$$j_1(M_1^n \setminus \partial M_1^n), \quad j_2(M_2^n \setminus \partial M_2^n), \quad \text{and} \quad U.$$

Since the differentiable structures on these sets which are induced by j_1, j_2, and g respectively, are compatible, this completes the proof. □

Proof of Lemma A.4. This follows from Lemma A.3 by taking $M_2^n = M_1^n$, with h the identity map of ∂M_1^n. The resulting manifold M^n is called the *double* of M_1^n.

Appendix B. C^r-triangulations.*

A triangulation (K, f) of a C^r-manifold M^n will be called a C^r-*triangulation* if

(1) for each closed n-simplex Δ of $|K|$ the map $f|_\Delta$ is a C^r-map[3]; and

(2) this map $f|_\Delta$ has Jacobian of rank n at all points.

For every point x of $|K|$ let N_x denote the open star neighborhood of x in $|K|$, and let $T_{f(x)} = T_{f(x)}M^n$ denote the tangent space of M^n at $f(x)$. Following Whitehead, we define a linear map $F_x : N_x \to T_{f(x)}$ as follows. For each $y \in N_x$ there is a unique linear map $g_y : [0,1] \to N_x$ satisfying $g_y(0) = x, \quad g_y(1) = y$. Define $F_x(y)$ as the velocity vector of the curve $f \circ g_y : [0,1] \to M^n$ at 0.

LEMMA B.1. *If (K, f) is a C^r-triangulation of M^n then for each x in $|K|$ the map F_x is a homeomorphism of N_x into the Euclidean space $T_{f(x)}$.*

The relationship of our definition of C^r-triangulation to that of Whitehead can now be described as follows. Whitehead considers only a bounded manifold M^n which is embedded in an unbounded manifold $\widehat{M}^n$. By Lemma A.4, this is no restriction on M^n. He calls a triangulation (K, f) of M^n a C^r-triangulation if it satisfies the following two conditions:

(1′) For each j-simplex τ of $|K|$, the map $f|_\tau$ can be extended to a C^r-map of a neighborhood of τ in Euclidean j-space into $\widehat{M}^n$.

(2′) For each point x of $|K|$, the map F_x is a homeomorphism of N_x into $T_{f(x)}$.

---

*For a more detailed treatment of C^r-triangulations, see J. Munkres, "Elementary Differential Topology", Annals of Math. Studies **54**, Princeton U. Press 1961.

[3]See Appendix A. The simplex Δ is to be considered as a rectilinear subset of a Euclidean n-space.

It follows easily from Lemmas A.1 and B.1 that this definition is equivalent to the definition given above.

Proof of Lemma B.1. We first consider the case of a manifold without boundary. Recall that the ***closed star neighborhood*** $\overline{N}_x$ of a point $x \in |K|$ is the subcomplex which is the union of all (closed) simplices Δ of K which contain x. The ***open star neighborhood*** N_x is the open set obtained from $\overline{N}_x$ by removing those simplexes Δ of this subcomplex which do not themselves contain x. The argument begins as follows.

> **(1)** *If Δ be any n-simplex of $|K|$ which contains x, then the map F_x maps Δ linearly and homeomorphically onto its image in the tangent space $T_{f(x)}$.*

This follows immediately from the fact that the Jacobian of the restriction $f|_\Delta$ has maximal rank n at x,

> **(2)** *Any point $y \in N_x$ has a neighborhood $N_y \subset N_x$ with the property that $F_x(w) \neq F_x(y)$ for every $w \neq y$ in N_y.*

In fact it follows easily from the definition of star neighborhood that the points x, y, and w are all contained in some common (closed) simplex on which F_x is one-to-one.

Now suppose that w is an interior point of some n-dimensional simplex Δ which contains x. The straight line segment $g_w([0,1])$ from x to w then maps to a curve $t \mapsto f \circ g_w(t)$ in $f(\Delta)$. The following is now easy to verify:

> **(3)** *If $c : [0,1] \to M^n$ is any differentiable curve which is tangent to this curve $f \circ g_w$ at $t = 0$, then $c(t) \in f(\Delta)$ for small values of t.*

As a consequence:

> **(4)** *With $w \in \mathrm{interior}(\Delta)$ as above, we have $F_x(z) \neq F_x(w)$ for any $z \neq w$ which belongs to N_x.*

In fact, this follows from **(1)** whenever $z \in \Delta$. But if $z \notin \Delta$, then the fact that $F_x(z) = F_x(w)$ would imply that the curves $f \circ g_w([0,1])$ and $f \circ g_z([0,1])$ were tangent at x, thus contradicting **(3)**.

> **(5)** *The map $F_x : N_x \to T_{f(x)}$ is an open mapping.*

To see this, note that **(2)** implies that F_x has a well defined local degree at any point y in the manifold N_x. (See for example [**2**, XII, §2].) But this local degree can only be ± 1. In fact, for any point w in the interior of a simplex Δ containing y, it follows from **(4)** that $F_x(w)$ has a unique pre-image in N_x. Since a map with local degree ± 1 everywhere must be open, the conclusion follows.

We are now ready to prove that the linear map F_x is one-to-one on N_x. Suppose $F_x(y) = F_x(z)$ with $y \neq z$. Since F_x is open, it would follow that suitable small neighborhoods of y and z have the same image in $T_{f(x)}$. Since this clearly contradicts **(4)**, this completes the proof of Lemma B.1 in the case of a manifold without boundary. The general case then follows easily using Lemma A.4, and extending the triangulation appropriately. □

References

[1] J. W. Alexander, *The combinatorial theory of complexes*, Ann. of Math. **31** (1930), 292–320.
[2] P. Alexandroff and H. Hopf, "Topologie," Berlin, Springer, 1935.
[3] S. Bochner, *Analytic mapping of compact Riemann spaces into Euclidean space*, Duke Math. J. **3** (1937), 339–354.
[4] S. S. Cairns, *Triangulated manifolds which are not Brouwer manifolds*, Ann. of Math. **41** (1940), 809–824.
[5] S. S. Cairns, *Introduction of a Riemannian geometry on a triangulable 4-manifold*, Ann. of Math. **45** (1944), 218–219.
[6] J. Milnor, *On manifolds homeomorphic to the 7-sphere*, Ann. of Math. **64** (1956), 399–405.
[7] M. H. A. Newman, *On the foundations of combinatory analysis situs*, Nederl. Akad. Wetensch. Proc. **29** (1926), 611–626.
[8] R. G. de Rham, "Variétés différentiables," Paris, Hermann, 1955.
[9] R. Thom, *Quelques propriétés globales des variétés différentiables,* Comm. Math. Helv. **28** (1954), 17–86.
[10] R. Thom, *Les classes caractéristiques de Pontrjagin des variétés triangulées*, to appear.‡
[11] J. H. C. Whitehead, *On C^1-complexes*, Ann. of Math. **41** (1940), 809–824.
[12] H. Whitney, *Differentiable functions defined in arbitrary subsets of Euclidean space*, Trans. Am. Math. Soc. **40** (1936), 309–318.
[13] H. Whitney, *Differentiable Manifolds*, Ann. of Math. **37** (1936), 645–680.

‡"Symposium Internacional de Topología Algebraíca (International Symposium on Algebraic Topology) pp. 54–67, Universidad Nacional Autónoma de México and UNESCO, 1958.

Colloques intern. C. N. R. S. [89, 1959, Lille].
Bull. Soc. math. France,
87, 1959, p. 439 à 444.

SOMMES DE VARIÉTES DIFFÉRENTIABLES ET STRUCTURES DIFFÉRENTIABLES DES SPHÈRES;

PAR

JOHN W. MILNOR (1)
(Princeton).

La somme (sous-entendu somme *connexe*) de deux n-variétés est obtenue en enlevant de chacune d'elles l'intérieur d'une boule, puis en identifiant les bords que cette opération fait apparaître. Les deux lemmes qui suivent permettent de rendre cette définition rigoureuse.

LEMME 1. (J. CERF). — *Soit M une n-variété connexe, orientée. Soit f_1, f_2 deux plongements de la boule unité (fermée) D^n dans l'intérieur de M, si f_1 et f_2 conservent l'orientation, il existe un difféomorphisme $g: M \to M$ tel que $gf_1 = f_2$.*

La preuve se trouvera dans un article à paraître de CERF.

LEMME 2. (R. THOM). — *Soit M l'espace obtenu à partir de la réunion disjointe de deux variétés M_1 et M_2 en « recollant les bords » c'est-à-dire en identifiant tout $x \in \partial M_1$ avec $fx \in \partial M_2$, où $f: \partial M_1 \to \partial M_2$ est un difféomorphisme (2). Alors M peut être munie d'une structure différentiable compatible avec celles de M_1 et M_2. En plus, la variété différentiable ainsi obtenue est unique à un difféomorphisme près.*

Une preuve est indiquée en [3], p. 35.

Définition de la *somme* de deux n-variétés différentiables orientées connexes, M_1 et M_2 : on choisit deux plongements (différentiables)

(1) L'auteur bénéficie d'un Sloan Fellowship.

(2) Si les bords ont plusieurs composantes connexes, on peut désirer identifier une composante connexe de ∂M_1 avec une composante connexe de ∂M_2. Le lemme 2 est encore valable dans ce cas.

$f_i : D^n \to$ (intérieur de M_i) tels que f_1 conserve l'orientation et que f_2 l'inverse. Soit $M_1 \# M_2$ la variété obtenue à partir de la réunion disjointe :

$$(M_1 - \text{intérieur } f_1(D^n)) \cup (M_2 - \text{intérieur } f_2(D^n))$$

en recollant les bords à l'aide du difféomorphisme :

$$f_2 f_1^{-1} : f_1(S^{n-1}) \to f_2(S^{n-1}).$$

(Il est essentiel de bien utiliser ce difféomorphisme-là.)

On munit $M_1 \# M_2$ d'une structure différentiable compatible avec celles de $M_1 -$ (intérieur de $f_1(D^n)$) et $M_2 -$ (intérieur de $f_2(D^n)$). Cette variété est munie d'une orientation compatible avec celles de M_1 et M_2.

LEMME 3. — *La somme $M_1 \# M_2$ est bien définie à un difféomorphisme conservant l'orientation près. Cette opération satisfait à* (3) :

a) $M_1 \# M_2 \approx M_2 \# M_1$

b) $M_1 \# (M_2 \# M_3) \approx (M_1 \# M_2) \# M_3$ et

c) $M \# S^n \approx M.$

En plus, $M \# R^n$ est difféomorphe à M privé d'un point. $M \# D^n$ est difféomorphe à M privé de l'intérieur d'une n-boule plongée.

Les preuves sont immédiates.

Le théorème suivant est dû principalement à B. MAZUR [1].

THÉORÈME 1. — *Pour une variété différentiable compacte M les trois conditions suivantes sont équivalentes* :

(1) *Il existe une variété N telle que $M \# N \approx S^n$*

(2) $M \# R^n \approx R^n$ (autrement dit M privé d'un point et difféomorphe à R^n)

(3) *M est la réunion de deux ouverts, tous deux difféomorphes à R^n*

La preuve que (1) entraîne (2) est basée sur le concept de somme infinie $M_1 \# M_2 \# \ldots$ de variétés différentiables. Une telle somme se définit comme dans le cas fini, et jouit de la même propriété d'associativité. Par exemple, la somme infinie $S^n \# S^n \# \ldots$ est difféomorphe à R^n.

Supposons maintenant $M \# N \approx S^n$, et considérons l'identité :

$$(M \# N) \# (M \# N) \# \ldots \approx M \# (N \# M) \# (N \# M) \# \ldots.$$

D'une part, ceci est difféomorphe à

$$S^n \# S^n \# \ldots \approx R^n.$$

(3) La notation $M_1 \approx M_2$ signifie qu'il existe un difféomorphisme, conservant l'orientation, de M_1 sur M_2. Les notations : $S^{n-1} \subset D^n \subset R^n$ désignent la sphère unité bordant la boule unité dans l'espace euclidien.

D'autre part, c'est difféomorphe à

$$M \# S^n \# S^n \ldots \approx M \# R^n.$$

Donc :

$$M \# R^n \approx R^n.$$

Il est clair que l'assertion (2) entraîne (3). Preuve que (3) entraîne (1).

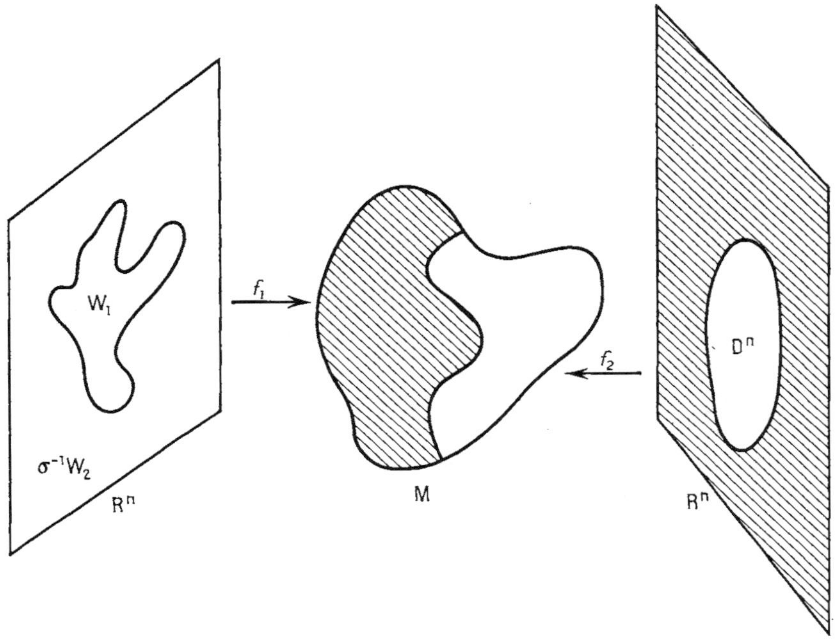

Supposons que $M = f_1(R^n) \cup f_2(R^n)$, où f_1 et f_2 sont des plongements, et considérons l'ensemble compact :

$$f_2^{-1}(M - f_1(R^n)) \subset R^n.$$

En faisant au besoin une homothétie, on peut supposer que cet ensemble est contenu dans l'intérieur de D^n. L'application : $x \to f_1^{-1} f_2(x)$ définit alors un plongement : $S^{n-1} \to R^n$. Soit $W_1 \subset R^n$ la variété compacte dont $f_1^{-1} f_2(S^{n-1})$ est le bord. On peut obtenir une variété difféomorphe à M à partir de la réunion disjointe de W_1 et D^n en recollant les bords à l'aide du difféomorphisme :

$$f_2^{-1} f_1 = \partial W_1 \to \partial D^n.$$

Soit maintenant σ un difféomorphisme : $R^n \to (S^n -$ un point), par exemple la projection stéréographique; et soit W_2 l'adhérence du complémentaire de $S^n - \sigma W_1$. Définissons N à partir de la réunion disjointe de D^n et W_2

en recollant les bords à l'aide du difféomorphisme :

$$\sigma f_1^{-1} f_2 : \partial D^n \to \partial W_2.$$

Alors $M \# N$ peut être obtenue à partir de la réunion disjointe de W_1 et W_2 en recollant les bords à l'aide du difféomorphisme :

$$(\sigma f_1^{-1} f_2)(f_2^{-1} f_1) = \sigma : \partial W_1 \to \partial W_2.$$

Autrement dit : $M \# N$ est difféomorphe à $\sigma W_1 \cup W_2 = S^n$.

Ceci complète la preuve du théorème 1.

Corollaire 1. — *Une variété qui vérifie l'une des conditions* (1), (2), (3) *est nécessairement homéomorphe à* S^n.

Preuve : Cela résulte de la condition (2).

Corollaire 2. — *Soit* A^n *l'ensemble de toutes les variétés orientées (définies à un difféomorphisme conservant l'orientation près) qui vérifient* (1), (2) *ou* (3). *Alors* A^n *est un groupe abélien relativement à l'opération de composition.*

Preuve : Cela résulte de la condition (1).

Thom a montré que les deux conditions suivantes relatives à une variété M sont équivalentes (*cf.* [7], [4]).

a) M est obtenu à partir de deux exemplaires de D^n, en recollant les bords à l'aide d'un certain difféomorphisme (c'est-à-dire, $M \# D^n \approx D^n$).

b) M est combinatoirement équivalente à la sphère; c'est-à-dire qu'il existe une C^1-triangulation $f : K \to M$, où K est une subdivision rectiligne de la frontière d'un $(n+1)$-simplexe (*cf.* Whitehead [8]).

Les variétés satisfaisant à ces conditions définissent un sous-groupe $\Gamma^n \subset A^n$. Il serait extrêmement intéressant d'avoir des renseignements sur la structure de chacun de ces deux groupes. Voici un résumé de ce qu'on sait :

a) A^n est dénombrable (prouvé par Thom),
b) $\Gamma^n = 0$ pour $n \leq 4$, $A^n = 0$ pour $n \leq 3$ (*cf.* Smale [5]),
c) Γ^{4k-1} est non-trivial pour $k = 2$ ou $4 \leq k \leq 14$ (*cf.* [2]).

La démonstration, qui s'applique aussi bien à des valeurs de k plus grandes, fournit une borne inférieure spécifique pour l'ordre de Γ^{4k-1}. Par exemple, Γ^{31} contient au moins 16 931 177 éléments.

Il semble très difficile d'obtenir une borne supérieure pour la grandeur de ces groupes. (Par exemple, ont-ils un nombre fini de générateurs ?) La difficulté vient de ce que la question de décider si une variété donnée est difféomorphe à S^n est trop ardue. Le problème devient beaucoup plus facile si la relation de difféomorphisme est remplacée par le concept suivant, dû à THOM [6] :

DÉFINITION. — Deux variétés orientées sans bord M_1, M_2 sont *J-équivalentes*, s'il existe une variété différentiable à bord W telle que :

1° ∂W est la réunion disjointe de M_1 avec l'orientation donnée, et de M_2 avec l'orientation inverse de l'orientation donnée.

2° Les applications d'inclusion : $M_1 \to W$ et $M_2 \to W$ sont l'une et l'autre des homotopie-équivalences.

Avec cette définition, deux variétés J-équivalentes appartiennent à la même classe de cobordisme et ont le même type d'homotopie.

LEMME 4. — *Soit donnée une variété différentiable sans bord M; il existe N tel que $M \# N$ est J-équivalent à S^n si, et seulement si M a le type d'homotopie de S^n. Les classes de J-équivalence des sphères homotopiques forment un groupe abélien relativement à l'opération $\#$.*

On note ce groupe Θ^n.

THÉORÈME 2. — *Pour n impair et différent* ([4]) *de 3, le groupe Θ^n est fini.*

Pour n de la forme $4k-1$, on peut donner des renseignements plus détaillés. Soit B_k le k-ième nombre de Bernoulli.

THÉORÈME 3. — *Le groupe Θ^{4k-1}, $k > 1$, contient un sous-groupe cyclique d'ordre $2^{2k-2}(2^{2k-1}-1)$ (numérateur de B_k/k). L'indice de ce sous-groupe est un produit de nombres premiers qui sont petits au sens que $p(p-1)^2 < 2k$.*

Par exemple, Θ^7 est cyclique d'ordre 28, et Θ^{11} est cyclique d'ordre 992. Pour les plus grandes valeurs de k, je ne connais aucun Θ^{4k-1} entièrement.

D'autres précisions sur les groupes Θ^n sont données en [3].

BIBLIOGRAPHIE.

[1] MAZUR (B.). — *On embeddings of spheres*, (*Bull. Amer. math. Soc.*, t. 65, 1959, p. 59-65).

[2] MILNOR (J.). — *Differentiable structures on spheres*, (*Amer. J. of Math.*) (à paraître). *

([4]) Si la conjecture de Poincaré est exacte, alors $\Theta^3 = 0$.

* This volume pp. 35-45.

[3] MILNOR (J.). — *Differentiable manifolds which are homotopy spheres*. Princeton, Princeton University, 1959 (multigraphié). *
[4] MUNKRES (J.). — *Obstructions to the smoothing of piecewise differentiable homeomorphisms*, (*Bull. Amer. math. Soc.*, t. 65, 1959, p. 332-334).
[5] SMALE (S.). — *The space of diffeomorphisms of the 3-sphere* (à paraître). +
[6] THOM (R.). — *Les classes caractéristiques de Pontrjagin des variétés triangulées Symposium internacional de Topologia algebraica* [1956, Mexico]. Mexico, Universidad nacional autonomia, 1958; p. 54-67.
[7] THOM (R.). — *Les structures différentiables des boules et des sphères* (à paraître). ++
[8] WHITEHEAD (J. H. C.). — *On C^1-complexes* (*Annals of Math.*, t. 41, 1940, p. 809-824).

John W. MILNOR,
Princeton University,
Princeton, N. J. (États-Unis).

* This volume pp. 65-87.
\+ Correction: This should be 2-sphere; not 3-sphere. See SMALE [1959b] in the bibliography.
++ See THOM [1959].

DIFFERENTIABLE STRUCTURES ON SPHERES.*

By John Milnor.[1]

According to [5] the sphere S^7 can be given several differentiable structures which are essentially distinct. A corresponding result for the 15-sphere has been proved by Shimada [10] and Tamura [12]. The object of this note is to prove corresponding theorems for other dimensions of the form $4k-1$.

In § 1 certain differentiable manifolds $M(f_1, f_2)$ are constructed and studied; where $(f_1) \in \pi_m(SO_{n+1})$, $(f_2) \in \pi_n(SO_{m+1})$. In most cases these manifolds are topologically spheres. In § 2 an invariant λ is defined for differentiable $(4k-1)$-manifolds which are both homology spheres and boundaries. In § 3 the invariant $\lambda(M(f_1, f_2))$ is computed.

For $k \leqq 8$ the calculations are carried out explicitly. It is shown that there exist non-standard differentiable structures on S^{4k-1} for $k = 2, 4, 5, 6, 7, 8$. For example S^{31} has over sixteen million distinct differentiable structures. It is conjectured that the same argument works for all $k \geqq 4$; but I have not succeeded in solving the number theoretic problem which arises.

For $k = 1, 3$ the argument does not work. This is not surprising in the case $k = 1$, since J. Munkres, S. Smale, and J. H. C. Whitehead have shown (independently) that two differentiable 3-manifolds which are homeomorphic must necessarily be diffeomorphic.

The word *manifold* will always be used for a compact, oriented manifold, with or without boundary. The symbol D^k will stand for the unit disk in the euclidean space R^k.

1. Construction of manifolds homeomorphic to spheres. Given any diffeomorphism [2] $f\colon S^m \times S^n \to S^m \times S^n$, a manifold M of dimension $m+n+1$ is obtained by matching the boundaries of $D^{m+1} \times S^n$ and $S^m \times D^{n+1}$ under the correspondence f. That is: M is the identification space obtained from the disjoint union of $D^{m+1} \times S^n$ and $S^m \times D^{n+1}$ by identifying each point (x, y) in the boundary of $D^{m+1} \times S^n$ with $f(x, y)$, considered as a point in boundary of $S^m \times D^{n+1}$.

* Received February 10, 1959.

[1] The author holds a Sloan fellowship.

[2] A *diffeomorphism* is a differentiable homeomorphism with differentiable inverse.

An alternative definition of M, which makes it into a differentiable manifold, is the following. Let $f(x,y)=(x',y')$ and define $t'=1/t$. Start with disjoint spaces $R^{m+1}\times S^n$ and $S^m\times R^{n+1}$. Let M be obtained from these by matching (tx,y) with $(x',t'y')$ for every $x\in S^m$, $y\in S^n$, $0<t<\infty$.

[As an example suppose that f is the identity map of $S^m\times S^n$. Then M is diffeomorphic to the unit sphere $S^{m+n+1}\subset R^{m+1}\times R^{n+1}$. In fact the correspondence

$$(tx,y)\to(tx/(1+t^2)^{\frac{1}{2}},y/(1+t^2)^{\frac{1}{2}}),$$

$$(x',t'y')\to(x'/(1+t'^2)^{\frac{1}{2}},t'y'/(1+t'^2)^{\frac{1}{2}})$$

defines a diffeomorphism $M\to S^{m+n+1}$.]

If $y\in S^n$ has coordinates $(y_0,\cdots,y_n)$, define $h(y)=y_n$. The function $h\colon S^n\to[-1,1]$ has just two critical points.

LEMMA 1. *Suppose that the diffeomorphism*

$$(x,y)\xrightarrow{f}(x',y')$$

satisfies the restriction $h(y)=h(y')$ for all (x,y). Then the manifold M constructed above is homeomorphic to S^{m+n+1}.

Proof. A differentiable map $g\colon M\to[-1,1]$ is defined by the correspondence

$$(tx,y)\to h(y)/(1+t^2)^{\frac{1}{2}}\text{ (in the first coordinate system)},$$

$$(x',t'y')\to t'h(y')/(1+t'^2)^{\frac{1}{2}}\text{ (in the second)}.$$

It is easily verified that g has just two critical points, and that these are non-degenerate. Together with [5] Theorem 2, this completes the proof.

One way to construct such a diffeomorphism $(x,y)\to(x',y')$ is the following. Start with differentiable maps of spheres into rotation groups

$$f_1\colon S^m\to SO_{n+1},\qquad f_2\colon S^n\to SO_{m+1},$$

and let

$$y'=f_1(x)\cdot y,\qquad x'=f_2(y')^{-1}\cdot x=f_2(f_1(x)\cdot y)^{-1}\cdot x$$

for all $x\in S^m$, $y\in S^n$. This defines a diffeomorphism with inverse

$$x=f_2(y')\cdot x',\qquad y=f_1(x)^{-1}\cdot y'=f_1(f_2(y')\cdot x')^{-1}\cdot y'.$$

[More generally the rotation groups SO_{n+1} and SO_{m+1} could be replaced by the groups Diff S^n, Diff S^m consisting of all diffeomorphisms.] The condition $h(y)=h(y')$ is equivalent to the requirement that $f_1(S^m)$ be contained in the subgroup $SO_n \subset SO_{n+1}$. Whether this condition is satisfied or not, *the resulting* $(m+n+1)$*-manifold will be denoted by* $M(f_1, f_2)$.

Next we will show that $M(f_1, f_2)$ is a boundary. Start with three copies of the space $D^{m+1} \times D^{n+1}$. The notation (x_i, y_i) will be used for a point of $(D^{m+1} \times D^{n+1})_i$, $i = 1, 2, 3$. Identify $(S^m \times D^{n+1})_1$ with $(S^m \times D^{n+1})_2$ by the correspondence $(x_1, y_1) \to (x_2, y_2)$ where

$$x_2 = x_1, \qquad y_2 = f_1(x_1) \cdot y_1.$$

The resulting space $(D^{m+1} \times D^{n+1})_1 \cup (D^{m+1} \times D^{n+1})_2$ can be considered as a fibre bundle over the $(m+1)$-sphere $(D^{m+1})_1 \cup (D^{m+1})_2$ with fibre D^{n+1}, group SO_{n+1}, and characteristic map f_1. (See Steenrod [11] § 18.) It follows that this union can be given a differentiable structure in a natural way.

Next identify $(D^{m+1} \times S^n)_2$ with $(D^{m+1} \times S^n)_3$ by the correspondence $(x_2, y_2) \leftarrow (x_3, y_3)$, where $y_2 = y_3$, $x_2 = f_2(y_3) \cdot x_3$. Thus

$$(D^{m+1} \times D^{n+1})_2 \cup (D^{m+1} \times D^{n+1})_3$$

becomes a fibre bundle over an $(n+1)$-sphere with fibre D^{m+1} and characteristic map f_2.

Let W_1 denote the union of all three copies of $D^{m+1} \times D^{n+1}$. Clearly W_1 is a topological manifold with boundary:

$$\partial W_1 = (D^{m+1} \times S^n)_1 \cup (S^m \times D^{n+1})_3.$$

The intersection

$$(D^{m+1} \times S^n)_1 \cap (S^m \times D^{n+1})_3$$

of the two halves of the boundary is equal to $(S^m \times S^n)_1 = (S^m \times S^n)_3$. These two copies of $S^m \times S^n$ are identified under the composite correspondence

$$(x_1, y_1) \to (x_2, y_2) \to (x_3, y_3),$$

where

$$y_3 = y_2 = f_1(x_1) \cdot y_1, \qquad x_3 = f_2(y_3)^{-1} \cdot x_2 = f_2(y_3)^{-1} \cdot x_1.$$

But this is just the correspondence which was used to define the manifold $M(f_1, f_2)$. Thus ∂W_1 is homeomorphic to $M(f_1, f_2)$.

As it stands W_1 is not a differentiable manifold since there is an "angle" along the subset $(S^m \times S^n)_1$ of ∂W_1. Let W denote a differentiable manifold obtained by "straightening" this angle. (See the appendix to [7].) Clearly ∂W is diffeomorphic to $M(f_1, f_2)$.

2. The invariant $\lambda(M)$. First recall the index theorem of Hirzebruch [4]. If M_1 is a $4k$-manifold without boundary having Pontrjagin classes

$p_1, \cdots, p_k$, then the index $I(M_1)$ is equal to $L_k(p_1, \cdots, p_k)[M_1]$; where L_k is a certain polynomial.[3] For example

$$L_1 = p_1/3, \qquad L_2 = (7p_2 - p_1^2)/45, \cdots.$$

The coefficient s_k of p_k in L_k is particularly important. Hirzebruch expresses s_k in terms of the Bernoulli number B_k as follows (page 14):

$$s_k = 2^{2k}(2^{2k-1} - 1)B_k/(2k)!.$$

For example $s_2 = 7/45$, $s_3 = 62/945$, $s_4 = 127/4725$.

Now let M be a differentiable $(4k-1)$-manifold which

1) has the same rational homology groups as the $(4k-1)$-sphere, and

2) is a boundary: $M = \partial W$ with W differentiable.[4] Then a rational number modulo 1,

$$\lambda(M) \in Q/Z,$$

is defined as follows. The natural homomorphism

$$j \colon H^i(W, M; Q) \to H^i(W; Q)$$

is an isomorphism for $0 < i < 4k-1$. Hence the Pontrjagin classes $p_1, \cdots, p_{k-1}$ of W can be lifted back to $H^*(W, M; Q)$. Define $\lambda(M)$ as the residue class of

$$(I(W) - L_k(j^{-1}p_1, \cdots, j^{-1}p_{k-1}, 0)[W])/s_k$$

modulo 1. (Here the symbol $[W]$ stands for the homomorphism $H^{4k}(W, M; Q) \to Q$ associated with the orientation of W; and $I(W)$ denotes the index of the quadratic form $\alpha \to (\alpha \cup \alpha)[W]$, where $\alpha \in H^{2k}(W, M; Q)$.)

LEMMA 2. *This residue class $\lambda(M)$ is an invariant of M: that is it does not depend on the choice of W.*

The proof is completely analogous to that in [5], [10] or [12]. If M is the boundary of both W_1 and W_2, then an unbounded $4k$-manifold M_1 is obtained from W_1, W_2 by

---

[3] The symbol $[M_1]$ is used to denote the homomorphism of $H^{4k}(M_1; Q)$ into the rational numbers Q which is determined by an orientation for M_1. The index $I(M_1)$ is defined as the index of the quadratic form over $H^{2k}(M_1; Q)$ which is given by the formula $\alpha \to (\alpha \cup \alpha)[M_1]$.

[4] This second condition follows automatically if $H_*(M; Z)$ has no torsion. In fact every homology $(4k-1)$-sphere is a π-manifold (see [7]), and every π-manifold is a boundary (see [8]).

1) reversing the orientation of W_2;

2) matching W_1 and W_2 along the common boundary M;

3) constructing a differentiable structure in a neighborhood of $W_1 \cap W_2 = M$. (See [6] Lemma 4 or [7]). Then $I(M_1) = I(W_1) - (W_2)$; and each Pontrjagin number $p_{i_1} \cdots p_{i_r}[M_1]$ other than $p_k[M_1]$ is equal to the difference of corresponding Pontrjagin numbers for W_1, W_2. Now the index theorem for M_1 implies that the two definitions of $\lambda(M)$ differ only by the integer $p_k[M_1]$.

Example 1. For the $(4k-1)$-sphere it is clear that

$$\lambda(S^{4k-1}) \equiv 0 \pmod 1.$$

Example 2. For a 3-manifold the definition reduces to

$$\lambda(M^3) \equiv 3 \cdot I(W) \equiv 0 \pmod 1.$$

Example 3. For the 7-manifold $M_3{}^7$ of [5] the values

$$I(W) = 1, \quad (j^{-1}p_1)^2[W] = 36$$

give

$$\lambda(M_3{}^7) \equiv (45I(W) + (j^{-1}p_1)^2[W])/7 \equiv 4/7 \pmod 1.$$

Remark. If $H^*(M;Z)$ has no torsion, then the classes $j^{-1}p_i$ can be considered as integral cohomolgy classes, hence the Pontrjagin numbers of W are integers. This sharply restricts the denominator which $\lambda(M)$ can have. (For example $7\lambda(M^7)$ must be an integer.) On the other hand, if $H^*(M;Z)$ has torsion then arbitrarily large denominators may occur. (See the examples studied by Tamura.)

Example 4. In [7] § 4 certain homotopy spheres $M_0{}^{4k-1}$ are constructed for $k > 1$. These have the property that $M_0{}^{4k-1} = \partial W$, where W is parallelizable, and $I(W) = 8$. Thus

$$\lambda(M_0{}^{4k-1}) \equiv 8/s_k \pmod 1.$$

For $k = 2$ this gives $\lambda(M_0{}^7) \equiv 3/7$ with denominator 7. For $k = 3, 4, 5, 6, 7$ the denominator of $\lambda(M_0{}^{4k-1})$ is 31, 127, 73, 1414477, and 8191 respectively. (These numbers are prime, except for $1414477 = 23 \cdot 89 \cdot 691$.) I do not know whether the inequality $8/s_k \not\equiv 0 \pmod 1$ holds for all $k > 1$.

In conclusion, the following three properties of the invariant λ are easily verified.

1) If the orientation of M is reversed, then λ changes sign.

2) For the connected sum of manifolds (see [7]), λ satisfies

$$\lambda(M_1 \# M_2) \equiv \lambda(M_1) + \lambda(M_2) \quad (\text{mod } 1).$$

3) λ is an invariant of the J-equivalence class of M. (See Thom [13] or [7].)

3. Computation of $\lambda(M(f_1, f_2))$. Define the *Pontrjagin homomorphism*

$$p_r\colon \pi_{4r-1}(SO_q) \to Z$$

as follows. Every map $f\colon S^{4r-1} \to SO_q$ induces a bundle ξ over S^{4r} with Pontrjagin class $p_r(\xi) \in H^{4r}(S^{4r}; Z) \approx Z$. Define $p_r(f)$ as the corresponding integer $p_r(\xi)[S^{4r}]$.

Let $f_1\colon S^m \to SO_{n+1}$, $f_2\colon S^n \to SO_{m+1}$ be arbitrary differentiable maps, with $m + n + 1 = 4k - 1$. First suppose that $m \neq n$.

LEMMA 3. *If $m \neq n$ then $M(f_1, f_2)$ is a topological sphere. The invariant $\lambda(M(f_1, f_2))$ is zero if m, n are not of the form $4r - 1$. If $m = 4r - 1$, $n = 4(k-r) - 1$, then*

$$\lambda \equiv \pm p_r(f_1)\, p_{k-r}(f_2)\, s_r s_{k-r}/s_k \qquad (\text{mod } 1).$$

Proof. We may assume that $m < n$. The exact sequence

$$\pi_m(SO_n) \to \pi_m(SO_{n+1}) \to \pi_m(S^n) = 0$$

implies that f_1 is homotopic to a map f_1' which carries S^m into the subset $SO_n \subset SO_{n+1}$. According to Lemma 1 the manifold $M(f_1', f_2)$ is homeomorphic to S^{m+n+1}. But it can be verified that $M(f_1, f_2)$ is homeomorphic to $M(f_1', f_2)$, and therefore is also homeomorphic to the sphere.

Next consider the manifold W constructed in Section 1. Recall that W is the union of a fibre bundle over S^{m+1} with fibre D^{n+1} and a fibre bundle over S^{n+1} with fibre D^{m+1}. Call these sets W_2 and W_3 respectively, Thus $W_2 \cup W_3$ is W and $W_2 \cap W_3$ is a topological cell.

These bundles have canonical cross-sections corresponding to the center point of the disk. Hence S^{m+1} and S^{n+1} are imbedded in W. It follows easily that W has the same homology groups as $S^{m+1} \vee S^{n+1}$ (the union with a single point in common). That is $H_i(W; Z)$ is infinite cyclic for i equal to 0, $m+1$, or $n+1$, and zero otherwise.

The homology intersection ring of W (see Lefschetz [14]) is described as follows. Let a and b stand for generators in dimensions $n+1$, $m+1$ respectively. Clearly a and b have intersection number ± 1. The self-intersections $a \cdot a$ and $b \cdot b$ are zero. For example $a \cdot a$ is represented by a cycle of dimension

$$\dim a + \dim a - \dim W = n - m$$

which lies on the sphere $S^{n+1} \subset W_3$. Since $H_{n-m}(S^{n+1}; Z) = 0$, this cycle is homologous to zero.

Applying Poincaré duality it follows that *$H^*(W, M; Z)$ is free abelian on three generators, say α in dimension $m+1$, β in dimension $n+1$, and $\alpha\beta$ in dimension $m+n+2$. The cup products $\alpha\alpha$ and $\beta\beta$ are zero.* This implies that the index $I(W)$ is zero.

Computation of the Pontrjagin numbers of W. We may assume that $m = 4r - 1$, $n = 4k - 4r - 1$. (If the dimensions are not of this form, then the Pontrjagin numbers are certainly zero, hence $\lambda \equiv 0$.) First consider the tangent bundle of W_2. This splits into a Whitney sum $\xi \oplus \eta$, where ξ is the bundle of vectors tangent to the fibre and η is the bundle of vectors normal to the fibre. Restricting η to the sphere $S^{m+1} \subset W_2$ we obtain the tangent bundle of S^{m+1} with trivial Pontrjagin classes. Restricting ξ to S^{m+1} we obtain the bundle determined by $(f_1) \in \pi_m(SO_{n+1})$. Thus $p_r(W_2) = p_r(\xi)$ is equal to the integer $p_r(f_1)$ multiplied by a generator of the infinite cyclic group $H^{4r}(W_2; Z)$. Using the isomorphisms

$$H^{4r}(W_2; Z) \leftarrow H^{4r}(W; Z) \xleftarrow{\ j\ } H^{4r}(W, M; Z)$$

it follows that

$$p_r(W) = \pm\, p_r(f_1) j(\alpha).$$

Similarly

$$p_{k-r}(W) = \pm\, p_{k-r}(f_2) j(\beta).$$

Thus *the Pontrjagin number $(j^{-1}p_r)(j^{-1}p_{k-r})[W]$ is equal to $\pm\, p_r(f_1) p_{k-r}(f_2)$.* All other Pontrjagin numbers of W are zero (except $(j^{-1}p_k)[W]$ which is not defined).

Computation of the coefficients of $p_r p_{k-r}$ in the Hirzebruch polynomial L_k. Define the symmetric function $\sum t_1^{i_1} \cdots t_a^{i_a}$ in indeterminates $t_1, \cdots, t_N$ as the sum of all monomials which can be obtained from $t_1^{i_1} \cdots t_a^{i_a}$ by permuting $t_1, \cdots, t_N$. Each possible monomial should be included only once in the sum. (For example $\sum t_1^r = t_1^r + \cdots + t_N^r$.) Hirzebruch showed[5] that the coefficient of $p_{i_1} \cdots p_{i_a}$ in L_k can be expressed in the form $\sum t_1^{i_1} \cdots t_a^{i_a}$, where $t_1, \cdots, t_N$ are certain fixed complex numbers. (Here N stands for

[5] See [4] § 1. 4. 1.

some fixed integer greater than or equal to k.) In particular, the coefficient s_k of p_k is equal to $\sum t_1{}^k$.

The product rule

$$(\sum t_1{}^r)(\sum t_1{}^{k-r}) = \sum t_1{}^k + \sum t_1{}^r t_2{}^{k-r} \qquad \text{for } r \neq k-r$$

is easily verified. Hence *the coefficient* $\sum t_1{}^r t_2{}^{k-r}$ *of* $p_r p_{k-r}$ *in* L_k *is equal to* $s_r s_{k-r} - s_k$.

Thus we have $I(W) = 0$ and

$$L_k(j^{-1}p_1, \cdots, j^{-1}p_{k-1}, 0)[W] = \pm p_r(f_1) p_{k-r}(f_2)(s_r s_{k-r} - s_k).$$

Dividing by s_k and reducing modulo one, this yields the required formula

$$\lambda(M) \equiv \pm p_r(f_1) p_{k-r}(f_2) s_r s_{k-r}/s_k \pmod 1$$

Now consider the case $m = n$. Again it is necessary to assume that m has the form $4r-1$ in order to obtain a non-trivial λ.

LEMMA 4. *If the maps* f_1, f_2 *both carry* S^m *into the subgroup* $SO_m \subset SO_{m+1}$, *then the formula*

$$\lambda(M) \equiv p_r(f_1) p_r(f_2) s_r s_r / s_{2r}$$

holds, just as in Lemma 3.

Proof. Just as above, $H_*(W;Z)$ is isomorphic to $H_*(S^{m+1} \vee S^{m+1})$. If $b, a \in H_{m+1}(W;Z)$ are the generators corresponding to the two spheres, then the intersection number $a \cdot b$ is ± 1. The hypothesis $f_1(S^m) \subset SO_m$ implies that the normal bundle of the first $(m+1)$-sphere in W has a cross-section. Hence the self-intersection number $a \cdot a$ is zero. Similarly $b \cdot b = 0$. It follows that W has index zero.

The computation of Pontrjagin classes for W proceeds as before. Thus

$$p_r(W) = \pm p_r(f_1) j\alpha \pm p_r(f_2) j\beta.$$

However the Pontrjagin number $(j^{-1}p_r)^2[W]$ is now equal to $\pm 2 p_r(f_1) p_r(f_2)$. On the other hand, the coefficient of $p_r p_r$ in L_{2r} is equal[5] to $\frac{1}{2}(s_r s_r - s_{2r})$. Thus the factor of $\frac{1}{2}$ cancels the 2, so that

$$\lambda(M(f_1 f_2)) \equiv \pm p_r(f_1) p_r(f_2) s_r s_r / s_{2r}$$

as before.

In order to make use of Lemmas 3, 4 it is necessary to know what integers $p_r(f)$ can occur.

THEOREM OF BOTT [2], [3]. *In the stable range $q \geqq 4r$ the Pontrjagin homomorphism*

$$p_r\colon \pi_{4r-1}(SO_q) \to Z$$

has image generated by

$$(2r-1)! \text{ if } r \text{ is even}$$

$$2(2r-1)! \text{ if } r \text{ is odd.}$$

For smaller values of q this result can be augmented as follows.

LEMMA 5. *If $q \leqq 2r$ then the homomorphism p_r is zero. If $q > 2r$ then p_r is non-zero. In fact there exists an element*

$$(f) \in \pi_{4r-1}(SO_q)$$

such that the prime factors of $p_r(f)$ are all less than $2r$.

Proof by descending induction on q. Suppose that the assertion has been proved for $q+1$, and that $q > 2r$. In the exact sequence

$$\pi_{4r-1}(SO_q) \to \pi_{4r-1}(SO_{q+1}) \to \pi_{4r-1}(S^q),$$

the third group is stable. According to Serre [9] a prime π can divide the order of this group only if $2\pi - 3$ is less than or equal to the difference $4r - q - 1$. The inequalities $2\pi - 3 \leqq 4r - q - 1$, $q > 2r$, yield $\pi \leqq r$. Thus any element of $\pi_{4r-1}(SO_{q+1})$, after being multiplied by primes less than or equal to r, can be lifted back to $\pi_{4r-1}(SO_q)$. This completes the induction.

If $q < 2r$, then the Pontrjagin class p_r of any SO_q-bundle is zero. If $q = 2r$, then $p_r(\xi^{2r})$ is the square of the Euler class of ξ^{2r}. (See Borel and Serre [1].) Since our base space is S^{4r}, this implies that $p_r(\xi^{2r}) = 0$; which completes the proof of Lemma 5.

Combining Lemmas 3, 4, 5 this proves:

THEOREM 1. *Suppose that r is an integer satisfying*

$$k/3 < r \leqq k/2.$$

Then there exists a differentiable manifold M homeomorphic to S^{4k-1} for which $\lambda(M)$ is congruent modulo 1 *to $s_r s_{k-r}/s_k$ times some integer with prime factors all less than $2(k-r)$.*

The proof is straightforward. (The inequality $k/3 < r$ guarantees the existence of a map $f_2\colon S^{4(k-r)-1} \to SO_{4r}$ such that $p_{k-r}(f_2) \neq 0$.)

Note. Given k, the inequality $k/3 < r \leqq k/2$ has a solution r providing that $k = 2$ or $k \geqq 4$. It has no solution for $k = 1$ or 3.

THEOREM 2. *There exist at least 7 distinct differentiable structures on S^7; at least:*

$$\begin{aligned} 127 &\ \textit{on the 15-sphere,} \\ 73 &\ \textit{on the 19-sphere,} \\ 23 \cdot 89 \cdot 691 &\ \textit{on the 23-sphere,} \\ 8191 &\ \textit{on the 27-sphere, and at least} \\ 31 \cdot 151 \cdot 3617 &\ \textit{on the 31-sphere.} \end{aligned}$$

Proof. These results follow immediately from Theorem 1. As an example, for $k = 5$, taking $r = 2$, we have

$$s_2 s_3 / s_5 = 341/365.$$

Cancelling all prime factors less than 6 from the denominator, this leaves 73. But if M is a 19-manifold such that the denominator of $\lambda(M)$ is 73, then the first 73 manifolds

$$S^{19}, M, M \# M, M \# M \# M, \cdots$$

must be pairwise distinct. (Alternatively, if the homotopy class (f_1) is replaced by $q(f_1)$, $0 \leqq q < 73$, then we obtain 73 different values for the invariant λ.) Each one represent a possible differentiable structure for the 19-sphere.

In conclusion, here are two unsolved problems.

Problem 1. Does Theorem 1 imply the existence of non-standard differentiable structures on S^{4k-1} for all $k \geqq 4$? I have checked this only for k up to 14.

Problem 2. Is the invariant $\lambda(M^{4k-1})$ of a homotopy sphere always a multiple of the invariant

$$\lambda(M_0^{4k-1}) \equiv 8/s_k?$$

This question is of interest since, for any manifold M^{4k-1} which bounds a parallelizable manifold W, we have

$$\lambda(M^{4k-1}) \equiv I(W)/s_k \pmod 1,$$

and it can be shown that $I(W)$ is a multiple of 8. (Compare [7].)

REFERENCES.

[1] A. Borel and J. P. Serre, "Groupes de Lie et puissances réduites de Steenrod," *American Journal of Mathematics*, vol. 75 (1953), pp. 409-448.

[2] R. Bott, "The space of loops on a Lie group," *Michigan Mathematical Journal*, vol. 5 (1958), pp. 35-61.

[3] ——— and J. Milnor, "On the parallelizability of the spheres," *Bulletin of the American Mathematical Society*, vol. 64 (1958), pp. 87-89.

[4] F. Hirzebruch, *Neue topologische Methoden in der algebraischen Geometrie*, Springer, 1956.

[5] J. Milnor, "On manifolds homeomorphic to the 7-sphere," *Annals of Mathematics*, vol. 64 (1956), pp. 399-405.

[6] ———, "On the relationship between differentiable manifolds and combinatorial manifolds," (mimeographed), Princeton University, 1956. *

[7] ———, "Differentiable manifolds which are homotopy spheres," (mimeographed), Princeton University, 1959. +

[8] ———, "On the cobordism ring Ω^*, and a complex analogue," (in preparation). ++

[9] J. P. Serre, "Homologie singulière des espaces fibrés," *Annals of Mathematics*, vol. 54 (1951), pp. 425-505.

[10] N. Shimada, "Differentiable structures on the 15-sphere and Pontrjagin classes of certain manifolds," *Nagoya Mathematical Journal*, vol. 12 (1957), pp. 59-69.

[11] N. Steenrod, *The topology of fibre bundles*, Princeton, 1951.

[12] I. Tamura, "Homeomorphy classification of total spaces of sphere bundles over spheres," *Journal of the Mathematical Society of Japan*, vol. 10 (1958), pp. 29-43.

[13] R. Thom, "Les classes caractéristiques de Pontrjagin des variétés triangulées," *Topología Algebraica*, Mexico, 1958, pp. 54-67.

[14] S. Lefschetz, *Topology*, New York, Chelsea, 1956.

---

* This volume pp. 19-27.
\+ This volume pp. 65-87.
++ This volume pp. 257-273.

Reprinted from
Proceedings of Symposia in Pure Mathematics
volume 3, *Differential Geometry*
American Mathematical Society
1961

A PROCEDURE FOR KILLING HOMOTOPY GROUPS OF DIFFERENTIABLE MANIFOLDS

BY

JOHN MILNOR[1]

A "surgery" on a differentiable manifold W of dimension $n = p + q + 1$ has the effect of removing an imbedded sphere of dimension p from W, and replacing it by an imbedded sphere of dimension q (see §1). This construction is closely related to cobordism theory (§2). The main objective of this paper (§§3–6) is to study the extent to which this construction can be used to simplify the homotopy groups of a given manifold. A typical result is the following.

COROLLARY TO THEOREM 3. *Let W be a triangulated differentiable manifold of dimension $2m$. Suppose that W is compact and that its tangent bundle restricted to the $(m - 1)$-skeleton is a trivial bundle. Then by performing a series of surgeries on W, one can obtain a manifold W' which is $(m - 1)$-connected.*

CONVENTIONS. All manifolds are to be oriented and differentiable of class C^∞. The letter W is used for a manifold which may have boundaries. The letter V is reserved for a manifold without boundaries.

I am grateful to R. Thom for describing "surgery" to me, and for pointing out the possibility of using it to kill homotopy groups.

Added in proof. Much of the material below has been obtained independently by A. H. Wallace [**14**].

1. The construction.

Let D^{p+1} denote the unit disk in the euclidean space R^{p+1}, with boundary S^p and with center 0. The product manifold $S^p \times S^q$ can be considered either

(1) $$\text{as the boundary of } S^p \times D^{q+1},$$

or

(2) $$\text{as the boundary of } D^{p+1} \times S^q.$$

Given any imbedding of $S^p \times D^{q+1}$ in a manifold W of dimension $n = p + q + 1$, a new manifold W' can be formed by removing the interior of $S^p \times D^{q+1}$ and replacing it by the interior of $D^{p+1} \times S^q$. This procedure will be called surgery. To be more precise:

DEFINITION. Given a differentiable, orientation preserving imbedding

$$f : S^p \times D^{q+1} \to W$$

[1] The author holds a Sloan fellowship.

with $p + q + 1 = n$, let $\chi(W, f)$ denote the quotient manifold obtained from the disjoint sum

$$(W - f(S^p \times 0)) + (D^{p+1} \times S^q)$$

by identifying $f(u, \theta v)$ with $(\theta u, v)$ for each $u \in S^p$, $v \in S^q$, $0 < \theta \leq 1$. Thus $\chi(W, f)$ is an oriented differentiable manifold. The boundary of W (if any) is equal to the boundary of $\chi(W, f)$. If W' denotes any manifold which is diffeomorphic to $\chi(W, f)$ under an orientation preserving diffeomorphism, then we will say that *W' can be obtained from W by surgery of type $(p + 1, q + 1)$.*

This construction clearly makes sense in the range $0 \leq p < n$ (that is the range $p, q \geqq 0$). It will be convenient to extend it to the cases $p = -1$ or $p = n$ by defining

$$D^0 = R^0 = \text{the point } 0; \quad S^{-1} = \text{the vacuous set.}$$

With these conventions, a surgery of type $(0, n + 1)$ replaces W by the disjoint sum $W + S^n$, while a surgery of type $(n + 1, 0)$ replaces $W + S^n$ by W.

It is clear that W and W' play a symmetrical role in this construction. If $W' = \chi(W, f)$ is obtained from W by a surgery of type $(p + 1, q + 1)$, then W can be obtained from W' by a surgery of type $(q + 1, p + 1)$.

Given a sequence $W_1, \cdots, W_r$ of manifolds such that each W_{i+1} can be obtained from W_i by a surgery we will say that W_1 is χ-equivalent to W_r.

2. χ-equivalence and cobordism. Consider manifolds V without boundary which are compact and oriented.

THEOREM 1. *Two such manifolds are χ-equivalent if and only if they belong to the same cobordism class.*

PROOF. Let L denote the locus of points (x, y) in $R^{p+1} \times R^{q+1}$ which satisfy the inequalities

$$-1 \leqq ||x||^2 - ||y||^2 \leqq 1,$$

and

$$||x|| \, ||y|| < (\sinh 1)(\cosh 1).$$

Thus L is a differentiable manifold with two boundaries. The "upper" boundary, $||x||^2 - ||y||^2 = 1$, is diffeomorphic to $S^p \times (\text{Interior } D^{q+1})$ under the correspondence

$$(u, \theta v) \leftrightarrow (u \cosh \theta, v \sinh \theta), \qquad 0 \leqq \theta < 1.$$

The lower boundary, $||x||^2 - ||y||^2 = -1$, is diffeomorphic to (Interior D^{p+1}) $\times S^q$ under the correspondence

$$(\theta u, v) \leftrightarrow (u \sinh \theta, v \cosh \theta).$$

Consider the orthogonal trajectories of the surfaces $||x||^2 - ||y||^2 = \text{constant}$.

The trajectory which passes through the point (x, y) can be parametrized in the form

$$t \to (tx, t^{-1}y).$$

[Thus the product $||x|| \, ||y||$ is constant along any orthogonal trajectory.] If x or y is zero this trajectory is a straight line segment tending to the origin. For x and y different from zero it is a hyperbola which leads from some well defined point $(u \cosh \theta, v \sinh \theta)$ on the upper boundary of L to the corresponding point $(u \sinh \theta, v \cosh \theta)$ on the lower boundary.

Now let V be a differentiable manifold without boundary of dimension $n = p + q + 1$, and let f: $S^p \times D^{q+1} \to V$ be an imbedding. Construct an $(n + 1)$-dimensional manifold $\omega(V, f)$ as follows. Start with the disjoint sum

$$(V - f(S^p \times 0)) \times D^1 + L.$$

For each $u \in S^p$, $v \in S^q$, $0 < \theta < 1$, and $c \in D^1$ identify the point $(f(u, \theta v), c)$ in the first summand with the unique point $(x, y) \in L$ such that

(1) $||x||^2 - ||y||^2 = c$, and

(2) (x, y) lies on the orthogonal trajectory which passes through the point $(u \cosh \theta, v \sinh \theta)$.

It is not difficult to see that this correspondence defines a diffeomorphism

$$f(S^p \times (\text{Interior } D^{q+1} - 0)) \times D^1 \leftrightarrow L \cap (R^{p+1} - 0) \times (R^{q+1} - 0).$$

It follows from this that $\omega(V, f)$ is a well defined differentiable manifold.

This manifold $\omega(V, f)$ has two boundaries, corresponding to the values $c = ||x||^2 - ||y||^2 = +1$, and -1. The upper boundary, $c = +1$, can be identified with V, letting $z \in V$ correspond to:

$$\begin{cases} (z, 1) \in (V - f(S^p \times 0)) \times D^1 & \text{for } z \notin f(S^p \times 0). \\ (u \cosh \theta, v \sinh \theta) \in L & \text{for } z = f(u, \theta v). \end{cases}$$

The lower boundary can be indentified with $\chi(V, f)$: letting $z \in V - f(S^p \times 0)$ correspond to $(z, -1)$; and letting $(\theta u, v) \in D^{p+1} \times S^q$ correspond to $(u \sinh \theta, v \cosh \theta)$.

It is clear that $\omega(V, f)$ is compact if and only if V is compact. Since orientations can be chosen appropriately, this completes the proof that χ-equivalent manifolds belong to the same cobordism class.

An immediate consequence is the following.

COROLLARY (THOM). *The Stiefel-Whitney numbers, Pontrjagin numbers, and the index of a compact manifold V are invariant under surgery.*

Now suppose that V and V' are compact manifolds which belong to the same cobordism class. Thus V and V' together bound a compact manifold W.

LEMMA 1. *There exists a differentiable map $g : W \to [0, 1]$ with the following three properties:*

(1) $g^{-1}(0) = V', \quad g^{-1}(1) = V.$

(2) *The gradient of g vanishes only at isolated interior points $w_1, \cdots, w_k \in W$. Furthermore the matrix of second derivatives of g at each w_i is non-singular.*

(3) *The values $c_i = g(w_i)$ are distinct; say $0 < c_1 < \cdots < c_k < 1$.*

Using a partition of unity, it is easy to construct a function $g_0 : W \to [0, 1]$ which satisfies condition (1), and has non-zero gradient along the boundary of W. Hence g_0 has non-zero gradient along some compact neighborhood K_0 of the boundary.

Choose open sets $B_1, \cdots, B_r \subset$ Interior W which are diffeomorphic to open subsets of euclidean space, and which cover the compact set Closure$(W - K_0)$. Then g_0 restricted to B_1 can be approximated by a function g_0' which satisfies condition (2) throughout B_1. (See Morse [**6**, Theorem 16], or Whitney [**13**, Theorem 12A].)

Let $A_1, \cdots, A_r$ be an open covering of Closure$(W - K_0)$ such that Closure $A_i \subset B_i$. Let $\lambda_i : W \to [0, 1]$ be a differentiable function which takes the value of 1 on A_i, and has carrier contained in B_i. Now define

$$g_1(w) = (1 - \lambda_1(w))g_0(w) + \lambda_1(w)g_0'(w) \quad \text{for} \quad w \in B_1,$$

$$g_1(w) = g_0(w) \quad \text{for} \quad w \notin B_1.$$

Then $g_1 : W \to R$ will satisfy condition (2), at least throughout Closure A_1. Furthermore, if g_0' is sufficiently close to g_0, then

(a) the modified function g_1 will still have values between 0 and 1; and

(b) g_1 restricted to K_0 will still satisfy condition (2).

In other words g_1 will satisfy condition (2) throughout the compact set

$$K_1 = K_0 \cup \text{ Closure } A_1.$$

Now continue by induction, constructing functions $g_i : W \to [0, 1]$ which satisfy condition (1), and which satisfy (2) throughout the set

$$K_i = K_{i-1} \cup \text{ Closure } A_i.$$

Since K_r is equal to W, it follows that g_r satisfies condition (2) everywhere.

If g_r takes on the same value at two of its critical points, then by adding a function which is equal to ϵ in a small neighborhood of one point, and is zero outside a larger neighborhood, we can separate the two values. This completes the proof of Lemma 1.

The proof of Theorem 1 continues as follows. Define $c_0 = 0$, $c_{k+1} = 1$. If both c' and c'' lie between c_i and c_{i+1} note that the manifolds $g^{-1}(c')$ and $g^{-1}(c'')$ are diffeomorphic. This is proved by taking any Riemannian metric on W, and using the orthogonal trajectories of the surfaces $g^{-1}(c)$ to define the diffeomorphism. Call any one of these surfaces V_i. Thus V_0 is diffeomorphic to V' and V_k is diffeomorphic to V. We will prove that each V_{i-1} can be obtained from V_i by surgery.

Let $B(r)$ denote the open ball of radius r in R^{n+1}. According to Morse [**7**,

p. 172] there exists a neighborhood U of w_i in W, and a coordinate diffeomorphism

$$\phi : B(2\epsilon) \to U$$

so that the function $g\phi(z_0, \cdots, z_n)$ takes the form

$$c_i + z_0^2 + \cdots + z_p^2 - z_{p+1}^2 - \cdots - z_n^2,$$

for some $-1 \leqq p \leqq n$, and some $\epsilon > 0$. We will abbreviate this as

$$g\phi(x, y) - c_i = ||x||^2 - ||y||^2,$$

where $x \in R^{p+1}$, $y \in R^{q+1}$, $p + q + 1 = n$. Note that $\phi(0, 0) = w_i$.

We may assume that $g^{-1}(c_i + \epsilon^2) = V_i$ and that $g^{-1}(c_i - \epsilon^2) = V_{i-1}$. Define an imbedding

$$f : S^p \times D^{q+1} \to V_i$$

by $f(u, \theta v) = \phi(\epsilon u \cosh \theta, \epsilon v \sinh \theta)$. It will be shown that V_{i-1} is diffeomorphic to $\chi(V_i, f)$.

An imbedding $h: L \to W$ is defined by $h(x, y) = \phi(\epsilon x, \epsilon y)$. Choose a Riemannian metric on W which reduces to the euclidean metric

$$dx_0^2 + \cdots + dx_p^2 + dy_0^2 + \cdots + dy_q^2$$

throughout the closure of $h(L)$. The imbedding h carries the surfaces $||x||^2 - ||y||^2 =$ constant into surfaces $g^{-1}(c)$. Hence it carries orthogonal trajectories into orthogonal trajectories.

Define a diffeomorphism $k: \chi(V_i, f) \to V_{i-1}$ as follows. Starting at any point z of $V_i - f(S^p \times 0)$ the orthogonal trajectory is a non-singular curve which leads from z to some well defined point $k(z)$ in V_{i-1}. In the case $z = f(u, \theta v) = h(u \cosh \theta, v \sinh \theta)$ this orthogonal trajectory leads from z to the point $h(u \sinh \theta, v \cosh \theta)$ of V_{i-1}. Hence defining

$$k(\theta u, v) = \phi(\epsilon u \sinh \theta, \epsilon v \cosh \theta)$$

for all $(\theta u, v) \in D^{p+1} \times S^q$, we obtain a well defined diffeomorphism from $\chi(V_i, f)$ to V_{i-1}. This completes the proof that cobording manifolds are χ-equivalent.

REMARK. A similar proof shows that the region $g^{-1}[c_i - \epsilon^2, c_i + \epsilon^2]$ between V_i and V_{i-1} is diffeomorphic to $\omega(V_i, f)$.

3. Killing homotopy classes. Let W be a manifold of dimension n and let $\lambda \in \pi_p(W)$ be a homotopy class which we wish to kill. To simplify the discussion, assume that W is connected.

DEFINITION. An imbedding $f : S^p \times D^{n-p} \to W$ *represents* the homotopy class λ if $\lambda = f_*(\iota)$, where ι is a generator of the infinite cyclic group $\pi_p(S^p \times D^{n-p})$. [To be more precise one should specify base points for these homotopy groups, but this will be left to the reader.]

Suppose that λ is represented by such an imbedding $f : S^p \times D^{n-p} \to W$. Suppose further that $n \geqq 2p + 2$. Let $W' = \chi(W, f)$.

LEMMA 2. *Under these conditions, the homotopy groups* $\pi_i(W')$ *are isomorphic to* $\pi_i(W)$ *for* $i < p$; *and* $\pi_p(W')$ *is isomorphic to* $\pi_p(W)$ *modulo a subgroup which contains* λ.

Briefly: the effect of the surgery is to kill the homotopy class λ.

PROOF. As usual set $q = n - p - 1$. Note that $p < q$. Let X denote the space which is formed from the topological sum $W + (D^{p+1} \times D^{q+1})$ by identifying (u, y) with $f(u, y)$ for each $(u, y) \in S^p \times D^{q+1}$. The subset $W \cup (D^{p+1} \times 0)$ is clearly a deformation retract of X. This subset is formed from W by attaching a $(p + 1)$-cell; using the map $u \to f(u, 0)$ as attaching map. It follows immediately that the inclusion homomorphism $\pi_i(W) \to \pi_i(X)$ is an isomorphism for $i < p$, and is onto for $i = p$. Furthermore the homotopy class λ of the attaching map lies in the kernel of this homomorphism.

But W' is also imbedded topologically in X, and a similar argument shows that the homomorphism $\pi_i(W') \to \pi_i(X)$ is an isomorphism for $i < q$. In particular it is an isomorphism for $i \leq p$. Together with the preceding paragraph, this completes the proof.

REMARK. The manifold $\omega(W, f)$ could be used in place of the space X in this argument.

The above argument can also be applied to the case $p = 0$. In this case the appropriate statement is:

LEMMA 2′. *Let* W *be a manifold of dimension* $n \geqq 2$ *with* k *components,* $k \geqq 2$. *Let* $f : S^0 \times D^n \to W$ *be an orientation preserving*[2] *imbedding which carries the two components of* $S^0 \times D^n$ *into distinct components of* W. *Then* $\chi(W, f)$ *has only* $k - 1$ *components.*

The proof will be left to the reader.[3]

In order to make use of Lemma 2 it is necessary to answer the following question. For which homotopy classes $\lambda \in \pi_p(W)$ does there exist an imbedding $S^p \times D^{n-p} \to W$ which represents λ?

Let τ^n denote the tangent bundle of W and let $f_0 : S^p \to W$ denote any map in the homotopy class of λ. Let $f_0^*\tau^n$ denote the induced bundle over S^p.

LEMMA 3. *Assume that* $n \geqq 2p + 1$. *Then there exists an imbedding* $S^p \times D^{n-p} \to W$ *which represents* λ *if and only if the induced bundle* $f_0^*\tau^n$ *is trivial.*

The proof of Lemma 3 will be based on the following.

LEMMA 4. *Let* ξ^m *be an m-dimensional vector space bundle over a complex* K *of dimension* $p < m$. *Let* o^1 *denote the trivial line bundle over* K. *Then* ξ^m *is a trivial bundle if and only if the Whitney sum* $\xi^m \oplus o^1$ *is trivial.*

PROOF. Let $B(SO_{m+1})$ denote any classifying space for the rotation group. First observe that there exists a fibre bundle over $B(SO_{m+1})$ with fibre S^m and

[2] The sphere S^0 must be oriented as the boundary of D^1.

[3] Compare the definition of "connected sum" of manifolds, as given in Milnor [5].

total space $B(SO_m)$. [This is seen as follows. The group SO_{m+1} acts freely on the total space E of its universal bundle, with orbit space $E/SO_{m+1} = B(SO_{m+1})$. Hence the subgroup SO_m also acts freely on E. Define $B(SO_m)$ as the orbit space E/SO_m. It follows easily that $B(SO_m)$ is a bundle over $B(SO_{m+1})$ with fibre $SO_{m+1}/SO_m = S^m$.] Let $\pi : B(SO_m) \to B(SO_{m+1})$ denote the projection map.

Let $f : K \to B(SO_m)$ be a classifying map for the bundle ξ^m. Then $\pi f : K \to B(SO_{m+1})$ is a classifying map for the Whitney sum $\xi^m \oplus o^1$. If πf is null-homotopic then it follows from the covering homotopy theorem that f is homotopic to a map of K into the fibre S^m. Since K has dimension less than m, this implies that f is null-homotopic.[4]

PROOF OF LEMMA 3. If $n \geqq 2p + 1$ then any homotopy class of maps from S^p to W contains an imbedding

$$f_0 : S^p \to \text{Interior } W.$$

(See Whitney [**11**, Theorem 2].) Let τ^p denote the tangent bundle of S^p and let ν^{q+1} denote its normal bundle. Then the Whitney sum $\tau^p \oplus \nu^{q+1}$ can be identified with $f_0^*\tau^n$: the tangent bundle of W restricted to S^p.

If $f_0^*\tau^n$ is trivial then it follows that ν^{q+1} is trivial. To see this, let o^k denote the trivial k-dimensional vector space bundle over S^p. Then $o^1 \oplus \tau^p$ is known to be trivial. (Identify o^1 with the normal bundle of S^p in R^{p+1}.) Combining the relations

$$o^1 \oplus \tau^p \approx o^{p+1}, \qquad \tau^p \oplus \nu^{q+1} \approx o^n$$

it follows that $o^{p+1} \oplus \nu^{q+1} \approx o^{n+1}$. Together with Lemma 4 this implies that ν^{q+1} is trivial.

Now take a tubular neighborhood of $f_0(S^p)$ in W. This can be identified with the total space of the normal D^{q+1}-bundle. Hence this neighborhood is diffeomorphic to $S^p \times D^{q+1}$. The resulting imbedding

$$S^p \times D^{q+1} \to W$$

certainly represents the homotopy class of f_0.

Conversely suppose that such an imbedding is given. Since $S^p \times D^{q+1}$ is parallelizable, it follows immediately that $f_0^*\tau^n$ is a trivial bundle. This completes the proof of Lemma 3.

This lemma can also be formulated as follows. Let $B(SO_n)$ denote a classifying space for the rotation group, and let

$$T : W \to B(SO_n)$$

be a classifying map for the tangent bundle of W. Then there exists an imbedding $S^p \times D^{n-p} \to W$ which represents λ if and only if the homomorphism

$$T_* : \pi_p(W) \to \pi_p B(SO_n)$$

annihilates λ.

[4] This proof is due to A. Dold.

Combining this assertion with Lemmas 2 and 2′ it is easy to prove the following.

ASSERTION. *Every compact (oriented, differentiable) manifold is* χ*-equivalent to a manifold* W *which is connected (if* $n \geq 2$*), and which satisfies the following condition: the homomorphism*

$$T_* : \pi_p(W) \to \pi_p B(SO_n)$$

is a monomorphism for $1 \leq p \leq n/2 - 1$.

The proof proceeds by induction on p. First surgeries of type $(1, n)$ are used to connect the manifold. Then surgeries of type $(2, n-1)$ are used to kill its fundamental group, and so on. The kernel of T_* is finitely generated at each stage since W is compact and, for $p > 1$, simply connected.

4. π-manifolds. It is not possible to kill the entire homotopy group $\pi_p(W)$ by this construction, even if the dimension n is large, unless W satisfies some further restriction.

[As an example consider the second homotopy group of the complex projective space $P_{2m}(C)$. The Stiefel-Whitney number $w_2^{2m}[P_{2m}(C)]$ is non-zero; but for any 2-connected manifold V the number $w_2^{2m}[V]$ is zero. Hence $\pi_2(P_{2m})C^{44}$ cannot be killed by the surgery.]

As a first attempt one might try the hypothesis that W is parallelizable. However this condition is too easily destroyed by the surgery.

[For example consider the parallelizable manifold $S^1 \times S^3$. Killing the fundamental group of $S^1 \times S^3$ we obtain a manifold such as S^4 which has positive Euler characteristic; and therefore is not parallelizable.]

As a second attempt consider the following:

DEFINITION. W is a *π-manifold* if the Whitney sum of its tangent bundle and a trivial line bundle is trivial. The notation $(\tau^n \oplus o^1)W$ will be used for this Whitney sum. For example any sphere is a π-manifold.

REMARKS. A trivial vector space bundle of higher dimension could be used in place of the line bundle. (Compare Lemma 4.) If W is imbedded in a high dimensional euclidean space then W is π-manifold if and only if its normal bundle is trivial.[5]

THEOREM 2. *Let* W *be a* π*-manifold of dimension* $n \geq 2p + 1$. *Then any homotopy class in* $\pi_p(W)$ *is represented by an imbedding*

$$f : S^p \times D^{n-p} \to W$$

such that the new manifold $\chi(W, f)$ *is also a* π*-manifold.*

Together with Lemmas 2 and 2′ this clearly implies the following.

COROLLARY. *Any compact* π*-manifold of dimension* n *is* χ*-equivalent to a* π*-manifold which is* $[n/2 - 1]$*-connected.*

[5] This assertion is essentially due to J. H. C. Whitehead. It can be proved using Lemma 4 together with the relation $o^1 \oplus \tau^n \oplus \nu^k \approx o^{1+n+k}$.

REMARK 1. In the case of a compact π-manifold V without boundary one can prove the stronger statement that V is χ-equivalent to S^n. (See Milnor [**4**], Wall [**10**] together with Theorem 1.) This stronger statement is much more difficult to prove.

REMARK 2. It is definitely not true that $\chi(W, f)$ is a π-manifold for any imbedding f. If this were true then any manifold in the trivial cobordism class would have to be a π-manifold. But a counter-example is provided by the disjoint sum $P_2(C) + (-P_2(C))$ where the minus sign stands for reversal of orientation.

PROOF OF THEOREM 2. Since the boundary of W plays no role in this theorem, we may delete it. Then, given any imbedding $f: S^p \times D^{n-p} \to W$, the $(n + 1)$-dimensional manifold $\omega(W, f)$ is defined. (Compare the proof of Theorem 1.) If f is suitably chosen, we will prove that $\omega(W, f)$ is parallelizable. Since $\chi(W, f)$ is one of the two boundaries of $\omega(W, f)$, this will imply that $\chi(W, f)$ is a π-manifold.

Since $n \geq 2p + 1$, any element of $\pi_p(W)$ is represented by some imbedding $f_0: S^p \to W$. Since W is a π-manifold, there exist $n + 1$ linearly independent cross-sections of the bundle $(\tau^n \oplus o^1)W$. Restricting this bundle to $f_0(S^p)$ it splits up into the Whitney sum of $(\tau^p \oplus o^1)f_0(S^p)$ and the normal bundle ν^{n-p}. Note that the bundle $(\tau^p \oplus o^1)S^p$ can be identified with the tangent bundle of R^{p+1}, restricted to the unit sphere. Hence it has $p + 1$ canonical cross-sections, which will be denoted by $e_0, \cdots, e_p$. The corresponding sections of $(\tau^p \oplus o^1)f_0(S^p)$ will be denoted by $f_0.e_0, \cdots, f_0.e_p$.

LEMMA 5. *The $n + 1$ independent cross-sections $c_0, \cdots, c_n$ of $(\tau^n \oplus o^1)W$ can be chosen in such a way that, for each point $w = f_0(u)$ of $f_0(S^p)$, the first $p + 1$ of these vectors are given by*

$$c_i(w) = f_0.e_i(u), \qquad i = 0, \cdots, p.$$

Assuming this lemma for the moment, the proof of Theorem 2 proceeds as follows. We may assume that the remaining $n-p$ vectors, $c_{p+1}(W), \cdots, c_n(W)$ are normal to the submanifold $f_0(S^p) \subset W$. Hence these vector fields determine a specific product structure for the normal bundle ν^{n-p}. Making use of this product structure, a tubular neighborhood of $f_0(S^p)$ can be identified with $S^p \times D^{n-p}$. This gives the required imbedding $f: S^p \times D^{n-p} \to W$.

ASSERTION. If f is chosen in this way, then $\omega(W, f)$ is parallelizable.

To prove this assertion, identify $(\tau^n \oplus o^1)W$ with the restriction to W of the tangent bundle of $\omega(W, f)$. Thus $c_0, \cdots, c_n$ can be considered as independent cross-sections of $\tau^{n+1}\omega(W, f)$ which are defined only on W. We must extend these cross-sections throughout $\omega(W, f)$.

Consider first the problem of extending over the disk $D^{p+1} \times 0 \subset L \subset \omega(W, f)$. The boundary of this disk is $S^p \times 0 = f_0(S^p) \subset W$, and the sections $c_0, \cdots, c_p$ have been chosen so that, on $S^p \times 0$, they reduce to the standard cross-sections

$e_0, \cdots, e_p$ of $\tau^{p+1}(D^{p+1} \times 0)$. Thus the first $p+1$ cross-sections can certainly be extended over $D^{p+1} \times 0$.

But the remaining cross-sections $c_{p+1}, \cdots, c_n$, on $S^p \times 0$, are just the standard basis for the normal bundle of $D^{p+1} \times 0$ in L. Hence these can be extended over $D^{p+1} \times 0$. This extension can be carried out so that all $n+1$ vectors remain independent.

Thus the tangent bundle of $\omega(W, f)$, restricted to $W \cup (D^{p+1} \times 0)$, is a trivial bundle. To complete the argument it is only necessary to observe that $W \cup (D^{p+1} \times 0)$ is a deformation retract of $\omega(W, f)$. This can be proved in two steps as follows.

STEP 1. The manifold $\omega(W, f)$ can be deformed into the subset $W \cup$ (Closure L) by a deformation which leaves $W \cup (D^{n+1} \times 0)$ pointwise fixed. (The proof is not difficult.)

STEP 2. The space $W \cup$ (Closure L) has $W \cup (D^{n+1} \times 0)$ as deformation retract. In fact a retraction can be defined as follows, for $(x, y) \in$ Closure L:

$$(x, y) \rightarrow (x, 0) \quad \text{if} \quad ||x|| \leqq 1,$$

$$(x, y) \rightarrow (x, y(||x||^2 - 1)^{1/2}/||y||) \quad \text{otherwise.}$$

This retraction is clearly homotopic to the identity.

This proves that $W \cup (D^{n+1} \times 0)$ is a deformation retract of $\omega(W, f)$, and therefore proves that $\omega(W, f)$ is parallelizable.

To prove Theorem 2 observe that $\chi(W, f)$ is imbedded as a boundary of $\omega(W, f)$. Hence $(\tau^n \oplus o^1)\chi(W, f)$ can be identified with the tangent bundle of $\omega(W, f)$ restricted to $\chi(W, f)$. Since this is trivial it follows that $\chi(W, f)$ is a π-manifold. This completes the proof of Theorem 2, except for Lemma 5.

PROOF OF LEMMA 5. Start with any $n+1$ independent cross-sections $b_0, \cdots, b_n$ of $(\tau^n \oplus o^1)W$. For each $u \in S^p$ the required vectors $f_0.e_i$ at $w = f_0(u)$ can be expressed as linear combinations

$$f_0.e_i(u) = \sum z_{ij}(u) b_j(w).$$

The correspondence $u \rightarrow (z_{ij}(u))$ defines a map z from S^p to the Stiefel manifold consisting of all $(p+1) \times (n+1)$ matrices of rank $p+1$. But this Stiefel manifold is $(n-p-1)$-connected;[6] hence z is null-homotopic. Using the covering homotopy theorem it follows that z can be lifted to a null-homotopic map $\bar{z}$ which carries S^p into the Stiefel manifold of all non-singular $(n+1) \times (n+1)$ matrices. Let $\bar{z}_{ij}(u)$ denote the ijth component of $\bar{z}(u)$. Then $\bar{z}_{ij} = z_{ij}$ for $0 \leq i \leq p$.

Now define new sections $c_0, \cdots, c_n$ as follows. Let $c_i(w) = b_i(w)$ outside of some neighborhood of $f_0(S^p)$, and let

$$c_i(w) = \sum \bar{z}_{ij}(u) b_j(w) \qquad \text{for} \quad w = f_0(u).$$

[6] See Steenrod [8, §25.6] together with §7.7 and §12.9.

The null-homotopy of $\bar{z}$ can be used to extend throughout the neighborhood. This completes the proof of Lemma 5 and of Theorem 2.

5. k-parallelizable manifolds. The next step will be to extend the above procedure so that it applies to something more general than a π-manifold.

DEFINITION. A bundle ξ over a space B is *k-trivial* if for every complex k of dimension $\leq k$, and every map $f: K \to B$, the induced bundle $f^*\xi$ is trivial. In case B itself is a complex, this is equivalent to the requirement that the restriction of ξ to the k-skeleton of B should be trivial.

DEFINITION. A manifold is *k-parallelizable* if its tangent bundle is k-trivial. Thus every (orientable) manifold is 1-parallelizable. For large values of k, this condition implies that the manifold is actually parallelizable.

THEOREM 3. *Let W be a compact manifold which is k-parallelizable, $1 \leq k < n$. Then W is χ-equivalent to a k-parallelizable manifold W' which is c-connected where $c =$* Min $(k, [n/2 - 1])$.

PROOF. Suppose by induction that W is $(p - 1)$-connected with $1 \leq p \leq$ Min $(k, [n/2 - 1])$. Consider some triangulation of Interior W, together with its dual cell subdivision. Let W^r denote the r-skeleton of one of these cell subdivisions and W_r the r-skeleton of the other. Then the skeleton W^k is a deformation retract of (Interior W)$-W_{n-k-1}$. Therefore the open manifold (Interior W)$-W_{n-k-1}$ is parallelizable, and hence is a π-manifold. Since $p \leqq k$, any element of $\pi_p(W)$ is represented by a map of S^p into the skeleton.

$$W^k \subset (\text{Interior } W) - W_{n-k-1}.$$

Since $n \geq 2p + 1$ it follows from Theorem 2 that this homotopy class is represented by an imbedding

$$f : S^p \times D^{n-p} \to (\text{Interior } W) - W_{n-k-1}$$

such that the manifold $\chi((\text{Interior } W) - W_{n-k-1}, f)$ is again a π-manifold.

But f can also be considered as an imbedding of $S^p \times D^{n-p}$ in W. We assert that the manifold $\chi(W, f)$ is k-parallelizable. The above argument shows that

$$(\text{Interior } \chi(W, f)) - W_{n-k-1} = \chi((\text{Interior } W) - W_{n-k-1}, f)$$

is a π-manifold. But any map of a k-dimensional complex K into $\chi(W, f)$ can be deformed into a map

$$g : K \to (\text{Interior } \chi(W, f)) - W_{n-k-1},$$

since the set W_{n-k-1} is a countable union of imbedded simplexes, each having codimension $\geq k + 1$. Hence $g^*(\tau^n \oplus o^1) = (g^*\tau^n) \oplus o^1$ is a trivial bundle. Using Lemma 4 and the assumption that $k < n$, this implies that $g^*\tau^n$ is trivial. Therefore $\chi(W, f)$ is k-parallelizable.

Since W is compact and $(p - 1)$-connected it follows that $\pi_p(W)$ is finitely generated. Hence a finite number of iterations of this construction will kill the pth homotopy group. This completes the proof of Theorem 3.

6. Killing the middle homotopy group. Consider compact manifolds W of dimension $2m$. Using Theorem 2 or 3 it may be possible to kill the homotopy groups $\pi_p(W)$ in dimensions $p < m$. This section will study the possibility of killing $\pi_m(W)$.

REMARK. If one succeeds in obtaining a manifold W' which is m-connected, then the homotopy type of W' is almost determined by that of Boundary W. The Poincaré duality theorem implies that[7] $H_i(W', \text{Boundary } W) = 0$ for $m \leq i \leq 2m - 1$; and hence that $H_i(W')$ is isomorphic to H_i (Boundary W) for $m \leq i < 2m - 1$. As an example, if Boundary W is vacuous, then it follows easily that W' has the homotopy type of a $2m$-sphere.

In order to kill $\pi_m(W)$, the first problem is to represent a given homotopy class by an imbedded sphere.

LEMMA 6. *If W is a simply connected manifold of dimension $2m > 4$, then every element of $\pi_m(W)$ is represented by an imbedding $f_0 : S^m \to W$.*

More generally, if V is a compact connected manifold of dimension m, the proof will show that any homotopy class of maps $V \to W$ is represented by an imbedding. For the special case $W = R^{2m}$ this result is due to Whitney [**12**]. The present proof makes use of Whitney's method. I do not know whether the assertion is true in the case $m = 2$.

First any map can be approximated by an immersion $f : V \to W$ which has no singularities other than a finite number of double points:

$$f(a_1) = f(a_2), \qquad f(a_3) = f(a_4), \cdots,$$

where $a_1, \cdots, a_{2k}$ are distinct points of V. (See Whitney, [**11**, Theorem 2].) Since V is connected there exists an arc joining a_1 to a_2 in V which misses the points $a_3, \cdots, a_{2k}$. This arc projects into a simple closed curve C in W which is differentiable except for one angle.

Identify C with a simple closed curve in R^2, having one angle. Since W is simply connected of dimension ≥ 5, the imbedding of C in W can be extended to an imbedding g of R^2 in W. Now a tubular neighborhood N of $g(R^2)$ in W will be diffeomorphic to R^{2m}.

Consider the immersion $f^{-1}(N) \to N$. This has a double point, $f(a_1) = f(a_2)$, where a_1 and a_2 belong to the same component of $f^{-1}(N)$. Apply the construction described by Whitney [**12**, Theorems 3, 4] to this immersion. This has the effect of first adding one double point and then removing two double points. The map f is altered only in a compact subset of $f^{-1}(N)$. Therefore the modified immersion $f^{-1}(N) \to N$ gives rise to a new immersion $f' : V \to W$ which has only $k - 1$ double points.

[7] Integer coefficients are to be understood.

Since the set $N \subset W$ is homeomorphic to R^{2m} it follows easily that f' is homotopic to f. Iterating this construction k times we obtain an imbedding f_0 of V in W, which completes the proof.

The next problem is to decide whether the normal bundle ν^m of $f_0(S^m)$ is trivial. This turns out to be a difficult question for m odd. Hence we will concentrate on the case m even.

LEMMA 7. *Let W be m-parallelizable of dimension $2m$ with m even. Let $f_0 : S^m \to W$ be an imbedding which represents the homology class*[7] *$\beta \in H_m(W)$. Then the normal bundle ν^m of $f_0(S^m)$ is trivial if and only if the intersection number $\langle \beta, \beta \rangle$ is zero.*

[Here the notation $\langle \alpha, \alpha' \rangle$ stands for the intersection number of two homology classes $\alpha \in H_i(W)$, $\alpha' \in H_{2m-i}(W)$. See Lefschetz [**2**].]

PROOF. Let $[\nu^m] \in \pi_{m-1}(SO_m)$ denote the homotopy class which corresponds to the bundle ν^m. (See Steenrod [**8**, §18].) Just as in the proof of Lemma 3 it is seen that $\nu^m \oplus o^{m+1}$ is a trivial bundle and hence, by Lemma 4, that $\nu^m \oplus o^1$ is trivial. Therefore $[\nu^m]$ is annihilated by the homomorphism

$$i_* : \pi_{m-1}(SO_m) \to \pi_{m-1}(SO_{m+1}).$$

Now consider the exact sequence

$$\pi_m(S^m) \xrightarrow{\partial} \pi_{m-1}(SO_m) \xrightarrow{i_*} \pi_{m-1}(SO_{m+1}).$$

The homomorphism ∂ carries a generator of $\pi_m(S^m)$ into the class $[\tau^m]$ which corresponds to the tangent bundle of S_m. Hence $[\nu^m]$ must be equal to some multiple $k[\tau^m]$ where k is an integer.

The Euler class $X(\xi^m)$ of an m-dimensional vector space bundle can be defined as the first obstruction to the existence of a non-zero cross-section. Note that the correspondence

$$[\xi^m] \to X(\xi^m)$$

defines a homomorphism from $\pi_{m-1}(SO_m)$ to the infinite cyclic group $H^m(S^m)$. [Proof: Let $X \in H^m(B(SO_m))$ denote the universal Euler class and let $g : S^m \to B(SO_m)$ denote a classifying map for ξ^m. Thus $X(\xi^m) = g^*X$. This is clearly an additive function of the homotopy class $\{g\} \in \pi_m(B(SO_m)) \approx \pi_{m-1}(SO_m)$.]

For the tangent bundle τ^m of S^m the class $X(\tau^m)$ is known to be twice a generator of $H^m(S^m)$. (The hypothesis that m is even comes in here.) Since $[\nu^m] = k[\tau^m]$ it follows that $X(\nu^m)$ is equal to $2k$ times a generator.

This argument proves that the group Image ∂ = kernel i_* generated by $[\tau^m]$ is infinite cyclic. An element $[\xi^m]$ in this group is zero if and only if the Euler class $X(\xi^m)$ is zero.

But for the normal bundle ν^m, the class $X(\nu^m)$ can be interpreted as the intersection number $\langle \beta, \beta \rangle$ multiplied by a generator of $H^m(S^m)$. For given a normal vector field with only finitely many zeros, we can deform $f_0(S^m)$ along these vectors to obtain a new imbedding which intersects $f_0(S^m)$ at only finitely many places. The multiplicity of each such intersection is equal to the index of the corresponding zero of the normal vector field.

Thus ν^m is trivial if and only if $\langle\beta, \beta\rangle$ equals zero, which completes the proof.

REMARK. This argument also proves that $\langle\beta, \beta\rangle$ is equal to $2k$ and hence is always an even number. This fact will be important later.

Lemma 7 raises the following question: Does there exist a non-zero element β of $H_m(W)$ such that the intersection number $\langle\beta, \beta\rangle$ is zero?

Assume now that W is compact and $(m - 1)$-connected of dimension $2m$, and furthermore assume that the boundary of W has no homology in dimensions $m, m - 1$. Then the Poincaré duality theorem implies that $H_m(W)$ is a free abelian group. [The group $H_m(W)$ is isomorphic to $H_m(W$, Boundary $W)$ and hence, by duality, to $H^m(W)$. Since $H_{m-1}(W) = 0$, the universal coefficient theorem implies that $H^m(W)$ is free abelian.]

Since m is even the correspondence $\alpha \to \langle\alpha, \alpha\rangle$ for $\alpha \in H_m(W)$ defines a quadratic form with integer coefficients. The determinant of this quadratic form is ± 1. (Compare Lefschetz [**2**, p. 178] or Milnor [**3**, Lemma 1].) Our question now becomes: does the quadratic form of W have a non-trivial zero?

LEMMA 8. *A quadratic form φ with integer coefficients and with determinant ± 1 has a non-trivial zero if and only if it is indefinite.*

Both Lemma 8 and Lemma 9 (which follows) are immediate consequences of [**3**, Theorems 1, 2]. The more direct proofs given below are due to H. Sah. The basic reference for these proofs is B. Jones [**1**].

PROOF OF LEMMA 8. If φ is positive definite or negative definite then it clearly has only the trivial zero. Assume that φ is indefinite of rank r. That is, if ν denotes the number of negative terms when φ is diagonalized over the rational numbers, assume that $0 < \nu < r$.

CASE 1. $r \geqq 5$. According to [**1**, Corollary 27d], φ has a non-trivial zero over the rational numbers (that is in $H_m(W; Q)$). Clearing denominators it follows that φ has a non-trivial zero over the integers.

CASE 2. $r = 3$ or 4. According to [**1**, Theorem 14], φ has a non-trivial zero over the p-adic integers for p odd. It clearly has a non-trivial zero over the real numbers. The Hasse symbol $c_2(\varphi)$ can be computed as on pages 38–39 of [**1**], and turns out to be $+1$ for $\nu = 1, 2$ and -1 for $\nu = 3$. The determinant of φ is clearly equal to $(-1)^\nu$. Now using [**1**, Theorem 14] it follows that φ has a non-trivial zero over the 2-adic numbers. Together with the Hasse-Minkowski theorem, [**1**, Theorem 27], this implies that φ has a rational zero.

CASE 3. $r = 2$. Then the determinant of φ is -1, and the conclusion follows by an elementary argument. (See [**1**, Theorem 14a].) This completes the proof of Lemma 8

The *index* of a quadratic form is defined as $r - 2\nu$. (For a diagonalized form this is the number of positive terms minus the number of negative ones.) The index of the quadratic form $\alpha \to \langle \alpha, \alpha \rangle$ of W turns out to be invariant under the χ-construction. (Compare the Corollary to Theorem 1.) Thus in order to have any hope of killing $H_m(W)$ we must assume that this index is zero.

Now suppose that W is m-parallelizable. Then the quadratic form of W takes on only even values. This follows from the remark after Lemma 7, together with Lemma 6 and the Hurewicz theorem. (For an alternative proof, compare [**3**, Lemma 3].)

The conditions which we have obtained for the quadratic form of W turn out to be sufficient to characterize this form. Let U denote the matrix

$$\begin{pmatrix} 0 & 1 \\ 1 & 0 \end{pmatrix}.$$

LEMMA 9. *Consider a quadratic form over the integers with determinant* ± 1 *and index* 0, *which takes on only even values. The matrix of this form, with respect to a suitable basis, is*[8] diag $(U, \cdots, U)$.

PROOF. It will be convenient to use the notation $\alpha \to \langle \alpha, \alpha \rangle$, $\alpha \in H$, for this quadratic form. According to Lemma 8 there exists a non-zero element β_1 of H such that $\langle \beta_1, \beta_1 \rangle = 0$. We may assume that β_1 is indivisible. Since the determinant is ± 1, it follows that there exists an element α of H with $\langle \beta_1, \alpha \rangle = 1$. Define

$$\beta_2 = \alpha - \tfrac{1}{2}\langle \alpha, \alpha \rangle \beta_1 .$$

It follows that $\langle \beta_1, \beta_2 \rangle = 1$ and $\langle \beta_2, \beta_2 \rangle = 0$. Thus the matrix $\|\beta_i, \beta_j\|$ for $i, j = 1, 2$ is equal to U.

Let H' denote the set of α in H with $\langle \beta_1, \alpha \rangle = \langle \beta_2, \alpha \rangle = 0$. Since the determinant of U is ± 1 it follows that H splits into the direct sum of H' and the free abelian group generated by β_1 and β_2. By induction on the rank we may choose a basis $\beta_3, \cdots, \beta_r$ for H' which has the required form. This completes the proof of Lemma 9.

THEOREM 4. *Let W be m-parallelizable and $(m - 1)$-connected of dimension $2m$ where m is even $\neq 2$. Suppose that the quadratic form of W has index zero, and that the boundary of W has no homology in dimensions m, $m - 1$. Then W is χ-equivalent to an m-connected manifold.*

PROOF. According to Lemma 9 there exists a basis $\beta_1, \cdots, \beta_r$ for $H_m(W)$ so that the intersection matrix takes the form diag $(U, \cdots, U)$. By Lemma 6 and the Hurewicz theorem, there exists an imbedding

$$f_0 : S^m \to \text{Interior } W$$

which represents the homology class β_1. By Lemma 7 the normal bundle of

[8] Here diag$(U, \cdots, U)$ denotes the $r \times r$ matrix with $r/2$ copies of U along the diagonal, and zeros elsewhere.

$f_0(S^m)$ is trivial. Hence f_0 can be extended to an imbedding

$$f : S^m \times D^m \to W.$$

Let $W' = \chi(W, f)$, $W_0 = W - \text{Interior}\, f(S^m \times D^m)$.

First consider the exact sequence

$$0 \to H_m(W_0) \to H_m(W) \xrightarrow{j_*} H_m(W \bmod W_0) \xrightarrow{\partial} \cdots .$$

By excision it is seen that $H_m(W \bmod W_0)$ is infinite cyclic. A generator of this group has intersection number ± 1 with the cycle $f_0(S^m)$. Thus for $\alpha \in H_m(W)$ the image $j_*(\alpha)$ is equal to the intersection number $\langle \alpha, \beta_1 \rangle$ multiplied by a generator of $H_m(W \bmod W_0)$. Hence $H_m(W_0)$ is isomorphic to the subgroup of $H_m(W)$ generated by the elements $\beta_1, \beta_3, \beta_4, \cdots, \beta_r$; with β_2 omitted. It is easily verified that W_0 is $(m-1)$-connected.

Next consider the exact sequence

$$\cdots \to H_{m+1}(W' \bmod W_0) \xrightarrow{\partial} H_m(W_0) \to H_m(W') \to 0.$$

In this case the group $H_{m+1}(W' \bmod W_0)$ is infinite cyclic. The homomorphism ∂ carries a generator into the homology class represented by the imbedding

$$u \to f(u, v_0)$$

of S^m in W_0. But this is clearly the homology class β_1. Therefore $H_m(W')$ is a free abelian group with a basis corresponding to the elements $\beta_3, \beta_4, \cdots, \beta_r$. The manifold W' is $(m-1)$-connected.

The effect of this construction on W is to replace the sphere $f_0(S^m)$ by a sphere of dimension $m - 1$. Any map of an m-dimensional complex into W' can be deformed so as to miss this $(m-1)$-sphere, and hence can be deformed into the subspace W_0. This implies:

(1) that W' is also m-parallelizable and

(2) that the intersection number of any two elements of $H_m(W')$ is equal to the intersection number of two corresponding elements of $H_m(W)$. Thus the quadratic form of W' has matrix diag $(U, \cdots, U)$, with rank $r - 2$.

Now iterate this construction $r/2$ times. The resulting manifold W'' will still be $(m-1)$-connected and the group $H_m(W'')$ will be zero. Together with the Hurewicz theorem, this completes the proof of Theorem 4.

REMARK. If the hypothesis that W has index zero is replaced by the weaker hypothesis that the quadratic form of W is indefinite, then it is at least possible to reduce the rank of $H_m(W)$. The above argument shows that the intersection matrix of $H_m(W)$ has the form diag (U, X), for a suitable choice of basis. Applying surgery one obtains a manifold W' with intersection matrix X.

In conclusion we consider the corresponding problem for m odd. Then the quadratic form of W is replaced by a skew symmetric bilinear form. The following result is well known. (See Veblen [**9**, p. 183].) Let U' denote the matrix

$$\begin{pmatrix} 0 & 1 \\ -1 & 0 \end{pmatrix}.$$

ASSERTION. Every skew symmetric bilinear form with determinant ± 1 has matrix diag $(U', \cdots, U')$, for a suitable choice of basis.

This takes the place of Lemma 9, and is easier to prove. Unfortunately, however, Lemma 7 is not so easy to replace, except in two special cases.

If m is equal to 3 or 7 then the group $\pi_{m-1}(SO_m)$ is zero. Hence any m-sphere in a $2m$-dimensional manifold has trivial normal bundle. This makes it possible to prove the following analog of Theorem 4.

THEOREM 4'. *If W is compact and $(m - 1)$-connected of dimension $2m$, with $m = 3$ or 7, and if*

$$H_{m-1}(\text{Boundary } W) = H_m(\text{Boundary } W) = 0,$$

then W is χ-equivalent to an m-connected manifold.

The proof goes just as before, with the assertion concerning skew symmetric forms in place of Lemma 9.

For m odd but unequal to 1, 3, 7 the situation is more difficult, and will be considered in a later paper.

REFERENCES

1. B. Jones, *The arithmetic theory of quadratic forms*, New York, Wiley, 1950.
2. S. Lefschetz, *Topology*, 2nd ed., New York, Chelsea, 1956.
3. J. Milnor, *On simply connected 4-manifolds*, Symposium Topologia Algebraica, Mexico, 1958, pp. 122–128.
4. ———, *On the cobordism ring Ω^* and a complex analogue*. I, Amer. J. Math. vol. 82 (1960).
5. ———, *Sommes de variétés différentiables et structures différentiables des sphères*, Bull. Soc. Math. France vol. 87 (1959) pp. 439–444.
6. M. Morse, *The critical points of a function of n variables*, Trans. Amer. Math. Soc. vol. 33 (1931) pp. 72–91.
7. ———, *Calculus of variations in the large*, Amer. Math. Soc. Colloquium Publications, vol. 18, New York, 1934.
8. N. Steenrod, *The topology of fibre bundles*, Princeton, 1951.
9. O. Veblen, *Analysis situs*, 2nd ed., Amer. Math. Soc. Colloquium Publications, New York, 1931.
10. C. T. C. Wall, *Note on the cobordism ring*, Bull. Amer. Math. Soc. vol. 65 (1959) pp. 329–331.
11. H. Whitney, *Differentiable manifolds*, Ann. of Math. vol. 37 (1936) pp. 645–680.
12. ———, *The self-intersections of a smooth n-manifold in 2n-space*, Ann. of Math. vol. 45 (1944) pp. 220–246.
13. ———, *On singularities of mappings of euclidean spaces*. I. *Mappings of the plane into the plane*, Ann. of Math. vol. 62 (1955) pp. 374–410.
14. A. H. Wallace, *Modifications and cobounding manifolds*, Canadian J. Math. vol. 12 (1960) pp. 503–528.

PRINCETON UNIVERSITY,
PRINCETON, NEW JERSEY

Differentiable Manifolds Which Are Homotopy Spheres

John Milnor[1]

Princeton University, January 23, 1959

1. Introduction.

This paper will study the problem of classifying differentiable n-manifolds which are homotopy spheres, under the relation of J-equivalence. (See the "dictionary" below.) It is shown that the equivalence classes form an abelian group which is denoted by Θ^n. The only groups Θ^n which I have been able to determine completely are the following:

$$\Theta^1 = \Theta^2 = 0, \quad \Theta^5 = 0, \quad \Theta^7 = \mathbb{Z}_{28}, \quad \Theta^{11} = \mathbb{Z}_{992}.$$

However partial information is obtained in many other cases. For example (according to Lemmas 3.7, and Corollaries 5.11 and 6.9):

Theorem. *For $k > 1$ the group Θ^{4k-1} is finite but non-trivial.*

Section 2 of this paper will study a sum operation for connected manifolds of the same dimension. Section 3 defines an invariant λ' for certain $(4k-1)$-manifolds. Section 4 contains examples of homotopy spheres for which the invariant λ' takes on all possible values.

Section 5 describes a construction for simplifying manifolds, which was communicated to the author by R. Thom. Using this construction it is shown that the invariant $\lambda'(M)$ determines the J-equivalence class of M uniquely. A corresponding result for dimensions of the form $4k+1$ is stated without proof. Section 6 studies the following question: Is every homotopy sphere the boundary of a π-manifold?

Section 7 contains further discussion and a list of unsolved problems. Operations of "pasting together" manifolds and "straightening angles" are described in an appendix.

[1] The author holds a Sloan fellowship.

Dictionary of terms used. The word *manifold* will mean a compact, oriented, differentiable manifold, with or without boundaries. (The phrase "topological manifold" will be used in case the differentiable structure has not yet been specified.) The symbol $-M$ will be used for the manifold M with orientation reversed.

Two unbounded manifolds M_1, M_2 of the same dimension are J-*equivalent* if there exists a manifold W such that

(1) the boundary ∂W is the disjoint union of M_1 and $-M_2$, and
(2) both M_1 and M_2 are deformation retracts of W.

Thus J-equivalent manifolds belong to the same cobordism class and to the same homotopy type. This concept is due to [Thom, 3]. It is not known whether J-equivalent manifolds are necessarily diffeomorphic.

By a *homotopy sphere* we mean a (differentiable) manifold without boundary which has the homotopy type of a sphere. Similarly a *homology sphere* M must be unbounded and satisfy $H_*(M) \approx H_*(S^n)$. Here H_* denotes homology with integer coefficients, and S^n denotes the unit sphere in Euclidean space $\mathbb{R}^{n+1}$. The notation D^{n+1} will be used for the disk bounded by S^n.

LEMMA 1.1. *Let $M^n = \partial W^{n+1}$ where M^n is simply connected and W^{n+1} is contractible. Then M^n is J-equivalent to S^n.*

Proof. Choose an embedding of D^{n+1} in the interior of W^{n+1}. Then

$$W^{n+1} \setminus (\text{interior } D^{n+1})$$

has boundary equal to the disjoint union of M^n and S^n. It is not difficult to see that both boundaries are deformation retracts of $W^{n+1} \setminus (\text{interior } D^{n+1})$. □

A π-*manifold* W^n is characterized by the following property. If W^n is embedded in a high dimensional Euclidean space $\mathbb{R}^{n+q}$, then the normal bundle ν^q is trivial. This concept is due to [Whitehead, 2]. If W is a π-manifold, then clearly ∂W is also a π-manifold.

W^n will be called *almost parallelizable* if there exists a finite subset F so that $W^n \setminus F$ is parallelizable.

LEMMA 1.2. (J.H.C. Whitehead). *Every parallelizable manifold is a π-manifold. Every π-manifold is almost parallelizable.*

Proof. A field of tangent n-frames on $W^n \subset \mathbb{R}^{n+q}$ induces a map f from W^n to the Stiefel manifold $V_{n+q,n}$. Note that f is covered by a bundle map from ν^q to a corresponding $\mathrm{SO_q}$-bundle over $V_{n+q,n}$. But the space $V_{n+q,n}$ is $(q-1)$-connected. (See [Steenrod, 1, §25.6].) For $q > n$ this implies that f is homotopic to a constant; hence that ν^q is trivial.

Similarly a field of normal q-frames on W^n induces $f : W^n \to V_{n+q,q}$. Since $V_{n+q,q}$ is $(n-1)$-connected, the only obstruction to contracting f lies in

$$H^n(W^n;\ \pi_n(V_{n+q,q})).$$

But this cohomology group can be killed by removing a finite number of points from W^n. □

A similar argument shows the following.

LEMMA 1.3. *If every component of W^n has a non-vacuous boundary, then the three concepts: parallelizable, π-manifold, and almost parallelizable, are equivalent.*

The *J-homomorphism* of H. Hopf and G. Whitehead will be denoted by

$$J_n : \pi_n(\mathrm{SO_q}) \longrightarrow \pi_{\mathrm{n+q}}(\mathrm{S^q}).$$

(For a definition see [Kervaire, 4, §1.8]. Caution: this homomorphism has nothing to do with J-equivalence.) It will always be assumed that q is large. This homomorphism will play a fundamental role in what follows.

2. The connected sum of manifolds.

Let M_1, M_2 be connected differentiable manifolds of the same dimension n. The sum $M_1\#M_2$ is obtained by removing an n-cell from each, and then pasting the resulting boundaries together. There are three difficulties with this:

(1) The pasting must be done in such a way that $M_1\#M_2$ has an orientation compatible with that of both M_1 and M_2.

(2) Not every diffeomorphism between the boundaries will give rise to the same composite manifold, even when the orientations are properly matched. (According to [Milnor, 1] it is possible to past together the boundaries of two 7-cells, obtaining a manifold which is not diffeomorphic to S^7.)

(3) It is necessary to show that the result does not depend on which n-cell is chosen.

Definition. Choose an orientation preserving embedding $h_1 : \mathbb{R}^n \to M_1$ and an orientation reversing embedding $h_2 : \mathbb{R}^n \to M_2$. Let $M_1\#M_2$ be obtained from the disjoint union of $M_1 \setminus \{h_1(0)\}$ and $M_2 \setminus \{h_2(0)\}$ by identifying $h_1(x)$ with $h_2(x/||x||^2)$ for each $x \neq 0$ in $\mathbb{R}^n$.

Remark 2.1. It would be sufficient to specify $h_1(x)$ and $h_2(x)$ for $||x|| < 1+\epsilon$ in order to construct this manifold $M_1\#M_2$. In fact by removing all $h_i(x)$ with $||x|| \leq 1/(1+\epsilon)$ from each M_i, and then identifying $h_1(x)$ with $h_2(x/||x||^2)$ for $1+\epsilon > x > 1/(1+\epsilon)$, we obtain the identical manifold $M_1\#M_2$.

The following will be proved in a paper by J. Cerf.*

THEOREM 2.2. (Theorem of Cerf). *Let M be a connected n-manifold. Given two orientation preserving embeddings f, $f' : D^n \to$ (interior M), there exists a diffeomorphism $g : M \to M$ which satisfies $gf = f'$.*

COROLLARY 2.3. *The sum $M_1\#M_2$ is well defined up to orientation preserving diffeomorphism.*

Proof of the Corollary. The only choice which occurred in the definition was the choice of embeddings h_1, h_2. Given other embeddings h_1', h_2' there exist diffeomorphisms g_i of M_i so that

$$g_i h_i(x) = h_i'(x) \qquad \text{for} \quad ||x|| \leq 1+\epsilon.$$

*I have been unable to locate such a paper. However, this result is proved, for example, in the "Lectures on Differentiable Structures", on pages 177–190 below.

These g_i give rise to a diffeomorphism $g : M_1\#M_2 \to (M_1\#M_2)'$; which completes the proof. $\square$

LEMMA 2.4. *Suppose that the unbounded manifolds M_1, M_2 are J-equivalent to M_1' and M_2' respectively. Then the sum $M_1\#M_2$ is J-equivalent to $M_1'\#M_2'$.*

Proof. If the dimension n is ≤ 2, then the assertion is clear. Hence we may assume that $n \geq 3$. Choose manifolds W_i so that ∂W_i is the disjoint union of the deformation retracts M_i and $-M_i'$. Choose a differentiable arc a_i from $p_i \in M_i$ to $p_i' \in M_i'$ in W_i, so that the interior of a_i lies in the interior of W_i. We will see that the inclusion map

$$j : M_i \smallsetminus \{p_i\} \longrightarrow W_i \smallsetminus a_i$$

is a homotopy equivalence.

Since the codimension n of $\{p_i\}$ in M_i is ≥ 3, the homomorphisms

$$\pi_1(M_i \smallsetminus \{p_i\}) \longrightarrow \pi_1(M_i), \qquad \pi_1(W_i \smallsetminus a_i) \longrightarrow \pi_1(W_i)$$

are isomorphisms. Hence

$$j_* : \pi_1(M_i \smallsetminus \{p_i\}) \longrightarrow \pi_1(W_i \smallsetminus a_i)$$

is an isomorphism.

Let $\widehat{M}_i \subset \widehat{W}_i$ denote the universal covering spaces, and let $\{\widehat{p}_i\} \subset \widehat{a}_i$ denote the inverse images of $\{p_i\}$, a_i. The inclusion

$$(\widehat{M}_i,\ \widehat{M}_i \smallsetminus \{\widehat{p}_i\}) \longrightarrow (\widehat{W}_i,\ \widehat{W}_i \smallsetminus \widehat{a}_i)$$

gives rise to a homomorphism between exact sequences of homology groups. Using the Five Lemma it follows that

$$\widehat{\jmath}_* : H_k(\widehat{M}_i \smallsetminus \{\widehat{p}_i\}) \longrightarrow H_k(\widehat{W}_i \smallsetminus \widehat{a}_i)$$

is an isomorphism for all k. Therefore j is a homotopy equivalence. (Compare [Whitehead, 3].)

Choose tubular neighborhoods N_i of a_i, and let W be a manifold obtained from $W_1 \smallsetminus N_1$ and $W_2 \smallsetminus N_2$ by pasting together the boundaries in such a way that ∂W is the disjoint union of $M_1\#M_2$ and $-(M_1'\#M_2')$. Since the inclusions

$$M_i \smallsetminus (M_i \cap N_i) \longrightarrow W_i \smallsetminus N_i$$

are homotopy equivalences, it follows easily that the inclusion

$$M_1 \# M_2 \hookrightarrow W$$

is a homotopy equivalence. A corresponding argument takes care of the inclusion $(M_1'\#M_2') \hookrightarrow W$. This completes the proof of Lemma 2.4. $\square$

It is clear that the operation $\#$ is associative and commutative, providing that we do not distinguish between diffeomorphic manifolds. Furthermore the sphere acts as a zero element: $M \# S^n \approx M$.

LEMMA 2.5. *Suppose that M is a homotopy n-sphere. Then $M\#(-M)$ is J-equivalent to S^n.*

Proof. Let U denote the interior of a disk $D^n \subset M$. Consider the topological manifold $(M \smallsetminus U) \times [0,\, 1]$. This is differentiable, except along the "angles" $\partial U \times \{0\}$ and $\partial U \times \{1\}$. Let W be a differentiable manifold obtained from $(M \smallsetminus U) \times [0,\, 1]$ by straightening these angles. (See the Appendix.) Then W is a contractible manifold with boundary $M\#(-M)$. Together with Lemma 1.1 this completes the proof. □

Now combining Lemmas 2.4 and 2.5 this proves:

THEOREM 2.6. *The set of all J-equivalence classes of homotopy n-spheres forms an abelian group under the operation $\#$.*

This group will be denoted by Θ^n. It is clear that $\Theta^1 = 0$. Since Munkres has shown that a 2-manifold has an essentially unique differentiable structure, it follows that $\Theta^2 = 0$.

[Two subgroups of Θ^n will also be studied. $\Theta^n(\pi)$ will denote the subgroup formed by all π-manifolds in Θ^n, and $\Theta^n(\partial\pi)$ will denote the subgroup formed by all boundaries of π-manifolds.]

3. The invariant $\lambda'(M^{4k-1})$.

Let M be a $(4k-1)$-manifold which is **(1)** a homology sphere, and **(2)** the boundary of some π-manifold W. The intersection number of two homology class α, β of W will be denoted by $< \alpha,\, \beta >$. Let $I(W)$ denote the index of the quadratic form

$$\alpha \longrightarrow < \alpha,\, \alpha >,$$

where α varies over the Betti group $H_{2k}(W)/(\text{torsion})$. Integer coefficients are to be understood.

Define I_k as the greatest common divisor of $I(M')$ where M' ranges over all *almost parallelizable* manifolds of dimension $4k$ which have no boundary. This number has been studied by [Kervaire and Milnor, 1]. (See Lemma 3.7.)

LEMMA 3.1. *The residue class of $I(W)$ modulo I_k is an invariant of the homology sphere $M = \partial W$.*

Proof. If M is the boundary of two parallelizable manifolds W_1 and W_2, let N be the unbounded $4k$-manifold obtained from W_1 and $-W_2$ by pasting together the common boundary. Clearly

$$I(N) \;=\; I(W_1) \smallsetminus I(W_2).$$

Let p be a point of M. Then the complement $N \smallsetminus \{p\}$ is parallelizable. In fact $N \smallsetminus \{p\}$ is the union of parallelizable manifolds $W_1 \smallsetminus \{p\}$ and $W_2 \smallsetminus \{p\}$, having an intersection $M \smallsetminus \{p\}$ which is acyclic. Given a field of $4k$-frames on $W_1 \smallsetminus \{p\}$ and on $W_2 \smallsetminus \{p\}$, it is possible to deform one of the two so that they coincide along $M \smallsetminus \{p\}$. Therefore N is almost parallelizable; and

$$I(N) \;\equiv 0 \pmod{I_k}.$$

This completes the proof. □

Not every residue class can occur:

LEMMA 3.2. *The index $I(W)$ of an almost parallelizable manifold is always divisible by 8; providing that ∂W is a homology sphere.*

Proof. First observe that the intersection number $< \alpha, \alpha >$ is always an even integer. This is the homology translation of the statement that

$$\mathrm{Sq}^{2k} : H^{2k}(W, \partial W;\ \mathbb{Z}) \longrightarrow H^{4k}(W, \partial W;\ \mathbb{Z}_2)$$

is zero. If Sq^{2k} were not zero then the formulae of Wu (see [Wu Wen-Tsün, 1] and also [Kervaire, 2]) would imply that W had a non-trivial Stiefel-Whitney class in dimension $\leq 2k$.

Since ∂W is a homology sphere it follows by Poincaré duality that the matrix of intersection numbers has determinant ± 1. But a quadratic form with determinant ± 1 which takes on only even values must have index divisible by 8. (Compare [Milnor, 4]). This completes the proof. □

Definition. The residue class of $\frac{1}{8}I(W)$ modulo $\frac{1}{8}I_k$ will be denoted by $\lambda'(M)$.

LEMMA 3.3. *The properties of being* **(1)** *a homotopy n-sphere, and* **(2)** *the boundary of a π-manifold, are invariant under J-equivalence; and are preserved by the sum operation $\#$. Hence the manifolds which have these properties give rise to a subgroup of Θ^n.*

Definition. This subgroup will be denoted by $\Theta^n(\partial\pi)$.

LEMMA 3.4. *The invariant $\lambda'(M)$ depends only on the J-equivalence class of M. Furthermore*

$$\lambda'(M_1 \# M_2) = \lambda'(M_1) + \lambda'(M_2).$$

The proofs of Lemmas 3.3 and 3.4 are straightforward. Hence λ' gives rise to a homomorphism

$$\Lambda' : \Theta^{4k-1}(\partial\pi) \longrightarrow \mathbb{Z}_{\frac{1}{8}I_k}.$$

It will be proved in Sections 4 and 5 that Λ' is an isomorphism, at least for $k > 1$.

The principal difficulty with the invariant λ' is that it is extremely difficult to compute. For example it would be very interesting to evaluate λ' for the topological spheres which are constructed in [Milnor, 1, 5] and [Shimada, 1]. The invariant λ which is defined in these papers is somewhat weaker, but much easier to compute.

The numbers $\frac{1}{8}I_k$ can be described as follows. Let B_k denote the k-th Bernoulli number:

$$B_1 = \frac{1}{6}, \quad B_2 = \frac{1}{30}, \quad \ldots, \quad B_6 = \frac{691}{2730}, \quad \ldots.$$

Define j_k as the order of the image

$$J_{4k-1}(\mathrm{SO_q}) \subset \pi_{\mathrm{q+4k-1}}(\mathrm{S^q}) \qquad \text{for large q.}$$

Define

$$a_k = \begin{cases} 2, & \text{if } k \text{ is odd} \\ 1, & \text{if } k \text{ is even.} \end{cases}$$

Then according to [Kervaire and Milnor, 1]:

LEMMA 3.5. *I_k is equal to*

$$2^{2k-1}(2^{2k-1} - 1)B_k j_k a_k / k.$$

The only unknown quantity here is the integer j_k.

LEMMA 3.6. *j_k is a multiple of the denominator of $B_k/4k$.*

Proof. For k even, this is proved in [Kervaire and Milnor, 1]. For k odd this follows from the arguments of that paper, together with the following:

(Theorem of Hirzebruch [Not yet published].†) *If the unbounded manifold M^{4k} has Stiefel-Whitney class w_2 equal to zero, and if k is odd, then the $\widehat{A}$-genus $\widehat{A}[M^{4k}]$ is an even integer.*

This completes the proof of Lemma 3.6. □

On the other hand an upper bound for j_k is given by the order of the largest cyclic subgroup of $\pi_{q+4k-1}(S^q)$. The p-primary component of $\pi_{q+4k-1}(S^q)$ is known for $k < p^2(p-1)/2$ for any prime p. (See [Toda, 1, 2].) The full group is known (to me) only for $k = 1, 2, 3$. It turns out that the upper bound for the p-primary factor of j_k is exactly equal to the lower bound in each known case.

Combining the preceding information, we have:

LEMMA 3.7. *The number $\frac{1}{8}I_k$ is equal to*

$$a_k 2^{2k-2}(2^{2k-1}-1)\text{numerator}(B_k/k),$$

multiplied by an integer whose prime factors p satisfy $p^2(p-1) \le 2k$. In particular

$$\frac{1}{8}I_1 = 2, \quad \frac{1}{8}I_2 = 28, \quad \frac{1}{8}I_3 = 992, \quad \frac{1}{8}I_4 = 8128 \times \text{power of } 2.$$

4. Construction of $(4k-1)$-manifolds.

The following is perhaps the simplest example of a symmetric matrix with determinant ± 1, with only even elements on the diagonal, and with index different from zero. (Compare [Milnor, 4].)

$$\begin{pmatrix} 2 & 1 & 0 & 0 & 0 & 0 & 0 & 0 \\ 1 & 2 & 1 & 0 & -1 & 0 & 0 & 0 \\ 0 & 1 & 2 & 1 & 0 & 0 & 0 & 0 \\ 0 & 0 & 1 & 2 & 1 & 0 & 0 & 0 \\ 0 & -1 & 0 & 1 & 2 & 1 & 0 & 0 \\ 0 & 0 & 0 & 0 & 1 & 2 & 1 & 0 \\ 0 & 0 & 0 & 0 & 0 & 1 & 2 & 1 \\ 0 & 0 & 0 & 0 & 0 & 0 & 1 & 2 \end{pmatrix}$$

We will construct a manifold W^{4k} which has the above intersection matrix.

Let T be a tubular neighborhood of the diagonal in $S^{2k} \times S^{2k}$: say the set of all pairs (x, y) with distance $d(x, y) \le \epsilon$. Thus T is a $4k$-manifold having the

†See for example HIRZEBRUCH [1966, p.199].

homotopy type of S^{2k}. The intersection number of the fundamental $2k$-cycle with itself is $+2$.

Let $\alpha : S^{2k} \longrightarrow S^{2k}$ be the "twelve hour rotation" which leaves the north pole p fixed, and satisfies $\alpha(x) = -x$ for x on the equator. Let $T' = (1 \times \alpha)T$ be the set of pairs (x, y) with $d(x, \alpha y) \leq \epsilon$. Then T and T' intersect only in a small neighborhood of the pair (p, p), and a small neighborhood of the pair $(-p, -p)$.

The universal covering space of $T \cup T'$ consists of infinitely many disjoint copies of T and infinitely many disjoint copies of T'. Numbering these copies T_i and T'_i, we may assume that each T'_i intersects only T_i and T_{i+1}.

Define W_1 as the subset

$$T_1 \cup T'_1 \cup T_2 \cup T'_2 \cup T_3 \cup T'_3 \cup T_4 \cup T'_4$$

of this universal covering space. Thus W_1 is a topological $4k$-manifold, having the homotopy type of a union of eight $2k$-spheres with a single point in common. Choosing an appropriate basis for $H_{2k}(W_1)$, the intersection matrix is as follows:

$$\begin{pmatrix} 2 & 1 & \dots & \dots & \dots & \dots & \dots & \dots \\ 1 & 2 & 1 & \dots & \dots & \dots & \dots & \dots \\ \dots & 1 & 2 & 1 & \dots & \dots & \dots & \dots \\ \dots & \dots & 1 & 2 & 1 & \dots & \dots & \dots \\ \dots & \dots & \dots & 1 & 2 & 1 & \dots & \dots \\ \dots & \dots & \dots & \dots & 1 & 2 & 1 & \dots \\ \dots & \dots & \dots & \dots & \dots & 1 & 2 & 1 \\ \dots & \dots & \dots & \dots & \dots & \dots & 1 & 2 \end{pmatrix}$$

To correct this intersection matrix it is necessary to introduce an intersection between T'_1 and T_3, so as to obtain an intersection number -1. Choose a rotation of $S^{2k} \times S^{2k}$ which carries a region of T' near the "equator" onto a region of T near the "equator", so as to obtain an intersection number of -1. Matching the corresponding regions of T'_1 and T_3, we obtain a topological manifold W_2, with the required intersection matrix.

This manifold W_2 is differentiable except along eight "angles" which have been introduced in the boundary. Let W_3 be a differentiable manifold obtained by straightening these angles. (See the appendix.)

Unfortunately the transition from W_1 to W_2 changed the homotopy type. In fact the fundamental group $\pi_1(W_2) = \pi_1(W_3)$ is infinite cyclic. Next we will kill this fundamental group. A generator can be represented by a simple closed differentiable curve C lying on the boundary of W_3.

Choose an embedding $h : S^1 \times D^{4k-2} \longrightarrow \partial W_3$ which carries $S^1 \times \{0\}$ onto the given curve C. Le W_4 be the space obtained from the disjoint union

$$W_3 \bigcup D^2 \times D^{4k-2}$$

by identifying $S^1 \times D^{4k-2}$ with its image under h. Then W_4 is simply connected. In fact W_4 has the same homotopy type as W_1; but the same intersection matrix as W_2 or W_3.

This space W_4 is a differentiable manifold, except along the "angle" corresponding to $S^1 \times S^{4k-3}$. Let W_0 be a differentiable manifold obtained by "straightening" this angle.

THEOREM 4.1. *W_0 is a parallelizable $4k$-manifold, with boundary M_0 which is a homology $(4k-1)$-sphere. In fact for $k > 1$, M_0 is a homotopy sphere. The index $I(W_0)$ equals $+8$.*

Thus the invariant $\lambda'(M_0)$ is defined and equal to $+1$.

COROLLARY 4.2. *The homomorphism Λ' from $\Theta^{4k-1}(\partial\pi)$ to the cyclic group of order $\frac{1}{8}I_k$ is onto, providing that $k > 1$.*

COROLLARY 4.3. *The group Θ^{4k-1} is non-trivial, providing that $k > 1$.*

Proof that W_0 is parallelizable. The only obstruction to parallelizability lies in the group

$$H^{2k}(W_0;\ \pi_{2k-1}(\mathrm{SO}_{4k})).$$

But $H_{2k}(W_0)$ is generated by eight cycles, each of which is contained in a submanifold diffeomorphic to $T \subset S^{2k} \times S^{2k}$. Since S^{2k} is a π-manifold, it follows that T is a π-manifold, hence parallelizable. Therefore, W_0 is parallelizable.

Computation of $H_*(M_0)$. Since the groups $H_i(W_0)$ have no torsion, it follows by Poincaré-Lefschetz duality that $H_i(W_0, M_0) \approx H^{4k-1}(W_0)$ is isomorphic to $\mathrm{Hom}(H_{4k-i}(W_0), \mathbb{Z})$. The natural homomorphism

$$H_i(W_0) \longrightarrow H_i(W_0, M_0) \approx \mathrm{Hom}(H_{4k-i}(W_0), \mathbb{Z})$$

is determined by the matrix of intersection numbers.

Now recall that $H_0(W_0) \approx \mathbb{Z}$; that $H_i(W_0) = 0$ for $i \neq 0,\ 2k$; and that the matrix of intersection numbers in dimension $2k$ has determinant $+1$. Plugging this information into the exact sequence of the pair (W_0, M_0), it follows that M_0 has the homology of a $(4k-1)$-sphere.

Proof that M_0 is simply connected, providing that $k > 1$. (For k equal to 1 the group $\pi_1(M_0)$ depends on the choice of the curve C. If C could be chosen so that $\pi_1(M_0) = 0$, then M_0 would provide a counter-example to the Poincaré hypothesis.)

Let $K \subset W_3$ denote the union of 8 copies of S^{2k}, one in the center of each T_i and T_i'. Since the codimension $2k$ of K is greater than 2, it follows that

$$\pi_2(W_3,\ W_3 \setminus K) = 0.$$

But it is clear that ∂W_3 is a deformation retract of $W_3 \setminus K$. Hence $\pi_2(W_3, \partial W_3) = 0$. From the exact sequence of this pair it follows that $\pi_1(\partial W_3)$ is infinite cyclic, generated by the closed curve C.

The manifold ∂W_4 can be obtained from ∂W_3 in two steps, as follows. (Compare Lemma 5.5.)

(1) Remove a tubular neighborhood of $C \subset \partial W_3$. Since the codimension $4k-2$ of C in ∂W_3 is greater than 2, it follows that $\pi_1(\partial W_3 \smallsetminus C)$ is also infinite cyclic.

(2) Fill in the resulting hole with a copy of $D^2 \times S^{4k-3}$. The effect of this addition on the fundamental group is to kill the generator. Hence $\pi_1(\partial W_4) = 0$.

Since the differentiable manifold $M_0 = \partial W_0$ is homeomorphic to ∂W_4, this completes the proof that M_0 is a homotopy sphere. Since the index $I(W_0)$ is easily shown to be $+8$, this proves Theorem 4.1. □

5. Simplifying manifolds by surgery.

This section will describe an operation, suggested to the author by Thom, which can be used to kill off the lower homotopy groups of a manifold. To illustrate the method, the following will first be proved.

THEOREM 5.1. *Let M be an unbounded $4k$-manifold which is almost parallelizable. (That is there exists a finite subset F, so that $M \smallsetminus F$ is parallelizable.) Then there exists an unbounded $4k$-manifold M' which satisfies:*

(1) *the index $I(M')$ equals $I(M)$,*

(2) *M' is also almost parallelizable, and*

(3) *M' is $(2k-1)$-connected.*

Remark 5.2. It is not possible to kill off any further homotopy groups: If M' were $2k$-connected, then the index $I(M')$ would have to be zero.

Remark 5.3. The hypothesis that M is almost parallelizable is essential here. As an example, for $n = 12$, the complex projective space $\mathbb{P}_6(C)$ has index 1. But for any 5-connected 12-manifold, the index must be divisible by $I_3 = 7936$. (This follows since the only obstruction to almost parallelizability lies in $H^6(M; \pi_5(\mathrm{SO}_{12})) = 0$.)

The proof of Theorem 5.1 begins as follows. If M has several components, let M_1 denote the connected sum of these components. It is not hard to show that M_1 is almost parallelizable, and that $I(M_1) = I(M)$.

Suppose by induction that M is $(q-1)$-connected, where $0 < q < 2k$. Any given element

$$\alpha \in \pi_q(M)$$

can be represented by an embedding

$$f : S^q \longrightarrow M.$$

(This presents no difficulty since the dimension $4k$ is greater than $2q$. Compare [Whitney, 1].)

LEMMA 5.4. *Let $f : S^q \longrightarrow M^n$ be an embedding, with $q < \frac{n}{2}$; and suppose that the bundle $f^*(\tau^n)$ induced from the tangent bundle of M^n is trivial. Then the normal bundle ν^{n-q} is trivial.*

Proof. Let o^k denote the trivial $\mathrm{SO_k}$-bundle over S^q, and let τ^q denote the tangent bundle. It is well known that the Whitney sum $\tau^q \oplus o^1$ is trivial. We are assuming that the bundle $\nu^{n-q} \oplus \tau^q \sim f^*(\tau^n)$ is trivial. Therefore

$$\nu^{n-q} \oplus o^{q+1} \sim \nu^{n-q} \oplus \tau^q \oplus o^1 \sim o^n \oplus o^1 \sim o^{n+1}$$

is trivial. That is: the inclusion $\mathrm{SO_{n-q}} \hookrightarrow \mathrm{SO_{n+1}}$ carries the $\mathrm{SO_{n-q}}$-bundle ν^{n-q} into the trivial bundle. But the homomorphism

$$\pi_{q-1}(\mathrm{SO_{n-q}}) \longrightarrow \pi_{q-1}(\mathrm{SO_{n+1}})$$

is an isomorphism in the stable range $n - q > q$. This completes the proof. $\square$

Let T be a tubular neighborhood of $f(S^q)$. Then T can be identified with the total space of the D^{4k-q}-bundle which is associated with ν^{4k-q}. Choosing a specific product structure for ν^{4k-q}, it follows that T is homeomorphic to $S^q \times D^{4k-q}$. Let M_1 denote a differentiable manifold obtained from M by

(1) removing the interior of T, and
(2) pasting a copy of $D^{q+1} \times S^{4k-q-1}$ in its place, matching the common boundary $S^q \times S^{4k-q-1}$.

LEMMA 5.5. *The manifold M_1 is also $(q-1)$-connected. Furthermore $\pi_q(M_1)$ is isomorphic to $\pi_q(M)/(\alpha)$, where (α) denotes the normal subgroup generated by α.*

Proof. Since $f(S^q)$ has codimension $4k-q$ in M, it follows that $\pi_i(M \smallsetminus f(S^q))$ is isomorphic to $\pi_i(M)$ for $i < 4k - q - 1$. In particular

$$\begin{array}{lcll} \pi_i(M \smallsetminus f(S^q)) & = & 0 \quad \text{for} \quad i < q, & \text{and} \\ \pi_q(M \smallsetminus f(S^q)) & \approx & \pi_q(M). & \end{array}$$

The manifold $M \smallsetminus f(S^q)$ can be obtained from M_1 by removing the sphere $\{0\} \times S^{4k-q-1}$ of codimension $q + 1$.
Therefore

$$\pi_i(M_1) \approx \pi_i(M \smallsetminus f(S^q)) = 0 \quad \text{for} \quad i < q.$$

Case 1. $q = 1$. Then $\pi_1(M_1)$ can be computed as follows. The manifold M_1 can be obtained from $M \smallsetminus T$ by first adjoining a 2-cell $D^2 \times$ (constant) and then adjoining a 4k-cell. The first operation introduces the relation $\alpha = 0$ into the fundamental group; while the second operation leaves the group unchanged.

Case 2. $q > 1$. Then the group $\pi_{q+1}(M_1, M \smallsetminus f(S^q))$ is isomorphic to $H_{q+1}(M_1, M \smallsetminus f(S^q)) \approx \mathbb{Z}$. In the exact sequence

$$\mathbb{Z} \longrightarrow \pi_q(M \smallsetminus f(S^q)) \longrightarrow \pi_q(M_1) \longrightarrow 0,$$

it is clear that $\partial(1) = \alpha$, so that $\pi_q(M_1) \approx \pi_q(M)/(\alpha)$, as required.
This completes the proof of Lemma 5.5. $\square$

LEMMA 5.6. *If the product structure for the normal bundle ν^{4k-q} is correctly chosen, then the manifold M_1 will also be almost parallelizable.*

Before giving the proof, here is a description of some vector fields on $S^q \subset D^{q+1}$. Let $\epsilon_1, \ldots, \epsilon_{q+1}$ be the standard basis for the tangent vector space of D^{q+1}. The outward normal vector at a point $(t_1, \ldots, t_{q+1}) \in S^q$ is $\zeta = t_1\epsilon_1 + \ldots + t_{q+1}\epsilon_{q+1}$. Let ϵ'_i denote the projection of ϵ_i into the tangent bundle of S^q. Thus $\epsilon'_i = \epsilon_i - t_i\zeta$, so that $\epsilon_i = \epsilon'_i + t_i\zeta$.

Proof of Lemma 5.6. Choose some field φ_1 of vectors normal to $f(S^q)$. The "endpoints" of the vectors φ_1 sweep out a subset of ∂T which will be denoted by $S^q \times (1, 0, \ldots, 0)$. The outward normal vector to ∂T at a point of $S^q \times (1, 0, \ldots, 0)$ will also be denoted by φ_1 (since it can be considered as a translate of the vector φ_1 at the corresponding point of $f(S^q)$). Now consider the vector fields

$$\epsilon'_1 + t_1\varphi_1\,, \quad \epsilon'_2 + t_2\varphi_1\,, \quad \ldots, \quad \epsilon'_{q+1} + t_{q+1}\varphi_1\,,$$

along $S^q \times (1, 0, \ldots, 0)$. These are orthogonal unit vectors in the tangent bundle of M.

Since M is almost parallelizable, there exits a field $(\Psi_1, \ldots, \Psi_{4k})$ of $4k$-frames which is defined over $M \setminus F$. Here F denotes some finite set which we may assume is disjoint from T.

> **Assertion.** It is possible to deform this field $(\Psi_1, \ldots, \Psi_{4k})$ so as to obtain a field $(\Psi'_1, \ldots, \Psi'_{4k})$ of $4k$-frames such that along $S^q \times (1, 0, \ldots, 0)$:
>
> $$\Psi'_1 = \epsilon'_1 + t_1\varphi_1\,, \quad \ldots, \quad \Psi'_{q+1} = \epsilon'_{q+1} + t_{q+1}\varphi_1\,.$$

This is proved as follows. Define a matrix $a_{ij}(t_1, \ldots, t_{q+1})$ by the formulae

$$\epsilon'_i + t_i\varphi_1 = \sum_{j=1}^{4k} a_{ij}\Psi_j\,, \qquad i = 1, 2, \ldots, q+1.$$

This defines a map from S^q to the Stiefel manifold $V_{4k,q+1}$. This map is null-homotopic since $\pi_q(V_{4k,\,q+1}) = 0$. Hence it can be lifted to a null-homotopic map of S^q into $V_{4k,\,4k}$

$$(t_1, \ldots, t_{q+1}) \longrightarrow ||a_{ij}||, \qquad i = 1, \ldots, 4k.$$

Let $\Psi'_i = \Psi_i$ outside of a neighborhood of $S^q \times (1, 0, \ldots, 0)$ but let

$$\Psi'_i = \sum a_{ij}\Psi_j$$

for points in $S^q \times (1, 0, \ldots, 0)$, and for all i. The null-homotopy can now be used to define Ψ'_i throughout the neighborhood.

We may assume that the vectors Ψ'_i along $f(S^q)$ are translates of those along $S^q \times (1, 0, \ldots, 0)$.

Now choose the product structure for the normal bundle of $f(S^q)$ which is determined by the field

$$\varphi_1, \quad \Psi'_{q+2}, \quad \ldots, \quad \Psi'_{4k}$$

of normal $(4k-q)$-frames. In terms of this product structure, construct the manifold

$$M_1 = M \setminus (\text{interior } T) \bigcup D^{q+1} \times S^{4k-q-1}.$$

The $4k$-frame $(\Psi'_1, \ldots, \Psi'_{4k})$ in $M \setminus (\text{interior } T)$ can be extended throughout $D^{q+1} \times (1, 0, \ldots, 0)$ as follows. Note that the vector φ_1 along $S^q \times (1, 0, \ldots, 0)$ can be identified with the normal vector to S^q in D^{q+1}. Hence the vectors $\Psi'_i = \epsilon'_i + t_i\varphi_1$ $(1 \leq i \leq q+1)$ along $S^q \times (1, 0, \ldots, 0)$ can be identified with the standard basis for the tangent bundle of $D^{q+1} \times (1, 0, \ldots, 0)$. Hence these vectors $\Psi'_1, \ldots, \Psi'_{q+1}$ can be extended.

The remaining vectors $\Psi'_{q+2}, \ldots, \Psi'_{4k}$ are normal to D^{q+1}. The projection $S^q \times S^{4k-q-1} \longrightarrow S^{4k-q-1}$ carries these vectors onto a fixed $(4k-q-1)$-frame at the point $(1, 0, \ldots, 0) \in S^{4k-q-1}$. Hence it is certainly possible to extend $(\Psi'_{q+2}, \ldots, \Psi'_{4k})$ over $D^{q+1} \times (1, 0, \ldots, 0)$ as a field of normal $(4k-q-1)$-frames.

Thus a field of $4k$-frames has been defined over the subset

$$(M \setminus (\text{interior } T) \setminus F) \bigcup (D^{q+1} \times (1, 0, \ldots, 0))$$

of $M_1 \setminus F$. The complement of this set in $M_1 \setminus F$ consists of a single $4k$-cell: $(\text{interior } D^{q+1}) \times (S^{4k-q-1} \setminus \{\text{point}\})$. Let F' consist of F together with a single point in this cell. Then it is clearly possible to extend the $4k$-frame field throughout $M_1 \setminus F'$. This completes the proof that M_1 is almost parallelizable. □

Remark 5.7. It cannot be proved that M_1 is parallelizable, even assuming that M is parallelizable. As an example take $M = S^1 \times S^3$, $M_1 = S^4$.

LEMMA 5.8. (Thom). *The manifold M_1 belongs to the same cobordism class as M.*

Proof. Let W be the space obtained from the disjoint union of $M \times [0, 1]$ and $D^{q+1} \times D^{4k-q}$ by pasting together $T \times \{1\}$ and $S^q \times D^{4k-q}$, using the product structure for T constructed above. A differentiable manifold W_1 is obtained from W by "straightening" the angle $\partial T \times \{1\} \setminus S^q \times S^{4k-q-1}$. It is clear that ∂W_1 is the disjoint union of M_1 and $-M$, which completes the proof. □

Proof of Theorem 5.1. Suppose that M is $(q-1)$-connected, almost parallelizable, and $\pi_q(M)$ has r generators. The above construction yields a manifold M_1 which is $(q-1)$-connected, almost parallelizable, and such that $\pi_q(M_1)$ has $r-1$ generators. Iterating the construction r times, this yields a manifold M_r which is q-connected. Continue by induction on q until we obtain a manifold M' which is $(2k-1)$-connected. According to Lemma 5.5 the manifold M' has the same cobordism class as M. Therefore the index $I(M')$ is equal to $I(M)$. (Compare [Thom, 1].) This completes the proof of Theorem 5.1. □

THEOREM 5.9. *Let M be a homology sphere of dimension $4k-1$, $k > 1$, which bounds a π-manifold. If $\lambda'(M) = 0$ then M bounds a contractible manifold.*

COROLLARY 5.10. *If M is a homotopy $(4k-1)$-sphere with $\lambda'(M) = 0$, then M is J-equivalent to S^{4k-1}; providing that $k > 1$.*

COROLLARY 5.11. *For $k > 1$ the group $\Theta^{4k-1}(\partial\pi)$ is cyclic of order $\frac{1}{8}I_k$.*

The proof of Theorem 5.9 is similar to that of Theorem 5.1, but also used the following three results.

THEOREM 5.12. *Let W be a simply connected manifold of dimension $2n$, $n > 2$. Then every element of $\pi_n(W)$ is represented by an embedding $f : S^n \hookrightarrow W$.*

The proof is a modification of Whitney's proof that every n-manifold can be embedded in $2n$-space. (See [Whitney, 2].) Details will not be given. I do not know whether Theorem 5.12 is true for $n = 2$.[‡]

THEOREM 5.13. *Suppose that a quadratic form over the integers has determinant ± 1, index 0, and takes on only even values. Then it is equivalent to a quadratic form with matrix* diag$(U, U, \ldots, U)$, *where* $U = \begin{pmatrix} 0 & 1 \\ 1 & 0. \end{pmatrix}$

Proof.[§] This follows from Theorems 1, 2 of [Milnor, 4] (making use of Theorems of Eichler, etc).

[The following remark is due to H. Sah. In order to prove Theorem 5.13 it is sufficient to prove that there exists an isotropic vector: that is an $\alpha \neq 0$ such that the value $< \alpha, \alpha >$ of the quadratic form is zero. The existence of an isotropic vector is not difficult to prove; using the Hasse-Minkowski Theorem that such a vector exists if and only if **(1)** the form is indefinite and **(2)** for each prime p the corresponding form over the p-adic numbers has an isotropic vector. Given such an α let $\alpha_1 = \alpha/r$ be indivisible. Since the determinant is ± 1, there exists β with $< \alpha_1, \beta >= 1$. Define

$$\beta_1 = \beta - \frac{1}{2} < \beta, \beta > \alpha_1.$$

Then

$$< \alpha_1, \alpha_1 > = < \beta_1, \beta_1 > = 0, \quad < \alpha_1, \beta_1 > = 1.$$

Now consider the set of all γ which satisfy $< \alpha_1, \gamma > = < \beta_1, \gamma > = 0$. By induction on the rank we can choose a basis for this set so that the matrix has the required form.]

LEMMA 5.14. *Let $f : S^{2k} \to W^{4k}$ be an embedding, and suppose that*

(1) *the homology class β of $f(S^{2k})$ has self intersection number $< \beta, \beta > = 0$ and*

(2) *the induced bundle $f^*(\tau^{4k})$ over S^{2k} is trivial.*

Then the normal bundle ν^{2k} is trivial.

[‡]A counterexample is given in KERVAIRE AND MILNOR [1961].

[§]For a fuller treatment of relevant facts about quadratic forms, see MILNOR AND HUSEMOLLER [1973].

Proof. Just as in Lemma 5.4 it is seen that ν^{2k} corresponds to an element $\alpha \in \pi_{2k-1}(\mathrm{SO}_{2k})$ which is annihilated by the homomorphism

$$\pi_{2k-1}(\mathrm{SO}_{2k}) \longrightarrow \pi_{2k-1}(\mathrm{SO}_{4k+1}).$$

Since the group $\pi_{2k-1}(\mathrm{SO}_{2k+1})$ is already stable, it follows from the exact sequence

$$\pi_{2k}(S^{2k}) \approx \mathbb{Z} \xrightarrow{\partial} \pi_{2k-1}(\mathrm{SO}_{2k}) \longrightarrow \pi_{2k-1}(\mathrm{SO}_{2k+1})$$

that $\alpha = \partial n$ for some $n \in \mathbb{Z}$.

The element $\partial(1) \in \pi_{2k-1}(\mathrm{SO}_{2k})$ corresponds to the tangent bundle of S^{2k}, with Euler class equal to twice the generator of $H^{2k}(S^{2k})$. Therefore the Euler class of ν^{2k} is equal to $2n$ times a generator of $H^{2k}(S^{2k})$. But this Euler class can be interpreted as the self-intersection number $< \beta, \beta >$ times a generator. Therefore $2n = 0$, hence $\alpha = 0$. This completes the proof. □

Proof of Theorem 5.9. Let M be a $(4k-1)$-manifold with $\lambda'(M) = 0$. An argument similar to the proof of Theorem 5.1 shows that M bounds a manifold W which is almost parallelizable (hence parallelizable) and $(2k-1)$-connected. The index $I(W)$ is congruent to zero modulo I_k. Hence there exists an almost parallelizable $4k$-manifold N, without boundary, which satisfies $I(N) = -I(W)$. By Theorem 5.1 we may assume that N is also $(2k-1)$-connected.

Now consider the sum $W_1 = W \# N$. This is a parallelizable $4k$-manifold with index zero, and with boundary M. The self-intersection matrix of W_1 has determinant ± 1 by the Poincaré Duality Theorem, and has only even elements on the diagonal. (Compare the proof of Lemma 3.2.) Therefore, according to Theorem 5.13 it is possible to choose a basis $\{\alpha_1, \ldots, \alpha_r, \beta_1, \ldots, \beta_r\}$ for $H_{2k}(W_1)$ so that the intersection matrix is given by:

$$< \alpha_i, \alpha_j >= 0, \quad < \beta_i, \beta_j >= 0, \quad < \alpha_i, \beta_j >= \delta_{ij}.$$

(Here δ_{ij} is a Kronecker delta.)

According to Theorem 5.12 there exists an embedding $f : S^{2k} \to W_1$ which represents the homology class α_1. According to Lemma 5.14 the normal bundle of $f(S^{2k})$ is trivial. Hence we can remove a tubular neighborhood and replace it by $D^{2k+1} \times S^{2k-1}$, yielding a new manifold W_2.

From the pair $(W_1, W_1 \smallsetminus f(S^{2k}))$ we obtain an exact sequence

$$\cdots \longrightarrow 0 \xrightarrow{\partial} H_{2k}(W_1 \smallsetminus f(S^{2k})) \longrightarrow H_{2k}(W_1) \xrightarrow{j} \mathbb{Z} \longrightarrow \cdots$$

where $j(\gamma)$ is the intersection number of γ with the homology class α_1 of $f(S^{2k})$. Therefore $H_{2k}(W_1 \smallsetminus f(S^{2k}))$ is the subgroup of $H_{2k}(W_1)$ generated by the elements $\alpha_1, \ldots, \alpha_r, \beta_2, \ldots, \beta_r$ (with β_1 excluded).

From the pair $(W_2, W_1 \smallsetminus f(S^{2k}))$ we obtain an exact sequence

$$\longrightarrow \mathbb{Z} \xrightarrow{\partial} H_{2k}(W_1 \smallsetminus f(S^{2k})) \longrightarrow H_{2k}(W_2) \xrightarrow{j} 0 \longrightarrow \cdots,$$

where $\partial(1)$ is the class α_1. Hence $H_{2k}(W_2)$ is freely generated by the classes $\{\alpha_2, \ldots, \alpha_r, \beta_2, \ldots, \beta_r\}$. Note that the intersection numbers of these classes in

W_2 is the same as that in W_1. In fact any $2k$-cycle in W_2 can be deformed so that it does not intersect the submanifold $\{0\} \times S^{2k-1}$ which has codimension $2k+1$.

Now choose an embedding $f_2 : S^{2k} \to W_2$ which represents the class α_2. We may assume that $f_2(S^{2k})$ is contained in the parallelizable manifold

$$W_2 \setminus (\{0\} \times S^{2k-1}) = W_1 \setminus f(S^{2k}),$$

hence the normal bundle is trivial. Iterating this procedure r times, we obtain a manifold W_{r+1} which is $2k$-connected, and therefore contractible. This completes the proof of Theorem 5.9. □

This argument can be modified slightly to prove the following.

THEOREM 5.15. *The groups $\Theta^5(\partial\pi)$ and $\Theta^{13}(\partial\pi)$ are zero.*

Proof. Let $M^5 = \partial W^6$ where W^6 is parellelizable. Just as above, we may assume that W^6 is 2-connected. The self intersection matrix of $H_3(W^6)$ is skew symmetric with determinant ± 1. Hence it is equivalent to a matrix of the form $\operatorname{diag}(U', U', \ldots, U')$ where $U' = \begin{pmatrix} 0 & 1 \\ -1 & 0 \end{pmatrix}$. (See for example [Veblen, p.183].) The normal bundle of any 3-sphere in W^6 is trivial since $\pi_2(\mathrm{SO}_3) = 0$. Hence the argument above shows that we can kill $H_3(W^6)$.

The argument in dimension 13 is similar, using the fact that $\pi_6(\mathrm{SO}_7) = 0$. This completes the proof of Theorem 5.15. □

Remark 5.16. The following assertion will be proved in a later paper. For any n of the form $4k+1$, the group $\Theta^n(\partial\pi)$ is either zero or cyclic of order two. The proof will make use of the Arf invariant of a certain quadratic form over the field $\mathbb{Z}_2$.

THEOREM 5.17. *The groups $\Theta^6(\partial\pi)$ and $\Theta^{14}(\partial\pi)$ are zero.*

Outline of proof. Let $M^{2k} = \partial W^{2k+1}$, where W^{2k+1} is parallelizable. Just as above we may assume that W^{2k+1} is $(k-1)$-connected. Furthermore the group $H_k(W^{2k+1}; \mathbb{Q})$ with rational coefficients is not difficult to kill. Thus we may assume that $H_k(W^{2k+1})$ is a finite group. Any element of this group is represented by an embedded k-sphere with trivial normal bundle. Hence one can form W_1^{2k+1} as before. However the homology group $H_k(W_1^{2k+1})$ depends on the particular product structure which is chosen for the normal bundle. The following question arises: Given an arbitrary normal vector field φ_1, does there exist a field of normal $(k+1)$-frames $(\varphi_1, \ldots, \varphi_{k+1})$? For k equal to 1, 3 or 7 this is possible, since the homomorphism

$$\pi_k(\mathrm{SO}_{k+1}) \longrightarrow \pi_k(\mathrm{S}^k)$$

is onto. Hence it is possible to choose $(\varphi_1, \ldots, \varphi_{k+1})$ so that $H_k(W_1^{2k+1})$ is smaller than $H_k(W^{2k+1})$. However for other values of k this homomorphism is not onto, so that the proof does not go through.

6. The group $\Theta^n/\Theta^n(\partial\pi)$.

The main result of this section will be that the factor group $\Theta^n/\Theta^n(\partial\pi)$ is always finite. [This is the group of all homotopy n-spheres modulo those which bound π-manifolds.] Upper bounds for this group are given, but no lower bounds. It is possible that every homotopy sphere is the boundary of a π-manifold.

Let $M^n \subset \mathbb{R}^{n+q}$ be a homology sphere, with $q > n$. The only obstruction to triviality of the normal bundle is an element

$$\mathfrak{O} \in H^n(M^n; \pi_{n-1}(\mathrm{SO_q})) \approx \pi_{\mathrm{n}-1}(\mathrm{SO_q}).$$

This coefficient group has been computed as follows by [Bott, 1] :

n modulo 8	0	1	2	3	4	5	6	7
$\pi_{n-1}(\mathrm{SO_q})$	$\mathbb{Z}$	$\mathbb{Z}_2$	$\mathbb{Z}_2$	0	$\mathbb{Z}$	0	0	0

(This table is valid for $q > n \geq 2$.)

If n is congruent to 3, 5, 6 or 7 modulo 8, this clearly implies that $\mathfrak{O}$ is zero.

If n is equal to $4k$ then the obstruction class $\mathfrak{O}$ can be identified with a certain fraction of the Pontrjagin class $p_k(M^n)$. (See [Kervaire, 3] or [Kervaire and Milnor, 1].) But Hirzebruch's index formula ([Hirzebruch, 1, p.85]) implies that the Pontrjagin class of a homology sphere is zero. Again it follows that $\mathfrak{O} = 0$.

Finally suppose that n is congruent to 1 or 2 modulo 8, so that $\pi_{n-1}(\mathrm{SO_q}) \approx \mathbb{Z}_2$. A Theorem of Rohlin asserts that the obstruction class $\mathfrak{O}$ is annihilated by the homomorphism

$$J_{n-1} : \pi_{n-1}(\mathrm{SO_q}) \longrightarrow \pi_{\mathrm{n+q+1}}(\mathrm{S^q}).$$

(See [Rohlin, 1] or [Kervaire and Milnor, 1].) If J_{n-1} is non-trivial, it follows that $\mathfrak{O} = 0$. This proves:

THEOREM 6.1. *Every homology n-sphere is a π-manifold, unless*

(1) $n \equiv 1$ *or* 2 *modulo* 8, *and*

(2) *the homomorphism* J_{n-1} *is zero.*

For $n = 2$ it is well known that J_1 is an isomorphism. For $n = 9$, 10 we have:

LEMMA 6.2. (Lemma of [Kervaire, 5].) *The homomorphisms* J_8 *and* J_9 *are non-trivial.*

Therefore:

COROLLARY 6.3. *For* $n < 17$ *every homology n-sphere is a π-manifold.*

I do not know whether conditions **(1)** and **(2)** of Theorem 6.1 are ever satisfied. However, in any case the following is true.†

LEMMA 6.4. *For any n the homotopy n-spheres which are π-manifolds form a subgroup* $\Theta^n(\pi) \subset \Theta^n$ *which has index at most* 2.

†The following paper by Kervaire and Milnor shows that every homotopy sphere is a π-manifold. See Theorem 3.1 on page 93 below.

Proof. We may assume that $\pi_{n-1}(\mathrm{SO_q}) \approx \mathbb{Z}_2$. The obstruction correspondence $M^n \to \mathfrak{O}(M^n) \in \pi_{n-1}(\mathrm{SO_q})$ is easily seen to be additive, and invariant under J-equivalence. This completes the proof. □

Now let M^n be any π-manifold without boundary, and consider the question: *Is M^n the boundary of a π-manifold?* The theory of [Thom, 2] can be used to give an answer as follows.

Choose an embedding of M^n in the interior of a cube $[0, 1] \times \ldots \times [0, 1] = I^{n+q}$, and choose a field φ of normal q-frames. Then the Thom construction yields a map

$$(I^{n+q},\ \partial I^{n+q}) \longrightarrow (S^q,\ \text{base point}),$$

and hence a homotopy class

$$t(\varphi) \in \pi_{n+q}(S^q).$$

(See [Thom, 2, p.30], or [Kervaire, 1, p.223], or [Kervaire and Milnor, 1, proof of Lemma 1].) This class is zero if and only if there exists a π-manifold $W \subset I^{n+q+1}$ such that

(1) $\partial W = M^n \times \{0\}$, and

(2) the field φ of normal q-frames can be extended throughout W.

Now let φ range over all possible fields of normal q-frames. The set of all homotopy classes $t(\varphi)$ will be denoted by

$$t'(M^n) \subset \pi_{n+q}(S^q).$$

Evidently M^n bounds a π-manifold if and only if

$$0 \in t'(M^n).$$

LEMMA 6.5. *If M_1 and M_2 are π-manifolds; then*

$$t'(M_1 \# M_2) \supset t'(M_1) + t'(M_2).$$

(I do not know whether equality holds.)

Proof. Let W be a manifold formed from the disjoint union of $M_1 \times [0, 1]$, $M_2 \times [0, 1]$ and $D^n \times D^1$ by matching $D^n \times \{-1\}$ with a cell in $M_1 \times \{1\}$; matching $D^n \times \{1\}$ with a cell in $M_2 \times \{1\}$; and then straightening corners. If the orientations are correct, then ∂W will be the disjoint union of $M_1 \# M_2$, $-M_1$ and $-M_2$. Furthermore W has the homotopy type of the union of M_1 and M_2 with a single point in common.

Choose an embedding of W in $\mathbb{R}^{n+q} \times [0, 1]$ so that $-M_1$ and $-M_2$ go into $\mathbb{R}^{n+q} \times \{0\}$, while $M_1 \# M_2$ goes into $\mathbb{R}^{n+q} \times \{1\}$. Now given fields φ_1, φ_2 of normal q-frames on M_1 and M_2 respectively, there exists an extension Ψ which is defined throughout W.

If φ denotes the restriction of Ψ to $M_1 \# M_2$, then it is clear that $t(\varphi) = t(\varphi_1) + t(\varphi_2)$. This completes the proof of Lemma 6.5. □

Now consider the special case $M = S^n$. Every field φ of normal q-frames determines an element

$$\alpha \in \pi_n(\mathrm{SO_q}).$$

Kervaire has shown that $t(\varphi)$ is equal to $\pm J_n(\alpha)$. (See [Kervaire, 4].) Since any α may occur this proves:

LEMMA 6.6. *The set $t'(S^n)$ is equal to* (image J_n) $\subset \pi_{n+q}(S^q)$.

Applying Lemma 6.5 to the identity

$$M^n \# S^n = M^n$$

this shows that $t'(M^n) \supset t'(M^n) +$ (image J_n). In other words $t'(M^n)$ is a union of cosets of (image J_n). This suggests that we ***define*** $t(M^n)$ as the subset of

$$(\text{cokernel } J_n) = \pi_{n+q}(S^q)/(\text{image } J_n)$$

which corresponds to $t'(M^n)$.

THEOREM 6.7. *The Thom construction yields a correspondence*

$$M^n \longrightarrow t(M^n) \subset (\text{cokernel } J_n)$$

with the following properties:

- **(a)** *$t(M^n)$ is defined and non-vacuous for every unbounded π-manifold.*
- **(b)** *$t(M^n)$ contains 0 if and only if M^n bounds a π-manifold.*
- **(c)** *$t(M_1 \# M_2) \supset t(M_1) + t(M_2)$.*
- **(d)** *$t(S^n) = \{0\}$.*
- **(e)** *If M_1 is J-equivalent to M_2 then $t(M_1) = t(M_2)$.*
- **(f)** *If M^n is a homotopy sphere, then $t(M^n)$ consists of a single element.*

Proof. Properties **(a)** through **(d)** follow from the discussion above. Property **(e)** follows immediately from the definition. To prove **(f)** recall that $M^n\#(-M^n)$ is J-equivalent to S^n. Therefore $\{0\} \supset t(M^n)+t(-M^n)$. But this would be impossible if $t(M^n)$ contained more than one element. □

COROLLARY 6.8. *The factor group $\Theta^n(\pi)/\Theta^n(\partial\pi)$ is naturally isomorphic to a subgroup of* (cokernel J_n).

COROLLARY 6.9. *This factor group is finite for every n. Hence the subgroup $\Theta^n(\partial\pi) \subset \Theta^n$ has finite index.*

To conclude this section, here is a summary of what is known about the group (cokernel J_n). Toda has computed the p-primary component of the stable group $\pi_{n+q}(S^q)$ for the range $n < 2p^2(p-1) - 3$. (See [Toda, 2].) Combining this information with Lemma 3.6 the p-primary component of (cokernel J_n) is determined for the same range. As an example (compare [Milnor, 3]):

Assertion. The p-primary component of (cokernel J_n) is zero for $n < 2p(p-1)-2$, and is $\mathbb{Z}_p$ for $n = 2p(p-1)-2$.

The 2-primary component can be determined for $n \leq 13$, making use of [Toda, 1], together with Lemmas 6.2 and 3.6. The following is a tabulation of the first thirteen groups.

n	1	2	3	4	5	6	7	8	9	10	11	12	13	14–19
coker J_n	0	$\mathbb{Z}_2$	0	0	0	$\mathbb{Z}_2$	0	$\mathbb{Z}_2$	$\mathbb{Z}_2+\mathbb{Z}_2$	$\mathbb{Z}_6$	0	0	$\mathbb{Z}_3$	2-group

Since Θ^2 is known to be zero, the first unsolved case occurs for $n = 6$. Is the group $\Theta^6/\Theta^6(\partial\pi)$ non-trivial?

7. Discussion

Combining the results of the preceding sections, we have the following estimate of Θ^n for small values of n.

$\Theta^1 = \Theta^2 = \Theta^5 = 0$,

Θ^6 is either 0 or $\mathbb{Z}_2$,

Θ^7 is cyclic of order 28,

Θ^9 has order at most 8,

Θ^{11} is cyclic of order 992,

Θ^{13} is either 0 or $\mathbb{Z}_3$,

Θ^{14} is a 2-group,

Θ^{15} has order 127 times a power of 2. This group contains an element of order 8128.

Following are six questions which remain open.*

Problem 1. *Are the groups $\Theta^{2k}(\partial\pi)$ finite for $k \neq 1, 3, 7$? A solution would probably be based on a detailed study of $(2k+1)$-manifolds which are $(k-1)$-connected.* (Compare Theorem 5.17.)

Another outstanding problem is the decision as to whether every homotopy sphere bounds a π-manifold. (See §6.)

*In fact Problem 1 was solved positively and Problem 2 was solved negatively in the subsequent paper with Kervaire (pages 89–122). Problem 3 was solved negatively by Andrew Casson. (Compare Akbulut and McCarthy.) Problem 4, as well as problem 5 in the simply-connected case, were settled by Smale. (See pages 8, 9 above.) Problem 6 was disproved by Stallings. (Compare §11.1 and §12.4 of Milnor, "Whitehead Torsion", reprinted in Volume 2 of these Collected Papers, pp. 151-219.)

Problem 2. *Is there any theory which related the invariant* $t(M^n) \subset \text{cokernel}(J_n)$ *with the topology of* M^n*? In particular does this invariant vanish for a homotopy sphere?*

Another question would be the relationship between this paper and the Poincaré hypothesis.

Problem 3. *Does there exist a homotopy* 3*-sphere* M *such that* $\lambda'(M) \neq 0$*?* Such a manifold could not be homeomorphic to S^3. In fact it follows from work of E. Moise, J. Munkres, S. Smale and J.H.C. Whitehead that the differentiable structure of a topological 3-manifold is unique up to diffeomorphism.

Problem 4. *Are the homotopy spheres* M_0^{4k-1} *homeomorphic to* S^{4k-1}*?* (See §4. Note that k must be ≥ 2.)

The following seems to be a very deep question.

Problem 5. *Are* J*-equivalent manifolds necessarily diffeomorphic?*

An affirmative answer would imply the generalized Poincaré hypothesis for differentiable manifolds. For if M is a homotopy n-sphere then $M\#(-M)$ is J-equivalent to S^n. But if $M\#M_1$ is diffeomorphic to S^n then an argument due to Mazur implies that M itself is homeomorphic to S^n.

Most known invariants of differentiable manifolds depend only on the J-equivalence class. For example:

Assertion. If M_1 is J-equivalent to M_2 then some homotopy equivalence $M_1 \to M_2$ is covered by a bundle map $\tau_1^n \to \tau_2^n$ between the tangent bundles.

Proof. Suppose that the boundaries M_1 and $-M_2$ are deformation retracts of W. Choose a non-singular vector field on W which points out of W along M_1 and into W along M_2. The orthogonal complement of this vector field in τ^{n+1} yields an $\mathrm{SO_n}$-bundle ξ^n over W. Now the bundle maps $\tau_1^n \to \xi^n \leftarrow \tau_2^n$ can be used to construct the required bundle map. □

Problem 6. *Is the "simple homotopy type" of* M *invariant under* J*-equivalence?* (See [Whitehead, 1, 4].)

Appendix: Pasting and Straightening.

Let $\mathbb{R}_+$ denote the set of real numbers t with $0 \leq t < \infty$.

Assertion. If W is a differentiable manifold with boundary, then there exists a neighborhood U of ∂W, and a diffeomorphism

$$h : \partial W \times \mathbb{R}_+ \longrightarrow U$$

which satisfies the identity $h(x, 0) = x$.

(A proof of this assertion is given in [Milnor, 2]. Alternatively this may be taken as part of the definition of "manifold with boundary".)

Given two manifolds W_1, W_2 and an orientation reversing diffeomorphism

$$f : \partial W_1 \longrightarrow \partial W_2,$$

let M denote the space obtained from the disjoint union of W_1 and W_2 by identifying each $x \in \partial W_1$ with $f(x)$.

LEMMA A.1. *The topological manifold M can be given a differentiable structure which is compatible with that of W_1 and W_2.*

LEMMA A.2. *If two such differentiable structures are given, then the resulting differentiable manifolds are diffeomorphic.*

Proof of Lemma A.1. Choose neighborhoods U_i of ∂W_i in W_i, and diffeomorphisms

$$h_i : \partial W_i \times \mathbb{R}_+ \longrightarrow U_i$$

as above. A homeomorphism

$$h : \partial W_1 \times \mathbb{R} \longrightarrow U_1 \cup U_2 \subset M$$

is defined by the formula

$$h(x,\, t) = \begin{cases} h_1(x,\, t), & \text{for} \quad t \geq 0 \\ h_2(f(x),\, -t), & \text{for} \quad t \leq 0\,. \end{cases}$$

Taking h to be a diffeomorphism, this defines the required differentiable structure. □

Proof of Lemma A.2. Let M and M' be the two differentiable manifolds. Choose a contravariant vector field φ_1 along the boundary of W_1 which points out of W_1. Considering W_1 and W_2 as submanifolds of M, this yields a vector field φ_2 along the boundary of W_2 which points into W_2. On the other hand, considering W_1 and W_2 as submanifolds of M', the field φ_1 corresponds to some other field φ_2' along the boundary of W_2.

Choose a diffeomorphism $g_2 : W_2 \to W_2$ which leaves ∂W_2 pointwise fixed, and carries the vector field φ_2 into φ_2'. (The construction is not difficult.) Then a homeomorphism $g : M \to M'$ is obtained by combining g_2 with the identity map of W_1. It is easily verified that g and g^{-1} are differentiable of class C^1.

Approximate g by a C^∞-differentiable map g'; where the approximation must be close enough so that the Jacobian of g' has rank n everywhere. (See [Whitney, 1].) Then $g' : M \to M'$ is the required diffeomorphism.

Several times in this paper it has been necessary to consider n-manifolds with boundary which are differentiable except along some $(n-2)$-dimensional submanifold of the boundary. The simplest example of such an object is the quadrant $\mathbb{R}_+ \times \mathbb{R}_+ \subset \mathbb{R}^2$. This example can be "straightened" by introducing new coordinates as follows. Map $\mathbb{R}_+ \times \mathbb{R}_+$ onto the half-plane $\mathbb{R} \times \mathbb{R}_+$ by the correspondence

$$(r \cos\theta,\; r \sin\theta) \xrightarrow{\;f\;} (r\cos(2\theta),\; r\sin(2\theta))$$

for $0 \le r$, $0 \le \theta \le \frac{\pi}{2}$. Thus f is a diffeomorphism, except at the singular point. Another example is provided by the three-quarter-plane $(\mathbb{R}_+ \times \mathbb{R}) \cup (\mathbb{R} \times \mathbb{R}_+)$. This can be straightened by the transformation

$$(r\cos\theta,\ r\sin\theta) \longrightarrow \big(r\cos\big(\frac{2\theta+\pi}{3}\big),\ r\sin\big(\frac{2\theta+\pi}{3}\big)\big),$$

for $0 \le r$, $-\frac{\pi}{2} \le \theta \le \pi$. $\square$

A higher dimensional example is given as follows. Let W_1 and W_2 be differentiable manifolds with boundary. The $W_1 \times W_2$ is differentiable except along $\partial W_1 \times \partial W_2$. Some neighborhood $U_1 \times U_2$ of this singular set is "diffeomorphic" to

$$(\partial W_1 \times \partial W_2) \times (\mathbb{R}_+ \times \mathbb{R}_+).$$

Form a new differentiable manifold W as follows. Take the disjoint union of $(W_1 \times W_2) \smallsetminus (\partial W_1 \times \partial W_2)$ and $(\partial W_1 \times \partial W_2) \times (\mathbb{R} \times \mathbb{R}_+)$, and identify

$$h_1(x_1,\ r\cos\theta) \times h_2(x_2,\ r\sin\theta) \in U_1 \times U_2$$

with

$$(x_1,\ x_2,\ r\cos 2\theta,\ r\sin 2\theta)$$

for each $x_1 \in \partial W_1$, $x_2 \in \partial W_2$, $0 < r$, $0 \le \theta \le \frac{\pi}{2}$. This construction will be referred to as "straightening the angle" . Note that the differentiable structure of $\partial W_1 \times W_2$, and of $W_1 \times \partial W_2$ is left fixed, so that Lemma A.2 applies to their union ∂W.

A similar construction works for each of the examples considered in this paper.

References

Bott, R.: [1] *The stable homotopy of the classical groups,* Proc. Nat. Acad. Sci. U.S.A., **43** (1957), 933–935.

Hirzebruch, F.: [1] "Neue topologische Methoden in der algebraischen Geometrie," Springer (1956).

Kervaire, M.: [1] *Courbure intégrale généralisée et homotopie,* Math. Annalen, **131** (1956), 219–252.

———— : [2] *Relative characteristic classes,* Amer. J. Math., **79** (1957), 517–558.

————: [3] *On the Pontrjagin classes of certain* SO(n)*-bundles over manifolds,* Amer J. Math., **80** (1958), 632–638.

———— : [4] *An interpretation of G. Whitehead's generalization of Hopf's invariant,* Annals of Math., **69** (1959), 345–365.†

———— : [5] *Some non stable homotopy groups of Lie groups*, Illinois J. Math., **4** (1960), 161–169.†

———— and Milnor, J.: [1] *Bernoulli numbers, homotopy groups and a theorem of Rohlin*, Proc. Internat. Congress Math., (1958); Cambridge University Press, (1960), 454–458.†

Mazur, B.: [1] *On embeddings of spheres*, Thesis, Princeton University, (1959).

†Several of these citations were incomplete in the original manuscript.

Milnor, J.: [1] *On manifolds homeomorphic to the 7-sphere,* Annals of Math., **64** (1956), 399–405.

——— : [2] *On the relationship between differentiable manifolds and combinatorial manifolds*, Princeton University (1956), (mimeographed). [See this volume pp. 19–27.]

——— : [3] *On the Whitehead homomorphism,* J. Bull. Amer. Math. Soc., **64** (1958), 79–82.

——— : [4] *On simply connected 4-manifolds,* Topología Algebraíca, Mexico, Symposium Internacional de Topología Algebraíca, Univ. Nac. Autónoma de México and UNESCO, Mexico City, (1958), 122–128.†

——— : [5] *Differentiable structures on spheres*, Amer. J. Math. **81** (1959), 962–972.†

Munkres, J.: [1] *Some applications of triangulation theorems,* Thesis, University of Michigan, (1955).

Rohlin, V. A.: [1] *Classification of mappings of S^{n+3} onto S^n,* (Russian) Dokl. Acad. Nauk. SSSR **81** (1951), 19–92.

Shimada, N.: [1] *Differentiable structures on the 15-sphere and Pontrjagin classes of certain manifolds,* Nagoya Math. J. **12** (1957), 59–69.

Steenrod, N.: [1] "The Topology of Fibre Bundles," Princeton 1951.

Thom, R.: [1] *Espaces fibrés en sphères et carrés de Steenrod,* Ann. Sci. Ecole Norm. Sup. **69** (1952), 109–181.

——— : [2] *Quelques propriétés globales des variétés différentiables,* Comment. Math. Helv. **28** (1954), 17–86.

——— : [3] *Les classes caractéristiques de Pontrjagin des variétés triangulées*, Symposium Internacional de Topología Algebraíca, Univ. Nac. Autónoma de México and UNESCO, Mexico City, (1958), 54–67.†

Toda, H. [1] *On exact sequences in Steenrod algebra mod 2,* Memoire College of Sci., Kyoto **31** (1958), 33–64.

——— : [2] *p-primary components of homotopy groups III,* Memoirs College of Sci., Kyoto **31** (1958), 191–210.

Veblen, O. "Analysis situs" (2nd-edition), Amer. Math. Soc. Colloquium Publications, New York, 1931.

Whitehead, J.H.C.: [1] *On C^1-complexes,* Annals of Math. **41** (1940), 809–824.

——— : [2] *On the homotopy type of manifolds,* Annals of Math. **41** (1940), 825–832.

——— : [3] *On the homotopy type of ANR's,* Bull. Amer. Math. Soc. **54** (1948), 1133–1145.

——— : [4] *Simple homotopy types,* Amer. J. Math. **72** (1950), 1–57.

Whitney, H.: [1] *Differentiable manifolds,* Annals of Math. **37**(1936), 645–680.

——— : [2] *The self intersections of a smooth n-manifold in 2n-space,* Annals of Math. **45** (1944), 220–246.

Wu, W.T.: [1] *Classes caractéristiques et i-carrés d'une variété,* C. R. Acad. Sci. Paris **230** (1950), 508–511.

ANNALS OF MATHEMATICS
Vol. 77, No. 3, May, 1963
Printed in Japan

GROUPS OF HOMOTOPY SPHERES: I

BY MICHEL A. KERVAIRE AND JOHN W. MILNOR

(Received April 19, 1962)

1. Introduction

All manifolds, with or without boundary, are to be compact, oriented, and differentiable of class C^∞. The boundary of M will be denoted by bM. The manifold M with orientation reversed is denoted by $-M$.

DEFINITION. The manifold M is a *homotopy n-sphere* if M is closed (that is, $bM = \varnothing$) and has the homotopy type of the sphere S^n.

DEFINITION. Two closed n-manifolds M_1 and M_2 are *h-cobordant*[1] if the disjoint sum $M_1 + (-M_2)$ is the boundary of some manifold W, where both M_1 and $(-M_2)$ are deformation retracts of W. It is clear that this is an equivalence relation.

The *connected sum* of two connected n-manifolds is obtained by removing a small n-cell from each, and then pasting together the resulting boundaries. Details will be given in § 2.

THEOREM 1.1. *The h-cobordism classes of homotopy n-spheres form an abelian group under the connected sum operation.*

This group will be denoted by Θ_n, and called the nth *homotopy sphere cobordism group*. It is the object of this paper (which is divided into 2 parts) to investigate the structure of Θ_n.

It is clear that $\Theta_1 = \Theta_2 = 0$. On the other hand these groups are not all zero. For example, it follows easily from Milnor [14] that $\Theta_7 \neq 0$.

The main result of the present Part I will be:

THEOREM 1.2. *For $n \neq 3$ the group Θ_n is finite.*

(Our methods break down for the case $n = 3$. However, if one assumes the Poincaré hypothesis, then it can be shown that $\Theta_3 = 0$.)

More detailed information about these groups will be given in Part II. For example, for $n = 1, 2, 3, \cdots, 18$, it will be shown that the order of the group Θ_n is respectively:

n	1	2	3	4	5	6	7	8	9	10	11	12	13	14	15	16	17	18
$[\Theta_n]$	1	1	?	1	1	1	28	2	8	6	992	1	3	2	16256	2	16	16.

Partial summaries of results are given also at the end of § 4 and of § 7.

[1] The term "J-equivalent" has previously been used for this relation. Compare [15], [16], [17].

REMARK. S. Smale [25] and J. Stallings [27], C. Zeeman [33] have proved that every homotopy n-sphere, $n \neq 3, 4$, is actually homeomorphic to the standard sphere S^n. Furthermore, Smale has proved [26] that two homotopy n-spheres, $n \neq 3, 4$, are h-cobordant if and only if they are diffeomorphic. Thus for $n \neq 3, 4$ (and possibly for all n) the group Θ_n can be described as the set of all diffeomorphism classes of differentiable structures on the topological n-sphere. These facts will not be used in the present paper.

2. Construction of the group Θ_n

First we give a precise definition of the connected sum $M_1 \# M_2$ of two connected n-manifolds M_1 and M_2. (Compare Seifert [22] and Milnor [15], [16].) The notation D^n will be used for the unit disk in euclidean n-space. Choose imbeddings

$$i_1\colon D^n \to M_1\,, \qquad i_2\colon D^n \to M_2$$

so that i_1 preserves orientation and i_2 reverses orientation. Now obtain $M_1 \# M_2$ from the disjoint sum

$$(M_1 - i_1(0)) + (M_2 - i_2(0))$$

by identifying $i_1(tu)$ with $i_2((1-t)u)$ for each unit vector $u \in S^{n-1}$ and each $0 < t < 1$. Choose the orientation for $M_1 \# M_2$ which is compatible with that of M_1 and M_2. (This makes sense since the correspondence $i_1(tu) \to i_2((1-t)u)$ preserves orientation.)

It is clear that the sum of two homotopy n-spheres is a homotopy n-sphere.

LEMMA 2.1. *The connected sum operation is well defined, associative, and commutative up to orientation preserving diffeomorphism. The sphere S^n serves as identity element.*

PROOF. The first assertions follow easily from the lemma of Palais [20] and Cerf [5] which asserts that any two orientation preserving imbeddings $i, i'\colon D^n \to M$ are related by the equation $i' = f \circ i$, for some diffeomorphism $f\colon M \to M$. The proof that $M \# S^n$ is diffeomorphic to M will be left to the reader.

LEMMA 2.2. *Let M_1, M_1' and M_2 be closed and simply connected.*[2] *If M_1 is h-cobordant to M_1' then $M_1 \# M_2$ is h-cobordant to $M_1' \# M_2$.*

PROOF. We may assume that the dimension n is ≥ 3. Let $M_1 + (-M_1') = bW_1$, where M_1 and $-M_1'$ are deformation retracts of W_1. Choose a differentiable arc A from a point $p \in M_1$ to a point $p' \in -M_1'$ within W_1 so that

[2] This hypothesis is imposed in order to simplify the proof. It could easily be eliminated.

a tubular neighborhood of this arc is diffeomorphic to $R^n \times [0, 1]$. Thus we obtain an imbedding

$$i: R^n \times [0, 1] \to W_1$$

with $i(R^n \times 0) \subset M_1$, $i(R^n \times 1) \subset M_1'$, and $i(0 \times [0, 1]) = A$. Now form a manifold W from the disjoint sum

$$(W_1 - A) + \big(M_2 - i_2(0)\big) \times [0, 1]$$

by identifying $i(tu, s)$ with $i_2((1 - t)u) \times s$ for each $0 < t < 1$, $0 \leqq s \leqq 1$, $u \in S^{n-1}$. Clearly W is a compact manifold bounded by the disjoint sum

$$M_1 \# M_2 + \big(-(M_1' \# M_2)\big) .$$

We must show that both boundaries are deformation retracts of W.

First it is necessary to show that the inclusion map

$$M_1 - p \xrightarrow{j} W_1 - A$$

is a homotopy equivalence. Since $n \geqq 3$, it is clear that both of these manifolds are simply connected. Mapping the homology exact sequence of the pair $(M_1, M_1 - p)$ into that of the pair $(W_1, W_1 - A)$, we see that j induces isomorphisms of homology groups, and hence is a homotopy equivalence. Now it follows easily, using a Mayer-Vietoris sequence, that the inclusion

$$M_1 \# M_2 \to W$$

is a homotopy equivalence; hence that $M_1 \# M_2$ is a deformation retract of W. Similarly $M_1' \# M_2$ is a deformation retract of W, which completes the proof of Lemma 2.2.

LEMMA 2.3. *A simply connected manifold M is h-cobordant to the sphere S^n if and only if M bounds a contractible manifold.*

(Here the hypothesis of simple connectivity cannot be eliminated.)

PROOF. If $M + (-S^n) = bW$ then filling in a disk D^{n+1} we obtain a manifold W' with $bW' = M$. If S^n is a deformation retract of W, then it clearly follows that W' is contractible.

Conversely if $M = bW'$ with W' contractible, then removing the interior of an imbedded disk we obtain a simply connected manifold W with $bW = M + (-S^n)$. Mapping the homology exact sequence of the pair (D^{n+1}, S^n) into that of the pair (W', W), we see that the inclusion $S^n \to W$ induces a homology isomorphism; hence S^n is a deformation retract of W. Now applying the Poincaré duality isomorphism

$$H_k(W, M) \simeq H^{n+1-k}(W, S^n) ,$$

we see that the inclusion $M \to W$ also induces isomorphisms of homology groups. Since M is simply connected, this completes the proof.

LEMMA 2.4. *If M is a homotopy sphere, then $M \# (-M)$ bounds a contractible manifold.*

PROOF. Let $H^2 \subset D^2$ denote the half-disk consisting of all $(t \sin \theta, t \cos \theta)$ with $0 \leqq t \leqq 1$, $0 \leqq \theta \leqq \pi$, and let $\frac{1}{2}D^n \subset D^n$ denote the disk of radius $\frac{1}{2}$. Given an imbedding $i\colon D^n \to M$, form W from the disjoint union

$$\left(M - i\left(\frac{1}{2}D^n\right)\right) \times [0, \pi] + S^{n-1} \times H^2$$

by identifying $i(tu) \times \theta$ with $u \times ((2t-1)\sin\theta, (2t-1)\cos\theta)$ for each $\frac{1}{2} < t \leqq 1$, $0 \leqq \theta \leqq \pi$. (Intuitively we are removing the interior of $i(\frac{1}{2}D^n)$ from M and then "rotating" the result through 180° around the resulting boundary.) It is easily verified that W is a differentiable manifold with $bW = M \# (-M)$. Furthermore W contains $M-$Interior $i(\frac{1}{2}D^n)$ as deformation retract, and therefore is contractible. This proves Lemma 2.4.

PROOF OF THEOREM 1.1. Let Θ_n denote the collection of all h-cobordism classes of homotopy n-spheres. By Lemmas 2.1 and 2.2 there is a well defined, associative, commutative addition operation in Θ_n. The sphere S^n serves as zero element. By Lemmas 2.3, 2.4, each element of Θ_n has an inverse. Therefore Θ_n is an additive group.

Clearly Θ_1 is zero. For $n \leqq 3$, Munkres [19] and Whitehead [31] have proved that a topological n-manifold has a differentiable structure which is unique up to diffeomorphism. It follows that $\Theta_2 = 0$. If the Poincaré hypothesis were proved, it would follow that Θ_3 is zero; but at present the structure of Θ_3 remains unknown. For $n > 3$ the structure of Θ_n will be studied in the following sections.

Addendum. There is a slight modification of the connected sum construction which is frequently useful. Let W_1 and W_2 be $(n+1)$-manifolds with connected boundary. Then the sum $bW_1 \# bW_2$ is the boundary of a manifold W constructed as follows. Let H^{n+1} denote the half-disk consisting of all $x = (x_0, x_1, \cdots, x_n)$ with $|x| \leqq 1$, $x_0 \geqq 0$ and let D^n denote the subset $x_0 = 0$. Choose imbeddings

$$i_q\colon (H^{n+1}, D^n) \to (W_q, bW_q)\,, \qquad q = 1, 2\,,$$

so that $i_2 \circ i_1^{-1}$ reverses orientation. Now form W from

$$\left(W_1 - i_1(0)\right) + \left(W_2 - i_2(0)\right)$$

by identifying $i_1(tu)$ with $i_2((1-t)u)$ for each $0 < t < 1$, $u \in S^n \cap H^{n+1}$.

It is clear that W is a differentiable manifold with $bW = bW_1 \# bW_2$.

Note that W has the homotopy type of $W_1 \vee W_2$: the union with a single point in common.

W will be called the *connected sum along the boundary* of W_1 and W_2. The notation $(W, bW) = (W_1, bW_1) \# (W_2, bW_2)$ will be used for this sum.

3. Homotopy spheres are S-parallelizable

Let M be a manifold with tangent bundle $\tau = \tau(M)$, and let ε^1 denote a trivial line bundle over M.

DEFINITION. M will be called S-*parallelizable* if the Whitney sum $\tau \oplus \varepsilon^1$ is a trivial bundle.[3] The bundle $\tau \oplus \varepsilon^1$ will be called the *stable tangent bundle* of M. It is a stable bundle in the sense of [10]. (The expression S-parallelizable stands for stably parallelizable.)

THEOREM 3.1. *Every homotopy sphere is* S-*parallelizable.*

In the proof, we will use recent results of J. F. Adams [1], [2].

PROOF. Let Σ be a homotopy n-sphere. Then the only obstruction to the triviality of $\tau \oplus \varepsilon^1$ is a well defined cohomology class

$$\mathfrak{o}_n(\Sigma) \in H^n(\Sigma; \pi_{n-1}(\mathrm{SO}_{n+1})) = \pi_{n-1}(\mathrm{SO}_{n+1}) .$$

The coefficient group may be identified with the stable group $\pi_{n-1}(\mathrm{SO})$. But these stable groups have been computed by Bott [4], as follows, for $n \geqq 2$:

residue class of n mod 8:	0	1	2	3	4	5	6	7
$\pi_{n-1}(\mathrm{SO})$	Z	Z_2	Z_2	0	Z	0	0	0 .

(Here Z, Z_2, 0 denote the cyclic groups of order ∞, 2, 1 respectively.)

Case 1. $n \equiv 3, 5, 6$, or 7 (modulo 8). Then $\pi_{n-1}(\mathrm{SO}) = 0$, so that $\mathfrak{o}_n(\Sigma)$ is trivially zero.

Case 2. $n \equiv 0$ or 4 (modulo 8). Say that $n = 4k$. According to [18], [10], some non-zero multiple of the obstruction class $\mathfrak{o}_n(\Sigma)$ can be identified with the Pontrjagin class $p_k(\tau \oplus \varepsilon^1) = p_k(\tau)$. But the Hirzebruch signature[4] theorem implies that $p_k[\Sigma]$ is a multiple of the signature $\sigma(\Sigma)$ which is zero since $H^{2k}(\Sigma) = 0$. Therefore every homotopy $4k$-sphere is S-parallelizable.

Case 3. $n \equiv 1$ or 2 (modulo 8), so that $\pi_{n-1}(\mathrm{SO})$ is cyclic of order 2. For

[3] The authors have previously used the term "π-manifold" for an S-parallelizable manifold.

[4] We will substitute the word "signature" for "index" as used in [7; 14; 17; 18; 28] since this is more in accord with the usage in other parts of mathematics. The *signature* of the form $x_1^2 + \cdots + x_k^2 - x_{k+1}^2 - \cdots - x_{k+l}^2$ is defined to be $\sigma = k - l$.

each homotopy sphere Σ the residue modulo 2

$$\mathfrak{o}_n[\Sigma] \in \pi_{n-1}(\mathrm{SO}) \simeq Z_2$$

is well defined. It follows from an argument of Rohlin that

$$J_{n-1}(\mathfrak{o}_n) = 0 ,$$

where J_{n-1} denotes the Hopf-Whitehead homomorphism

$$J_{n-1}: \pi_{n-1}(\mathrm{SO}_k) \rightarrow \pi_{n+k-1}(S^k)$$

in the stable range $k > n$. (Compare [18, Lemma 1].) But J_{n-1} is a monomorphism for $n \equiv 1$ or 2 (modulo 8). For the case $n = 2$ this fact is well known, and for $n = 9, 10$ it has been proved by Kervaire [11]. For $n = 17, 18$, it has been verified by Kervaire and by Toda in unpublished computations. A proof that J_{n-1} is injective for all $n \equiv 1$ or 2 (modulo 8) has recently been given by J.F. Adams [1], [2]. Now the relation $J_{n-1}(\mathfrak{o}_n) = 0$ together with the information that J_{n-1} is a monomorphism implies that $\mathfrak{o}_n = 0$. This finishes the proof of Theorem 3.1.

In conclusion, here are two lemmas which clarify the concept of s-parallelizability. The first is essentially due to J. H. C. Whitehead [32].

LEMMA 3.3. *Let M be an n-dimensional submanifold of S^{n+k}, $n < k$. Then M is s-parallelizable if and only if its normal bundle is trivial.*

LEMMA 3.4. *A connected manifold with non-vacuous boundary is s-parallelizable if and only if it is parallelizable.*

The proofs will be based on the following lemma. (Compare Milnor [17, Lemma 4].)

Let ξ be a k-dimensional vector space bundle over an n-dimensional complex, $k > n$.

LEMMA 3.5. *If the Whitney sum of ξ with a trivial bundle ε^r is trivial then ξ itself is trivial.*

PROOF. We may assume that $r = 1$, and that ξ is oriented. An isomorphism $\xi \oplus \varepsilon^1 \approx \varepsilon^{k+1}$ gives rise to a bundle map f from ξ to the bundle γ^k of oriented k-planes in $(k + 1)$-space. Since the base space of ξ has dimension n, and since the base space of γ^k is the sphere S^k, $k > n$, it follows that f is null-homotopic; and hence that ξ is trivial.

PROOF OF LEMMA 3.3. Let τ, ν denote the tangent and normal bundles of M. Then $\tau \oplus \nu$ is trivial hence $(\tau \oplus \varepsilon^1) \oplus \nu$ is trivial. Applying Lemma 3.5 the conclusion follows.

PROOF OF LEMMA 3.4. This follows by a similar argument. The hypothesis on the manifold guarantees that every map into a sphere of the same dimension is null-homotopic.

4. Which homotopy spheres bound parallelizable manifolds?

Define a subgroup $bP_{n+1} \subset \Theta_n$ as follows. A homotopy n-sphere M represents an element of bP_{n+1} if and only if M is the boundary of a parallelizable manifold. We will see that this condition depends only on the h-cobordism class of M, and that bP_{n+1} does form a subgroup. The object of this section will be to prove the following

THEOREM 4.1. *The quotient group Θ_n/bP_{n+1} is finite.*

PROOF. Given an s-parallelizable closed manifold M of dimension n, choose an imbedding

$$i\colon M \to S^{n+k}$$

with $k > n + 1$. Such an imbedding exists and is unique up to differentiable isotopy. By Lemma 3.3 the normal bundle of M is trivial. Now choose a specific field φ of normal k-frames. Then the Pontrjagin-Thom construction yields a map

$$p(M, \varphi)\colon S^{n+k} \to S^k .$$

(See Pontrjagin [21, pp. 41–57], Thom [28].) The homotopy class of $p(M, \varphi)$ is a well defined element of the stable homotopy group

$$\Pi_n = \pi_{n+k}(S^k) .$$

Allowing the normal frame field φ to vary, we obtain a set of elements

$$p(M) = \{p(M, \varphi)\} \subset \Pi_n .$$

LEMMA 4.2. *The subset $p(M) \subset \Pi_n$ contains the zero element of Π_n if and only if M bounds a parallelizable manifold.*

PROOF. If $M = bW$ with W parallelizable then the imbedding $i\colon M \to S^{n+k}$ can be extended to an imbedding $W \to D^{n+k+1}$, and W has a field ψ of normal k-frames. We set $\varphi = \psi \mid M$. Now the Pontrjagin-Thom map $p(M, \varphi)\colon S^{n+k} \to S^k$ extends over D^{n+k+1}, hence is null-homotopic.

Conversely if $p(M, \varphi) \simeq 0$, then M bounds a manifold $W \subset D^{n+k+1}$, where φ extends to a field ψ of normal frames over W. It follows from Lemmas 3.3 and 3.4 that W is parallelizable. This completes the proof of Lemma 4.2.

LEMMA 4.3. *If M_0 is h-cobordant to M_1, then $p(M_0) = p(M_1)$.*

PROOF. If $M_0 + (-M_1) = bW$, we choose an imbedding of W in $S^{n+k} \times [0, 1]$ so that $M_q \to S^{n+k} \times (q)$ for $q = 0, 1$. Then a normal frame field φ_q on M_q extends to a normal frame field ψ on W which restricts to some normal frame field φ_{1-q} on M_{1-q}. Clearly (W, ψ) gives rise to a homotopy between $p(M_0, \varphi_0)$ and $p(M_1, \varphi_1)$.

LEMMA 4.4. *If M and M' are* s-*parallelizable then*

$$p(M) + p(M') \subset p(M \# M') \subset \Pi_n .$$

PROOF. Start with the disjoint sum

$$M \times [0, 1] + M' \times [0, 1] ,$$

and join the boundary components $M \times 1$ and $M' \times 1$ together, as described in the addendum at the end of § 2. Thus we obtain a manifold W bounded by the disjoint sum

$$(M \# M') + (-M) + (-M') .$$

Note that W has the homotopy type of $M \vee M$, the union with a single point in common.

Choose an imbedding of W in $S^{n+k} \times [0, 1]$ so that $(-M)$ and $(-M')$ go into well separated submanifolds of $S^{n+k} \times 0$, and so that $M \# M'$ goes into $S^{n+k} \times 1$. Given fields φ and φ' of normal k-frames on $(-M)$ and $(-M')$, it is not hard to see that there exists an extension defined throughout W. Let ψ denote the restriction of this field to $M \# M'$. Then clearly $p(M, \varphi) + p(M', \varphi')$ is homotopic to $p(M \# M', \psi)$. This completes the proof.

LEMMA 4.5. *The set $p(S^n) \subset \Pi_n$ is a subgroup of the stable homotopy group Π_n. For any homotopy sphere Σ the set $p(\Sigma)$ is a coset of this subgroup $p(S^n)$. Thus the correspondence $\Sigma \to p(\Sigma)$ defines a homomorphism p' from Θ_n to the quotient group $\Pi_n/p(S^n)$.*

PROOF. Combining Lemma 4.4 with the identities

(1) $S^n \# S^n = S^n$,

(2) $S^n \# \Sigma = \Sigma$,

(3) $\Sigma \# (-\Sigma) \sim S^n$

we obtain

$$(1) \qquad p(S^n) + p(S^n) \subset p(S^n) ,$$

which shows that $p(S^n)$ is a subgroup of Π_n;

$$(2) \qquad p(S^n) + p(\Sigma) \subset p(\Sigma) ,$$

which shows that $p(\Sigma)$ is a union of cosets of this subgroup; and

$$(3) \qquad p(\Sigma) + p(-\Sigma) \subset p(S^n) ,$$

which shows that $p(\Sigma)$ must be a single coset. This completes the proof of Lemma 4.5.

By Lemma 4.2 the kernel of $p'\colon \Theta_n \to \Pi_n/p(S^n)$ consists exactly of all h-cobordism classes of homotopy n-spheres which bound parallelizable manifolds. Thus these elements form a group which we will denote by $bP_{n+1} \subset \Theta_n$. It follows that Θ_n/bP_{n+1} is isomorphic to a subgroup of $\Pi_n/p(S^n)$. Since Π_n is finite (Serre [24]), this completes the proof of Theorem 4.1.

REMARKS. The subgroup $p(S^n) \subset \Pi_n$ can be described in more familiar terms as the image of the Hopf-Whitehead homomorphism

$$J_n: \pi_n(\mathrm{SO}_k) \to \pi_{n+k}(S^k) .$$

(See Kervaire [9, p. 349].) Hence $\Pi_n/p(S^n)$ is the cokernel of J_n. The actual structure of these groups for $n \leq 8$ is given in the following table. For details, and for higher values of n, the reader is referred to Part II of this paper.

n	1	2	3	4	5	6	7	8
Π_n	Z_2	Z_2	Z_{24}	0	0	Z_2	Z_{240}	$Z_2 + Z_2$
$\Pi_n/p(S^n)$	0	Z_2	0	0	0	Z_2	0	Z_2
Θ_n/bP_{n+1}	0	0	0	0	0	0	0	Z_2

The prime $q \geq 3$ first divides the order of Θ_n/bP_{n+1} for $n = 2q(q-1) - 2$.

Using Theorem 4.1, the proof of the main theorem (Theorem 1.2), stating that Θ_n is finite for $n \neq 3$, reduces now to proving that bP_{n+1} is finite for $n \neq 3$.

We will prove that the group bP_{n+1} is zero for n even (§§ 5, 6), and is finite cyclic for n odd, $n \neq 3$, (see §§ 7, 8). The first few groups can be given as follows:

n	1	3	5	7	9	11	13	15	17	19
order of bP_{n+1}	1	?	1	28	2	992	1	8128	2	261,632*

(Again see Part II for details.) The cyclic group bP_{n+1} has order 1 or 2 for $n \equiv 1 \pmod 4$, but the order grows more than exponentially for $n \equiv 3 \pmod 4$.

5. Spherical modifications

This section, and § 6 which follows, will prove that the groups bP_{2k+1} are zero.[5] That is:

THEOREM 5.1. *If a homotopy sphere of dimension $2k$ bounds an* s-*parallelizable manifold M, then it bounds a contractible manifold M_1.*

[5] An independent proof of this theorem has been given by C.T.C. Wall [29].

*Corrected 2006.

For the case $k = 1$, this assertion is clear since every homotopy 2-sphere is actually diffeomorphic to S^2. The proof for $k > 1$ will be based on the technique of "spherical modifications." (See Wallace [30], Milnor [15; 17][6].)

DEFINITION. Let M be a differentiable manifold of dimension $n = p + q + 1$ and let

$$\varphi: S^p \times D^{q+1} \to M$$

be a differentiable imbedding. Then a new differentiable manifold $M' = \chi(M, \varphi)$ is formed from the disjoint sum

$$(M - \varphi(S^p \times 0)) + D^{p+1} \times S^q$$

by identifying $\varphi(u, tv)$ with (tu, v) for each $u \in S^p$, $v \in S^q$, $0 < t \leqq 1$. We will say that M' is obtained from M by the *spherical modification* $\chi(\varphi)$. Note that the boundary of M' is equal to the boundary of M.

In order to prove Theorem 5.1 we will show that the homotopy groups of M can be completely killed by a sequence of such spherical modifications. The effect of a single modification $\chi(\varphi)$ on the homotopy groups of M can be described as follows.

Let $\lambda \in \pi_p M$ denote the homotopy class of the map $\varphi \mid S^p \times 0$ from $S^p \times 0$ to M.

LEMMA 5.2. *The homotopy groups of M' are given by*

$$\pi_i M' \simeq \pi_i M \qquad \textit{for } i < \operatorname{Min}(p, q),$$

and

$$\pi_p M' \simeq \pi_p M / \Lambda ,$$

provided that $p < q$; where Λ denotes a certain subgroup of $\pi_p M$ containing λ.

The proof is straightforward. (Compare [17, Lemma 2].)

Thus if $p < q$ (that is, if $p \leqq n/2 - 1$), the effect of the modification $\chi(\varphi)$ is to kill the homotopy class λ.

Now suppose that some homotopy class $\lambda \in \pi_p M$ is given.

LEMMA 5.3. *If M^n is s-parallelizable and if $p < n/2$, then the class λ is represented by some imbedding $\varphi: S^p \times D^{n-p} \to M$.*

PROOF. (Compare [17, Lemma 3].) Since $n \geqq 2p + 1$ it follows from a well known theorem of Whitney that λ can be represented by an imbedding

$$\varphi_0: S^p \to M .$$

[6] The term "surgery" is used for this concept in [15; 17].

It follows from Lemma 3.5 that the normal bundle of $\varphi_0 S^p$ in M is trivial. Hence φ_0 can be extended to the required imbedding $S^p \times D^{n-p} \to M$.

Thus Lemmas 5.2 and 5.3 assert that spherical modifications can be used to kill any required element $\lambda \in \pi_p M^n$ provided that $p \leqq n/2 - 1$. There is one danger however. If the imbedding φ is chosen badly then the modified manifold $M' = \chi(M, \varphi)$ may no longer be s-parallelizable. However the following was proven in [17]. Again let $n \geqq 2p + 1$.

LEMMA 5.4. *The imbedding φ: $S^p \times D^{n-p} \to M$ can be chosen within its homotopy class so that the modified manifold $\chi(M, \varphi)$ will also be* s-*parallelizable.*

For the proof, the reader may either refer to [17, Theorem 2], or make use of the sharper Lemma 6.2 which will be proved below.

Now combining Lemmas 5.2, 5.3, 5.4, one obtains the following. (Compare [17, p. 46].)

THEOREM 5.5. *Let M be a compact, connected* s-*parallelizable manifold of dimension $n \geqq 2k$. By a sequence of spherical modifications on M one can obtain an* s-*parallelizable manifold M_1 which is $(k-1)$-connected.*

Recall that $bM_1 = bM$.

PROOF. Choosing a suitable imbedding φ: $S^1 \times D^{n-1} \to M$, one can obtain an s-parallelizable manifold $M' = \chi(M, \varphi)$ such that $\pi_1 M'$ is generated by fewer elements than $\pi_1 M$. Thus after a finite number of steps, one can obtain a manifold M'' which is 1-connected. Now, after a finite number of steps, one can obtain an s-parallelizable manifold M''' which is 2-connected, and so on until we obtain a $(k-1)$-connected manifold. This proves Theorem 5.5.

In order to prove 5.1, where $\dim M = 2k + 1$, we must carry this argument one step further obtaining a manifold M_1 which is k-connected. It will then follow from the Poincaré duality theorem that M_1 is contractible.

The difficulty in carrying out this program is that Lemma 5.2 is no longer available. Thus if $M' = \chi(M, \varphi)$ where φ imbeds $S^k \times D^{k+1}$ in M, the group $\pi_k M'$ may actually be larger than $\pi_k M$. It is first necessary to describe in detail what happens to $\pi_k M$ under such a modification. Since we may assume that M is $(k-1)$-connected with $k > 1$, the homotopy group $\pi_k M$ may be replaced by the homology group $H_k M = H_k(M; Z)$.

LEMMA 5.6. *Let $M' = \chi(M, \varphi)$ where φ imbeds $S^k \times D^{k+1}$ in M, and let*

$$M_0 = M - (\text{interior } \varphi(S^k \times D^{k+1})) .$$

Then there is a commutative diagram

$$\begin{array}{ccccccccc} & & & & H_{k+1}M' & & & & \\ & & & & \downarrow{\scriptstyle \cdot\lambda'} & & & & \\ & & & & Z & & & & \\ & & & & \downarrow{\scriptstyle \varepsilon} & \searrow{\scriptstyle \lambda} & & & \\ H_{k+1}M & \xrightarrow{\cdot\lambda} & Z & \xrightarrow{\varepsilon'} & H_kM_0 & \xrightarrow{i} & H_kM & \longrightarrow & 0 \\ & & & {\scriptstyle \lambda'}\searrow & \downarrow{\scriptstyle i'} & & & & \\ & & & & H_kM' & & & & \\ & & & & \downarrow & & & & \\ & & & & 0 & & & & \end{array}$$

such that the horizontal and vertical sequences are exact. It follows that the quotient group $H_kM/\lambda(Z)$ *is isomorphic to* $H_kM'/\lambda'(Z)$.

Here the following notations are to be understood. The symbol λ denotes the element of H_kM which corresponds to the homotopy class $\varphi \mid S^k \times 0$, and λ also denotes the homomorphism $Z \to H_kM$ which carries 1 into λ. On the other hand $\cdot\lambda$: $H_{k+1}M \to Z$ denotes the homomorphism which carries each $\mu \in H_{k+1}M$ into the intersection number $\mu \cdot \lambda$. The symbols λ' and $\cdot\lambda'$ are to be interpreted similarly. The element $\lambda' \in H_kM'$ corresponds to the homotopy class $\varphi' \mid 0 \times S^k$ where

$$\varphi' \colon D^{k+1} \times S^k \to M'$$

denotes the canonical imbedding.

PROOF OF LEMMA 5.6. As horizontal sequences take the exact sequence

$$H_{k+1}M \longrightarrow H_{k+1}(M, M_0) \xrightarrow{\varepsilon'} H_kM_0 \xrightarrow{i} H_kM \longrightarrow H_k(M, M_0)$$

of the pair (M, M_0). By excision, the group $H_j(M, M_0)$ is isomorphic to

$$H_j(S^k \times D^{k+1}, S^k \times S^k) \simeq \begin{cases} Z & \text{for } j = k+1 \\ 0 & \text{for } j < k+1 \, . \end{cases}$$

Thus we obtain

$$H_{k+1}M \longrightarrow Z \xrightarrow{\varepsilon'} H_kM_0 \xrightarrow{i} H_kM \longrightarrow 0 \, ,$$

as asserted. Since a generator of $H_{k+1}(M, M_0)$ clearly has intersection number ± 1 with the cycle $\varphi(S^k \times 0)$ which represents λ, it follows that the homomorphism $H_{k+1}M \to Z$ can be described as the homomorphism $\mu \to \mu \cdot \lambda$. The element $\varepsilon' = \varepsilon'(1) \in H_kM_0$ can clearly be described as the homology class corresponding to the "meridian" $\varphi(x_0 \times S^k)$ of the torus

$\varphi(S^k \times S^k)$, where x_0 denotes a base point in S^k.

The vertical exact sequence is obtained in a similar way. Thus $\varepsilon = \varepsilon(1) \in H_k M_0$ is the homology class of the "parallel" $\varphi(S^k \times x_0)$ of the torus. Clearly $i(\varepsilon) \in H_k M$ is equal to the homology class λ of $\varphi(S^k \times 0)$. Similarly $i'(\varepsilon') = \lambda'$.

From this diagram the isomorphisms

$$H_k M/\lambda(Z) \simeq H_k M_0/\varepsilon(Z) + \varepsilon'(Z) \simeq H_k M'/\lambda'(Z)$$

are apparent. This completes the proof of Lemma 5.6.

As an application, suppose that one chooses an element $\lambda \in H_k M$ which is *primitive* in the sense that $\mu \cdot \lambda = 1$ for some $\mu \in H_{k+1} M$. It follows that

$$i: H_k M_0 \to H_k M$$

is an isomorphism, and hence that

$$H_k M' \simeq H_k M/\lambda(Z) .$$

Thus:

ASSERTION. *Any primitive element of $H_k M$ can be killed by a spherical modification.*

In order to apply this assertion we assume the following:

Hypothesis. M is a compact, s-parallelizable manifold of dimension $2k+1$, $k > 1$, and is $(k-1)$-connected. The boundary bM is either vacuous or a homology sphere.

This hypothesis will be assumed for the rest of § 5 and for § 6.

LEMMA 5.7. *Subject to this hypothesis, the homology group $H_k M$ can be reduced to its torsion subgroup by a sequence of spherical modifications. The modified manifold M_1 will still satisfy the hypothesis.*

PROOF. Suppose that $H_k M \simeq Z \oplus \cdots \oplus Z \oplus T$ where T is the torsion subgroup. Let λ generate one of the infinite cyclic summands. Using the Poincaré duality theorem one sees that $\mu_1 \cdot \lambda = 1$ for some element $\mu_1 \in H_{k+1}(M, bM)$. But the exact sequence

$$H_{k+1} M \to H_{k+1}(M, bM) \to H_k(bM) = 0$$

shows that μ_1 can be lifted back to $H_{k+1} M$. Therefore λ is primitive, and can be killed by a modification. After finitely many such modifications, one obtains a manifold M_1 with $H_k M_1 \simeq T \subset H_k M$. This completes the proof of Lemma 5.7.

Let us specialize to the case k even. Let M be as above, and let $\varphi: S^k \times D^{k+1} \to M$ be any imbedding.

LEMMA 5.8. *If k is even then the modification $\chi(\varphi)$ necessarily changes* *the kth Betti number of M.*

The proof will be based on the following lemma. (See Kervaire [8, Formula (8.8)].)

Let F be a fixed field and let W be an orientable homology manifold of dimension $2r$. Define the *semi-characteristic* $e^*(bW; F)$ to be the following residue class modulo 2:

$$e^*(bW; F) \equiv \sum_{i=0}^{r-1} \operatorname{rank} H_i(bW; F) \pmod 2 .$$

LEMMA 5.9. *The rank of the bilinear pairing*

$$H_r(W; F) \otimes H_r(W; F) \to F ,$$

given by the intersection number, is congruent modulo 2 to $e^*(bW; F)$ plus the Euler characteristic $e(W)$.

[For the convenience of the reader, here is a proof. Consider the exact sequence

$$H_r W \xrightarrow{h} H_r(W, bW) \longrightarrow H_{r-1}(bW) \longrightarrow \cdots \longrightarrow H_0(W, bW) \longrightarrow 0 ,$$

where the coefficient group F is to be understood. A counting argument shows that the rank of the indicated homomorphism h is equal to the alternating sum of the ranks of the vector spaces to the right of h in this sequence. Reducing modulo 2 and using the identity

$$\operatorname{rank} H_i(W, bW) = \operatorname{rank} H_{2r-i} W ,$$

this gives

$$\begin{aligned}\operatorname{rank} h &\equiv \textstyle\sum_{i=0}^{r-1} \operatorname{rank} H_i(bW) + \sum_{i=0}^{2r} \operatorname{rank} H_i W \\ &\equiv e^*(bW; F) + e(W) \pmod 2 .\end{aligned}$$

But the rank of

$$h\colon H_r W \to H_r(W, bW) \simeq \operatorname{Hom}_F(H_r W, F)$$

is just the rank of the intersection pairing. This completes the proof.]

PROOF OF LEMMA 5.8. First suppose that M has no boundary. As shown in [15] or [17] the manifolds M and $M' = \chi(M, \varphi)$, suitably oriented, together bound a manifold $W = W(M, \varphi)$ of dimension $2k + 2$. For the moment, since no differentiable structure on W is needed, we can simply define W to be the union

$$(M \times [0, 1]) \cup (D^{k+1} \times D^{k+1}) ,$$

where it is understood that $S^k \times D^{k+1}$ is to be pasted onto $M \times 1$ by the imbedding φ. Clearly W is a topological manifold with

$$bW = M \times 0 + M' \times 1 .$$

Note that W has the homotopy type of M with a $(k+1)$-cell attached. Since the dimension $2k+1$ of M is odd, this means that the Euler characteristic

$$e(W) = e(M) + (-1)^{k+1} = (-1)^{k+1} .$$

Since k is even, the intersection pairing

$$H_{k+1}(W; Q) \otimes H_{k+1}(W; Q) \to Q$$

is skew symmetric, hence has even rank. Therefore Lemma 5.9 (with rational coefficients) asserts that

$$e^*(M + M'; Q) + (-1)^{k+1} \equiv 0 \pmod 2 ,$$

and hence that

$$e^*(M; Q) \not\equiv e^*(M'; Q) .$$

But $H_i M \simeq H_i M' \simeq 0$ for $0 < i < k$, so this implies that

$$\operatorname{rank} H_k(M; Q) \not\equiv \operatorname{rank} H_k(M'; Q) .$$

This proves Lemma 5.8 provided that M has no boundary.

If M is bounded by a homology sphere, then attaching a cone over bM, one obtains a homology manifold M_* without boundary. The above argument now shows that

$$\operatorname{rank} H_k(M_*; Q) \not\equiv \operatorname{rank} H_k(M'_*; Q) .$$

Therefore the modification $\chi(\varphi)$ changes the rank of $H_k(M; Q)$ in this case also. This completes the proof of Lemma 5.8.

It is convenient at this point to insert an analogue of 5.8 which will only be used later. (See the end of § 6.) Let M be as above, with k even or odd, and let $W = W(M, \varphi)$.

LEMMA 5.10. *Suppose that every* mod 2 *homology class*

$$\xi \in H_{k+1}(W; Z_2)$$

has self-intersection number $\xi \cdot \xi = 0$. *Then the modification* $\chi(\varphi)$ *necessarily changes the rank of the* mod 2 *homology group* $H_k(M; Z_2)$.

The proof is completely analogous to that of 5.8. The hypothesis, $\xi \cdot \xi = 0$ for all ξ, guarantees that the intersection pairing

$$H_{k+1}(W; Z_2) \otimes H_{k+1}(W; Z_2) \to Z_2$$

will have even rank.

We now return to the case k even.

PROOF OF THEOREM 5.1, *for k even*. According to 5.6, we can assume that $H_k M$ is a torsion group. Choose

$$\varphi\colon S^k \times D^{k+1} \to M$$

as in 5.4 so as to represent a non-trivial $\lambda \in H_k M$. According to 5.6 we have

$$H_k M/\lambda(Z) \simeq H_k M'/\lambda'(Z) .$$

Since the group $\lambda(Z)$ is finite, it follows from 5.8 that $\lambda'(Z)$ must be infinite. Thus the sequence

$$0 \longrightarrow Z \xrightarrow{\lambda'} H_k M' \longrightarrow H_k M'/\lambda'(Z) \longrightarrow 0$$

is exact. It follows that the torsion subgroup of $H_k M'$ maps monomorphically into $H_k M'/\lambda'(Z)$; and hence is definitely smaller than $H_k M$. Now according to 5.7, we can perform a modification on M' so as to obtain a new manifold M'' with

$$H_k M'' \simeq \text{Torsion subgroup of } H_k M' < H_k M .$$

Thus in two steps one can replace $H_k M$ by a smaller group. Iterating this construction a finite number of times, the group $H_k M$ can be killed completely. This completes the proof of Theorem 5.1 for k even.

6. Framed spherical modifications

This section will complete the proof of Theorem 5.1 by taking care of the case k odd. This case is somewhat more difficult than the case k even (which was handled in § 5), since it is necessary to choose the imbeddings φ more carefully, taking particular care not to lose s-parallelizability in the process. Before starting on the proof, it is convenient to sharpen the concepts of s-parallelizable manifold, and of spherical modification.

DEFINITION. A *framed manifold* (M, f) will mean a differentiable manifold M together with a fixed trivialization f of the stable tangent bundle $\tau_M \oplus \varepsilon_M$.

Now consider a spherical modification $\chi(\varphi)$ of M. Recall that M and $M' = \chi(M, \varphi)$ together bound a manifold

$$W = (M \times [0, 1]) \cup (D^{p+1} \times D^{q+1})$$

where the subset $S^p \times D^{q+1}$ of $D^{p+1} \times D^{q+1}$ is pasted onto $M \times 1$ by the imbedding φ. (Compare Milnor [17].) It is easy to give W a differentiable structure, except along the "corner" $S^p \times S^q$. A neighborhood of this corner will be "diffeomorphic" with $S^p \times S^q \times Q$ where

$$Q \subset R^2$$

denotes the three-quarter disk consisting of all $(r\cos\theta, r\sin\theta)$ with $0 \leq r < 1$, $0 \leq \theta \leq 3\pi/2$. In order to "straighten" this corner, map Q onto the half-disk H, consisting of all $(r\cos\theta', r\sin\theta')$ with $0 \leqq r < 1$, $0 \leqq \theta' \leqq \pi$; by setting $\theta' = 2\theta/3$. Now carrying the differentiable structure of H back to Q, this makes Q into a differentiable manifold. Carrying out the same transformation on the neighborhood of $S^p \times S^q$, this makes $W = W(M, \varphi)$ into the required differentiable manifold. Note that both boundaries of W get the correct differential structures.

Now identify M with $M \times 0 \subset W$, and identify the stable tangent bundle $\tau_M \oplus \varepsilon_M$ with the restriction $\tau_W \mid M$. Thus a framing f of M determines a trivialization of $\tau_W \mid M$.

DEFINITION. A *framed spherical modification* $\chi(\varphi, F)$ of the framed manifold (M, f) will mean a spherical modification $\chi(\varphi)$ of M together with a trivialization F of the tangent bundle of W, satisfying the condition

$$F \mid M = f\ .$$

Note that the modified manifold $M' = \chi(M, \varphi)$ automatically acquires a framing

$$f' = F \mid M'\ .$$

It is only necessary to identify $\tau_W \mid M'$ with the stable tangent bundle $\tau_{M'} \oplus \varepsilon_{M'}$. To do this, we identify the positive direction in $\varepsilon_{M'}$ with the outward normal direction in $\tau_W \mid M'$.

The following question evidently arises. Given a modification $\chi(\varphi)$ of M and a framing f of M, does f extend to a trivialization F of τ_W? The obstructions to such an extension lie in the cohomology groups

$$H^{r+1}(W, M; \pi_r(\mathrm{SO}_{n+1})) \simeq \begin{cases} \pi_p(\mathrm{SO}_{n+1}) & \text{for } r = p \\ 0 & \text{for } r \neq p\ . \end{cases}$$

Thus the only obstruction to extending f is a well defined class

$$\gamma(\varphi) \in \pi_p(\mathrm{SO}_{n+1})\ .$$

The modification $\chi(\varphi)$ can be framed if and only if this obstruction $\gamma(\varphi)$ is zero.

Now consider the following alteration of the imbedding φ. Let

$$\alpha\colon S^p \to \mathrm{SO}_{q+1}$$

be a differentiable map, and define

$$\varphi_\alpha\colon S^p \times D^{q+1} \to M$$

by

$$\varphi_\alpha(u, v) = \varphi(u, v\cdot\alpha(u))\ ,$$

where the dot denotes the usual action of SO_{q+1} on D^{q+1}. Clearly φ_α is an imbedding which represents the same homotopy class $\lambda \in \pi_p M$ as φ.

LEMMA 6.1. *The obstruction $\gamma(\varphi_\alpha)$ depends only on $\gamma(\varphi)$ and on the homotopy class (α) of α. In fact*

$$\gamma(\varphi_\alpha) = \gamma(\varphi) + s_*(\alpha)$$

where $s_*\colon \pi_p(SO_{q+1}) \to \pi_p(SO_{n+1})$ is induced by the inclusion $s\colon SO_{q+1} \to SO_{n+1}$.

PROOF. (Compare [17], proof of Theorem 2.) Let W_α be the manifold constructed as W above, now using φ_α. There is a natural differentiable imbedding

$$i_\alpha\colon D^{p+1} \times \operatorname{int} D^{q+1} \to W_\alpha ,$$

and $i_\alpha \mid S^p \times D^{q+1}$ coincides with $\varphi_\alpha\colon S^p \times D^{q+1} \to M$ followed by the inclusion $M \to M \times 1 \subset W_\alpha$.

$\gamma(\varphi_\alpha)$ is the obstruction to extending $f \mid \varphi_\alpha(S^p \times 0)$ to a trivialization of $\tau(W_\alpha)$ restricted to $i_\alpha(D^{p+1} \times 0)$ Let $t^{n+1} = e^{p+1} \times e^{q+1}$ be the standard framing on $D^{p+1} \times D^{q+1}$. Then $i'_\alpha(t^{n+1})$ is a trivialization of the tangent bundle of W_α restricted to $i_\alpha(D^{p+1} \times D^{q+1})$, and $\gamma(\varphi_\alpha)$ is the homotopy class of the map $g\colon S^p \to SO_{n+1}$, where $g(u)$ is the matrix $\langle f^{n+1}, i'_\alpha(t^{n+1})\rangle$ at $\varphi_\alpha(u, 0)$.

Since $i_\alpha \mid D^{p+1} \times 0$ is independent of α, and $i_\alpha \mid S^p \times D^{q+1} = \varphi_\alpha$, we have

$$i'_\alpha(t^{n+1}) = i'(e^{p+1}) \times \varphi'_\alpha(e^{q+1})$$

at every point $(u, 0) \in S^p \times D^{q+1}$.

Since

$$\varphi'_\alpha(e^{q+1}) = \varphi'(e^{q+1}) \cdot \alpha(u)$$

at $(u, 0)$, it follows that

$$i'_\alpha(t^{n+1}) = i'(t^{n+1}) \cdot s(\alpha) .$$

Hence

$$\langle f^{n+1}, i'_\alpha(t^{n+1})\rangle = \langle f^{n+1}, i'(t^{n+1})\rangle \cdot s(\alpha)$$

and the lemma follows.

Now suppose (as usual) that $p \leqq q$. Then the homomorphism

$$s_*\colon \pi_p(SO_{q+1}) \to \pi_p S(O_{n+1})$$

is onto. Hence α can be chosen so that

$$\gamma(\varphi_\alpha) = \gamma(\varphi) + s_*(\alpha)$$

is zero. Thus we obtain:

LEMMA 6.2. *Given $\varphi\colon S^p \times D^{q+1} \to M$ with $p \leqq q$, a map α can be chosen* *so that the modification $\chi(\varphi_a)$ can be framed.*

In particular, it follow that the manifold $\chi(M, \varphi_\alpha)$ will be s-parallelizable. Thus we have proved Lemma 5.4 in a sharpened form.

We note however that α is not always uniquely determined. In the case $p = q = k$ odd, the homomorphism

$$s_*: \pi_k(\mathrm{SO}_{k+1}) \to \pi_k(\mathrm{SO}_{n+1})$$

has an infinite cyclic kernel. This freedom in the choice of α will be the basis of the proof of 5.1 for k odd.

Let us study the homology of the manifold

$$M'_\alpha = \chi(M, \varphi_\alpha) ,$$

where φ is now chosen, by Lemma 6.1, so that the spherical modification $\chi(\varphi)$ can be framed. Clearly the deleted manifold

$$M_0 = M - \left(\text{interior } \varphi_\alpha(S^k \times D^{k+1})\right)$$

does not depend on the choice of α. Furthermore the meridian $\varphi_\alpha(x_0 \times S^k)$ of the torus $\varphi_\alpha(S^k \times S^k) \subset M_0$ does not depend on the choice of α; hence the homology class

$$\varepsilon' \in H_k M_0$$

does not depend on α. On the other hand the parallel $\varphi_\alpha(S^k \times x_0)$ does depend on α. In fact it is clear that the homology class $\varepsilon_\alpha \in H_k M_0$ of this parallel is given by

$$\varepsilon_\alpha = \varepsilon + j(\alpha)\varepsilon'$$

where the homomorphism

$$j_*: \pi_k(\mathrm{SO}_{k+1}) \to Z \simeq \pi_k(S^k)$$

is induced by the canonical map

$$\rho \xrightarrow{j} x_0 \cdot \rho$$

from SO_{k+1} to S^k.

The spherical modification $\chi(\varphi_\alpha)$ can still be framed provided α is an element of the kernel of

$$s_*: \pi_k(\mathrm{SO}_{k+1}) \to \pi_k(\mathrm{SO}_{n+1}) .$$

Identifying the stable group $\pi_k(\mathrm{SO}_{n+1})$ with the stable group $\pi_k(\mathrm{SO}_{k+2})$, there is an exact sequence

$$\pi_{k+1}(S^{k+1}) \xrightarrow{\partial} \pi_k(\mathrm{SO}_{k+1}) \xrightarrow{s_*} \pi_k(\mathrm{SO}_{k+2})$$

associated with the fibration $\mathrm{SO}_{k+2}/\mathrm{SO}_{k+1} = S^{k+1}$. It is well known that the composition

$$\pi_{k+1}(S^{k+1}) \xrightarrow{\partial} \pi_k(\mathrm{SO}_{k+1}) \xrightarrow{j_*} \pi_k(S^k)$$

carries a generator of $\pi_{k+1}(S^{k+1})$ onto twice a generator of $\pi_k(S^k)$, provided that k is odd. Therefore the integer $j_*(\alpha)$ can be any multiple of 2.

Let us study the effect of replacing ε by $\varepsilon_\alpha = \varepsilon + j(\alpha)\varepsilon'$ on the homology of the modified manifold. Consider the exact sequence

$$0 \longrightarrow Z \xrightarrow{\varepsilon'} H_k M_0 \xrightarrow{i} H_k M \longrightarrow 0$$

of 5.6, where i carries ε into an element λ of order $l > 1$. Evidently $l\varepsilon$ must be a multiple of ε', say:

$$l\varepsilon + l'\varepsilon' = 0 .$$

Since ε' is not a torsion element, these two elements can satisfy no other relation. Since $\varepsilon_\alpha = \varepsilon + j_*(\alpha)\varepsilon'$ it follows that

$$l\varepsilon_\alpha + (l' - lj(\alpha))\varepsilon' = 0 .$$

Now using the sequence

$$Z \xrightarrow{\varepsilon_\alpha} H_k M_0 \xrightarrow{i'_\alpha} H_k M'_\alpha \longrightarrow 0 ,$$

we see that the inclusion homomorphism i'_α carries ε' into an element

$$\lambda'_\alpha \in H_k M'_\alpha$$

of order $|l' - lj(\alpha)|$. Since $H_k M'_\alpha / \lambda'_\alpha(Z)$ is isomorphic to $H_k M/\lambda(Z)$, we see that the group $H_k M'_\alpha$ is smaller than $H_k M_\alpha$ if and only if

$$0 < |l' - lj(\alpha)| < l .$$

But $j(\alpha)$ can be any even integer. Thus $j(\alpha)$ can be chosen so that

$$-l < l' - lj(\alpha) \leqq l .$$

This choice of $j(\alpha)$ will guarantee an improvement except in the special case where l' happens to be divisible by l.

Our progress so far can be summarized as follows.

LEMMA 6.3. *Let M be a framed $(k-1)$-connected manifold of dimension $2k+1$ with k odd, $k > 1$, such that $H_k M$ is finite. Let $\chi(\varphi, F)$ be a framed modification of M which replaces the element $\lambda \in H_k M$ of order $l > 1$ by an element $\lambda' \in H_k M'$ of order $\pm l'$. If $l' \not\equiv 0 \bmod l$ then it is possible to choose $(\alpha) \in \pi_k(\mathrm{SO}_{k+1})$ so that the modification $\chi(\varphi_\alpha)$ can still be framed, and so that the group $H_k M'_\alpha$ is definitely smaller than $H_k M$.*

Thus one must study the residue class of l' modulo l. Recall the definition of linking numbers. (Compare Seifert-Threlfall [23, § 77].) Let

$\lambda \in H_pM$, $\mu \in H_qM$ be homology classes of finite order, with $\dim M = p + q + 1$. Consider the homology sequence

$$\cdots \longrightarrow H_{p+1}(M; Q/Z) \xrightarrow{\beta} H_pM \xrightarrow{i_*} H_p(M; Q) \longrightarrow \cdots$$

associated with the coefficient sequence

$$0 \longrightarrow Z \xrightarrow{i} Q \longrightarrow Q/Z \longrightarrow 0 .$$

Since λ is of finite order, $i_*\lambda = 0$ and $\lambda = \beta(\nu)$ for some $\nu \in H_{p+1}(M; Q/Z)$. The pairing

$$Q/Z \otimes Z \to Q/Z$$

defined by multiplication induces a pairing

$$H_{p+1}(M; Q/Z) \otimes H_qM \to Q/Z$$

defined by the intersection of homology classes. We denote this pairing by a dot.

DEFINITION. The *linking number* $L(\lambda, \mu)$ is the rational number modulo 1 defined by

$$L(\lambda, \mu) = \nu \cdot \mu .$$

This linking number is well defined, and satisfies the symmetry relation

$$L(\mu, \lambda) + (-1)^{pq} L(\lambda, \mu) = 0 .$$

(Compare Seifert and Threlfall.)

LEMMA 6.4. *The ratio l'/l modulo 1 is, up to sign, equal to the self-linking number $L(\lambda, \lambda)$.*

PROOF. Since

$$l\varepsilon + l'\varepsilon' = 0$$

in H_kM_0, we see that the cycle $l\varepsilon + l'\varepsilon'$ on bM_0 bounds a chain c in M_0. Let $c_1 = \varphi(x_0 \times D^{k+1})$ denote the cycle in $\varphi(S^k \times D^{k+1}) \subset M$ with boundary ε'. Then the chain $c - l'c_1$ has boundary $l\varepsilon$; hence $(c - l'c_1)/l$ has boundary ε, representing the homology class λ in H_kM. Taking the intersection of this chain with $\varphi(S^k \times 0)$, representing λ, we obtain $\pm l'/l$, since c is disjoint and c_1 has intersection number ∓ 1. Thus $L(\lambda, \lambda) = \pm l'/l \bmod 1$.

Now if $L(\lambda, \lambda) \neq 0$, then $l' \not\equiv 0 \pmod{l}$, hence the class λ can be replaced by an element of smaller order under a spherical modification. Hence, unless $L(\lambda, \lambda) = 0$ for all $\lambda \in H_kM$, this group can be simplified.

LEMMA 6.5. *If H_kM is a torsion group, with $L(\lambda, \lambda) = 0$ for every*

$\lambda \in H_k M$, and if k is odd, then this group $H_k M$ must be a direct sum of cyclic groups of order 2.

PROOF. The relation

$$L(\eta, \xi) + (-1)^{pq} L(\xi, \eta) = 0$$

with $p = q \equiv 1 \pmod 2$ implies that

$$L(\eta, \xi) = L(\xi, \eta) .$$

Now if self-linking numbers are all zero, the identity

$$L(\xi + \eta, \xi + \eta) = L(\xi, \xi) + L(\eta, \eta) + L(\xi, \eta) + L(\eta, \xi)$$

implies that

$$2L(\xi, \eta) = 0$$

for all ξ and η. But, according to the Poincaré duality theorem for torsion groups (see [23, p. 245]), L defines a completely orthogonal pairing

$$T_p M \otimes T_q M \to Q/Z .$$

Hence the identity $L(2\xi, \eta) = 0$ for all η implies that $2\xi = 0$. This proves Lemma 6.5.

It follows that, by a sequence of modifications, one can reduce $H_k M$ to a group of the form $Z_2 \oplus \cdots \oplus Z_2 = sZ_2$.

Now let us apply Lemma 5.10. Since the modification $\chi(\varphi_\alpha)$ is framed, the corresponding manifold $W = W(M, \varphi_\alpha)$ is parallelizable. It follows from the formulas of Wu that the Steenrod operation

$$\mathrm{Sq}^{k+1}\colon H^{k+1}(W, bW; Z_2) \to H^{2k+2}(W, bW; Z_2)$$

is zero. (See Kervaire [8, Lemma (7.9)].) Hence every $\xi \in H_{k+1}(W; Z_2)$ has self-intersection number $\xi \cdot \xi = 0$. Thus, according to 5.10, the modification $\chi(\varphi_\alpha)$ changes the rank of $H_k(M; Z_2)$.

But the effect of $\chi(\varphi_\alpha)$ on $H_k(M; Z)$, provided that α is chosen properly, will be to replace the element λ of order $l = 2$ by an element λ'_α of order l'_α where

$$-2 < l'_\alpha \leqq 2 , \qquad l'_\alpha \equiv 0 \qquad (\mathrm{mod}\ 2) .$$

Thus l'_α must be 0 or 2. Now using the sequence

$$0 \to Z_{l'_\alpha} \to H_k M'_\alpha \to H_k M_\alpha / \lambda_\alpha(Z) \to 0 ,$$

where the right hand group is isomorphic to $(s-1)Z_2$, we see that $H_k M'_\alpha$ is given by one of the following:

$$H_k M'_\alpha \simeq \begin{cases} Z + (s-1)Z_2 , \\ Z_2 + (s-1)Z_2 , \\ Z + (s-2)Z_2 , \quad \text{or} \\ Z_4 + (s-2)Z_2 . \end{cases}$$

But the first two possibilities cannot occur, since they do not change the rank of $H_k(M; Z_2)$. In the remaining two cases, a further modification will replace $H_k M'_\alpha$ by a group which is definitely smaller than $H_k M$. Thus in all cases $H_k M$ can be replaced by a smaller group by a sequence of framed modifications.

This completes the proof of Theorem 5.1. Actually we have proved the following result which is slightly sharper.

THEOREM 6.6. *Let M be a compact, framed manifold of dimension $2k+1$, $k>1$, such that bM is either vacuous or a homology sphere. By a sequence of framed modifications, M can be reduced to a k-connected manifold M_1.*

If bM is vacuous then the Poincaré duality theorem implies that M_1 is a homotopy sphere. If bM is a homology sphere, then M_1 is contractible.

The proof of 6.6 is contained in the above discussion, provided that M is connected. But using [17, Lemma 2′] it is easily seen that a disconnected manifold can be connected by framed modifications. This completes the proof.

7. The groups bP_{2k}

The next two sections will prove that the groups bP_{2k} are finite cyclic for $k \neq 2$. In fact for k odd, the group bP_{2k} has at most two elements. For $k = 2m \neq 2$ we will see in Part II that bP_{4m} is a cyclic group of order[7]

$$\varepsilon_m 2^{2m-2}(2^{2m-1}-1) \text{ numerator } (4B_m/m) ,$$

where B_m denotes the mth Bernoulli number, and ε_m equals 1 or 2.

The proofs will be based on the following.

LEMMA 7.1. *Let M be a $(k-1)$-connected manifold of dimension $2k$, $k \geqq 3$, and suppose that $H_k M$ is free abelian with basis $\{\lambda_1, \cdots, \lambda_r, \mu_1, \cdots, \mu_r\}$ where*

$$\lambda_i \cdot \lambda_j = 0 , \qquad \lambda_i \cdot \mu_j = \delta_{ij}$$

for all i, j (where δ_{ij} denotes a Kronecker delta). Suppose further that every imbedded sphere in M which represents a homology class in the subgroup generated by $\lambda_1, \cdots, \lambda_r$ has trivial normal bundle. Then $H_k M$

[7] This expression for the order of bP_{4m} relies on recent results of J. F. Adams [1].

can be killed by a sequence of spherical modifications.

PROOF. According to [17, Lemma 6] or Haefliger [6] any homology class in H_kM can be represented by a differentiably imbedded sphere.

REMARK. It is at this point that the hypothesis $k \geqq 3$ is necessary. Our methods break down completely for the case $k = 2$ since a homology class in $H_2(M^4)$ need not be representable by a differentiably imbedded sphere. (Compare Kervaire and Milnor [13].)

Choose an imbedding $\varphi_0: S^k \to M$ so as to represent the homology class λ_r. Since the normal bundle is trivial, φ_0 can be extended to an imbedding $\varphi: S^k \times D^k \to M$. Let $M' = \chi(M, \varphi)$ denote the modified manifold, and let

$$M_0 = M - \text{Interior } \varphi(S^k \times D^k) = M' - \text{Interior } \varphi'(D^{k+1} \times S^{k-1}) .$$

The argument now proceeds just as in [17, p. 54]. There is a diagram

$$\begin{array}{ccccccccc} & & Z & & & & & & \\ & & \downarrow & \searrow^{\lambda_r} & & & & & \\ 0 & \longrightarrow & H_kM_0 & \longrightarrow & H_kM & \xrightarrow{\cdot\lambda_r} & Z & \longrightarrow & H_{k-1}M_0 \longrightarrow 0 \\ & & \downarrow & & & & & & \\ & & H_kM' & & & & & & \\ & & \downarrow & & & & & & \\ & & 0 & & & & & & \end{array}$$

where the notation and the proof is similar to that of Lemma 5.6. Since $\mu_r \cdot \lambda_r = 1$ it follows that $H_{k-1}M_0 = 0$. From this fact one easily proves that M_0 and M' are $(k-1)$-connected. The group H_kM_0 is isomorphic to the subgroup of H_kM generated by $\{\lambda_1, \cdots, \lambda_r, \mu_1, \cdots, \mu_{r-1}\}$. The group H_kM' is isomorphic to a quotient group of H_kM_0. It has basis $\{\lambda'_1, \cdots, \lambda'_{r-1}, \mu'_1, \cdots, \mu'_{r-1}\}$ where each λ'_i corresponds to a coset

$$\lambda_i + \lambda_r Z \subset H_kM ,$$

and each μ'_j corresponds to a coset $\mu_j + \lambda_r Z$.

The manifold M' also satisfies the hypothesis of 7.1. In order to verify that

$$\lambda'_i \cdot \lambda'_j = 0 , \qquad \lambda'_i \cdot \mu'_j = \delta_{ij} ,$$

note that each λ'_i or μ'_j can be represented by a sphere imbedded in M_0 and representing the homology class λ_i or μ_j of M. Thus the intersection numbers in M' are the same as those in M. In order to verify that any imbedded sphere with homology class $n_1\lambda'_1 + \cdots + n_{r-1}\lambda'_{r-1}$ has trivial normal bundle, note that any such sphere can be pushed off $\varphi'(0 \times S^{k-1})$, and hence can be deformed into M_0. It will then represent a homology class

$$(n_1\lambda_1 + \cdots + n_{r-1}\lambda_{r-1}) + n_r\lambda_r \in H_k M ,$$

and thus will have trivial normal bundle.

Iterating this construction r times, the result will be a k-connected manifold. This completes the proof of Lemma 7.1.

Now consider an s-parallelizable manifold M of dimension $2k$, bounded by a homology sphere. By Theorem 5.5, we can assume that M is $(k-1)$-connected. Using the Poincaré duality theorem it follows that $H_k M$ is free abelian, and that the intersection number pairing

$$H_k M \otimes H_k M \to Z$$

has determinant ± 1. The argument now splits up into three cases.

Case 1. Let $k = 3$ or 7. (Compare [17, Theorem 4'].) Since k is odd the intersection pairing is skew symmetric. Hence there exists a "symplectic" basis for $H_k M$; that is, a basis $\{\lambda_1, \cdots, \lambda_r, \mu_1, \cdots, \mu_r\}$ with

$$\lambda_i \cdot \lambda_j = \mu_i \cdot \mu_j = 0 , \qquad \lambda_i \cdot \mu_j = \delta_{ij} .$$

Since $\pi_{k-1}(\mathrm{SO}_k) = 0$ for $k = 3, 7$, any imbedded k-sphere will have trivial normal bundle. Thus Lemma 7.1 implies that $H_k M$ can be killed. Since an analogous result for $k = 1$ is easily obtained, this proves:

LEMMA 7.2. *The groups bP_2, bP_6, and bP_{14} are zero.*

Case 2. k is odd, but $k \neq 1, 3, 7$. Again one has a symplectic basis; but the normal bundle of an imbedded sphere is not necessarily trivial. This case will be studied in § 8.

Case 3. k is even, say $k = 2m$. Then the following is true. (Compare [17, Theorem 4].)

LEMMA 7.3. *Let M be a framed manifold of dimension $4m > 4$, bounded by a homology sphere.*[8] *The homotopy groups of M can be killed by a sequence of framed spherical modifications if and only if the signature $\sigma(M)$ is zero.*

Since a proof of 7.3 is essentially given in [17] we will only give an outline here.

In one direction the lemma follows from the assertion that $\sigma(M)$ is invariant under spherical modifications. (See [17, p. 41]. The fact that M has a boundary does not matter here, since we can adjoin a cone over

[8] This lemma is of course also true if bM is vacuous. In this case the signature $\sigma(M)$ is necessarily zero, by Hirzebruch's signature theorem.

the boundary, thus obtaining a closed homology manifold with the same signature.)

Conversely suppose that $\sigma(M) = 0$. We may assume that M is $(k-1)$-connected. Since the quadratic form $\lambda \to \lambda \cdot \lambda$ has determinant ± 1 and signature zero, it is possible to choose a basis $\{\lambda_1, \cdots, \lambda_r, \mu_1, \cdots, \mu_r\}$ for $H_k M$ so that $\lambda_i \cdot \lambda_j = 0$, $\lambda_i \cdot \mu_j = \delta_{ij}$. The proof is analogous to that of [17, Lemma 9], but somewhat simpler since we do not put any restriction on $\mu_i \cdot \mu_j$. For any imbedded sphere with homology class $\lambda = n_1\lambda_1 + \cdots + n_r\lambda_r$, the self-intersection number $\lambda \cdot \lambda$ is zero. Therefore, according to [17, Lemma 7], the normal bundle is trivial.

Thus M satisfies the hypothesis of 7.1. It fcllows that $H_k M$ can be killed by spherical modifications. Since the homomorphism

$$\pi_k(\mathrm{SO}_k) \to \pi_k(\mathrm{SO}_{2k+1})$$

is onto for k even, it follows from Lemma 6.2 that we need use only framed spherical modifications. This completes the proof of Lemma 7.3.

LEMMA 7.4. *For each $k = 2m$ there exists a parallelizable manifold M_0 whose boundary bM_0 is the ordinary $(4m-1)$-sphere, such that the signature $\sigma(M_0)$ is non-zero.*

PROOF. According to Milnor and Kervaire [18, p. 457] there exists a closed "almost parallelizable" $4m$-manifold whose signature is non-zero. Removing the interior of an imbedded $4m$-disk from this manifold, we obtain the required parallelizable manifold M_0.

Now consider the collection of all $4m$-manifolds M_0 which are s-parallelizable, and are bounded by the $(4m-1)$-sphere. Clearly the corresponding signatures $\sigma(M_0) \in Z$ form a group under addition. Let $\sigma_m > 0$ denote the generator of this group.

THEOREM 7.5. *Let Σ_1 and Σ_2 be homotopy spheres of dimension $4m-1$, $m > 1$, which bound s-parallelizable manifolds M_1 and M_2 respectively. Then Σ_1 is h-cobordant to Σ_2 if and only if*

$$\sigma(M_1) \equiv \sigma(M_2) \qquad \bmod \sigma_m .$$

PROOF. First suppose that

$$\sigma(M_1) = \sigma(M_2) + \sigma(M_0) .$$

Form the connected sum along the boundary

$$(M, bM) = (-M_1, -bM_1) \# (M_2, bM_2) \# (M_0, bM_0)$$

as in § 2; with boundary

$$bM = -\Sigma_1 \# \Sigma_2 \# S^{4m-1} \approx -\Sigma_1 \# \Sigma_2 .$$

Since

$$\sigma(M) = -\sigma(M_1) + \sigma(M_2) + \sigma(M_0) = 0$$

it follows from 7.3 that $bM = -\Sigma_1 \# \Sigma_2$ belongs to the trivial h-cobordism class. Therefore Σ_1 is h-cobordant to Σ_2.

Conversely let W be an h-cobordism between $-\Sigma_1 \# \Sigma_2$ and the sphere S^{4m-1}. Pasting W onto $(-M_1, -bM_1)\#(M_2, bM_2)$ along the common boundary $-\Sigma_1 \# \Sigma_2$, we obtain a differentiable manifold M bounded by the sphere S^{4m-1}. Since M is clearly s-parallelizable, we have

$$\sigma(M) \equiv 0 \qquad (\mathrm{mod}\ \sigma_m)\ .$$

But

$$\sigma(M) = -\sigma(M_1) + \sigma(M_2)\ .$$

Therefore

$$\sigma(M_1) \equiv \sigma(M_2) \qquad (\mathrm{mod}\ \sigma_m)\ ,$$

which completes the proof.

COROLLARY 7.6. *The group bP_{4m}, $m > 1$, is isomorphic to a subgroup of the cyclic group of order σ_m. Hence bP_{4m} is finite cyclic.*

The proof is evident.

Discussion and computations. In Part II we will see that bP_{4m} is cyclic of order precisely $\sigma_m/8$. In fact a given integer σ occurs as $\sigma(M)$ for some s-parallelizable M bounded by a homotopy sphere if and only if

$$\sigma \equiv 0 \qquad (\text{modulo } 8)\ .$$

The following equality is proved in [18, p. 457]:

$$\sigma_m = 2^{2m-1}(2^{2m-1} - 1)B_m j_m a_m/m\ ,$$

where B_m denotes the mth Bernoulli number, j_m denotes the order of the cyclic group

$$J(\pi_{4m-1}(\mathrm{SO})) \subset \Pi_{4m-1}\ ,$$

and a_m equals 1 or 2 according as m is even or odd. Thus bP_{4m} is cyclic of order

$$(1) \qquad \sigma_m/8 = 2^{2m-4}(2^{2m-1} - 1)B_m j_m a_m/m\ .$$

According to recent work of J. F. Adams [1], the integer j_m is precisely equal to the denominator of $B_m/4m$, at least when m is odd. (Compare [18, Theorem 4].) Therefore

$$B_m j_m a_m/4m = a_m \text{ numerator } (B_m/4m) = \text{numerator } (4B_m/m)$$

where the last equality holds since the denominator of B_m is divisible by 2 but not 4. Thus bP_{4m} is cyclic of order

(2) $$\sigma_m/8 = 2^{2m-2}(2^{2m-1} - 1) \text{ numerator } (4B_m/m) ,$$

when m is odd.

One can also give a formula for the order of the full group Θ_{4m-1}. In Part II we will see that Θ_{4m-1}/bP_{4m} is isomorphic to $\Pi_{4m-1}/J(\pi_{4m-1}(\mathrm{SO}))$. (Compare § 4.) Together with formula (1) above this implies that:

$$\text{order } \Theta_{4m-1} = (\text{order } \Pi_{4m-1})2^{2m-4}(2^{2m-1} - 1)B_m a_m/m .$$

8. A cohomology operation

Let $2 \leqq k \leqq n - 2$ be integers and let (K, L) be a CW-pair which satisfies the following:

Hypothesis. The cohomology groups $H^i(K, L; G)$ vanish for $k < i < n$ and for all coefficient groups G.

Then a cohomology operation

$$\psi: H^k(K, L; Z) \to H^n(K, L; \pi_{n-1}(S^k))$$

is defined as follows[9]. Let $e^0 \in S^k$ denote a base point and let

$$s \in H^k(S^k, e^0; Z)$$

denote a generator. Then $\psi(c)$ will denote the first obstruction to the existence of a map

$$f: (K, L) \to (S^k, e^0)$$

satisfying the condition $f^*(s) = c$.

To be more precise let K^r denote the r-skeleton of K. Then given any class

$$x \in H^k(K, L; Z) \simeq H^k(K^{n-1} \cup L, L; Z) ,$$

it follows from standard obstruction theory that there exists a map

$$f_x: (K^{n-1} \cup L, L) \to (S^k, e^0)$$

with $f_x^* s = x$; and that the restriction

$$f_x | (K^{n-2} \cup L, L)$$

is well defined up to homotopy. The obstruction to extending f_x over $K^n \cup L$ is the required class

$$\psi(x) \in H^n(K, L; \pi_{n-1}(S^k)) .$$

[9] A closely related operation φ_0 has been studied by Kervaire [12]. The operation φ_0 would serve equally well for our purposes.

LEMMA 8.1. *The function*

$$\psi\colon H^k(K, L; Z) \to H^n(K, L; \pi_{n-1}(S^k))$$

is well defined, and is natural in the following sense. If the CW-*pair* (K', L') *also satisfies the hypothesis above, then for any map*

$$g\colon (K', L') \to (K, L)\ ,$$

and any $x \in H^k(K, L; Z)$ *the identity*

$$g^*\psi(x) = \psi g^*(x)$$

is satisfied.

The proof is straightforward. It follows that ψ does not depend on the particular cell structure of the pair (K, L).

Now let us specialize to the case $n = 2k$.

LEMMA 8.2. *The operator* ψ *satisfies the identity*

$$\psi(x + y) = \psi(x) + \psi(y) + [i, i](x \smile y)\ ,$$

where the last term stands for the image of the class $x \smile y \in H^{2k}(K, L; Z)$ *under that coefficient homomorphism*

$$Z \to \pi_{2k-1}(S^k)$$

which carries 1 *into the Whitehead product class* $[i, i]$.

PROOF. Let $U = e^0 \cup e^k \cup \{e_i^{2k}\} \cup \{e_j^{2k+1}\} \cup \cdots$ denote a complex formed from the sphere S^k by adjoining cells of dimensions $\geqq 2k$ so as to kill the homotopy groups in dimensions $\geqq 2k - 1$. Let

$$u \in H^k(U, e^0; Z)$$

be a standard generator. Evidently the functions

$$\psi\colon H^k U \to H^{2k}(U; \pi_{2k-1}(S^k))$$

and

$$\psi\colon H^k(U \times U) \to H^{2k}(U \times U; \pi_{2k-1}(S^k))$$

are defined. We will first evaluate $\psi(u \times 1 + 1 \times u)$.

The $(2k + 1)$-skeleton of $U \times U$ consists of the union

$$U^{2k+1} \times e^0 \cup e^0 \times U^{2k+1} \cup e^k \times e^k\ .$$

Therefore the cohomology class $\psi(u \times 1 + 1 \times u) \in H^{2k}(U \times U; \pi_{2k-1}(S^k))$ can be expressed uniquely in the form

$$a \times 1 + 1 \times b + \gamma(u \times u)$$

with $a, b \in H^{2k}(U; \pi_{2k-1}(S^k))$ and $\gamma \in \pi_{2k-1}(S^k)$. Applying 8.1 to the inclusion map

$$U \times e^0 \to U \times U$$

we see that a must be equal to $\psi(u)$. Similarly b is equal to $\psi(u)$. Applying 8.1 to the inclusion

$$S^k \times S^k \to U \times U$$

we see that $\psi(s \times 1 + 1 \times s) = \gamma(s \times s)$. But $\psi(s \times 1 + 1 \times s)$ is just the obstruction to the existence of a mapping

$$f: S^k \times S^k \to S^k$$

satisfying $f(e^0, x) = f(x, e^0) = x$. Therefore γ must be equal to the Whitehead product class $[i, i] \in \pi_{2k-1}(S^k)$. Thus we obtain the identity

$$\begin{aligned}\psi(u \times 1 + 1 \times u) &= \psi(u) \times 1 + 1 \times \psi(u) + [i, i](u \times u) \\ &= \psi(u \times 1) + \psi(1 \times u) + [i, i]\big((u \times 1) \smile (1 \times u)\big) .\end{aligned}$$

Now consider an arbitrary CW-pair (K, L), and two classes $x, y \in H^k(K, L)$. Choose a map

$$g: (K, L) \to (U \times U, e^0 \times e^0)$$

so that $g^*(u \times 1) = x$, $g^*(1 \times u) = y$. (Such a map can be constructed inductively over the skeletons of K since the obstruction groups $H^i(H, L; \pi_{i-1}(U \times U))$ are all zero.) Then by 8.1:

$$\begin{aligned}\psi(x + y) &= g^*\psi(u \times 1 + 1 \times u) \\ &= g^*\psi(u \times 1) + g^*\psi(1 \times u) + [i, i]g^*\big((u \times 1) + (1 \times u)\big) \\ &= \psi(x) + \psi(y) + [i, i](x \smile y) .\end{aligned}$$

This completes the proof of Lemma 8.2.

Now let M be a $2k$-manifold which is $(k - 1)$-connected. Then

$$\psi: H^k(M, bM) \to H^{2k}(M, bM; \pi_{2k-1}(S^k)) \simeq \pi_{2k-1}(S^k)$$

is defined.

LEMMA 8.3. *Let k be odd*[10] *and let M be* S*-parallelizable. Then an imbedded k-sphere in M has trivial normal bundle if and only if its dual cohomology class $v \in H^k(M, bM)$ satisfies the condition $\psi(v) = 0$.*

PROOF. Let N be a closed tubular neighborhood of the imbedded sphere, and let

$$M_0 = M - \text{Interior } N .$$

Then there is a commutative diagram

[10] This lemma is actually true for k even also.

$$\begin{array}{ccc} w \in H^k(N, bN) & \xrightarrow{\psi} & H^{2k}(N, bN; \pi_{2k-1}(S^k)) \\ \uparrow\simeq & & \uparrow\simeq \\ H^k(M, M_0) & \xrightarrow{\psi} & H^{2k}(M, M_0; \pi_{2k-1}(S^k)) \\ \downarrow & & \downarrow\simeq \\ v \in H^k(M, bM) & \xrightarrow{\psi} & H^{2k}(M, bM; \pi_{2k-1}(S^k)) \,; \end{array}$$

where a generator w of the infinite cyclic group $H^k(N, bN)$ corresponds to the cohomology class v under the left hand vertical arrows. Thus[11]

$$\psi(v)[M] = \psi(w)[N] \in \pi_{2k-1}(S^k) \,.$$

It is clear that the homotopy class $\psi(w)[N]$ depends only on the normal bundle of the imbedded sphere.

The normal bundle is determined by an element ν of the group $\pi_{k-1}(\mathrm{SO}_k)$. Since M is s-parallelizable, ν must belong to the kernel of the homomorphism

$$\pi_{k-1}(\mathrm{SO}_k) \to \pi_{k-1}(\mathrm{SO}) \,.$$

But this kernel is zero for $k = 1, 3, 7$, and is cyclic of order 2 for other odd values of k. The unique non-trivial element corresponds to the tangent bundle of S^k, or equivalently to the normal bundle of the diagonal in $S^k \times S^k$.

Thus if $\nu \neq 0$ then N can be identified with a neighborhood of the diagonal in $S^k \times S^k$. Then

$$\psi(w)[N] = \psi(s \times 1 + 1 \times s)[S^k \times S^k] = [i, i] \neq 0$$

(assuming that $k \neq 1, 3, 7$). On the other hand if $\nu = 0$ then $\psi(w)$ is clearly zero. This completes the proof of Lemma 8.3.

Henceforth we will assume that k is odd and $\neq 1, 3, 7$. The subgroup of $\pi_{2k-1}(S^k)$ generated by $[i, i]$ will be identified with the standard cyclic group Z_2. Thus a function

$$\psi_0\colon H_k M \to Z_2$$

is defined by the formula

$$\psi_0(\lambda) = \psi(x)[M]$$

where $x \in H^k(M, bM)$ denotes the Poincaré dual of the homology class λ. Evidently:

(1) $\psi_0(\lambda + \mu) \equiv \psi_0(\lambda) + \psi_0(\mu) + \lambda \cdot \mu \pmod 2$, and

[11] The symbol $[M]$ denotes the homomorphism $H^n(M, bM; G) \to G$ determined by the orientation homology class in $H_n(M, bM; Z)$.

(2) $\psi_0(\lambda)=0$ if and only if an imbedded sphere representing the homology class λ has trivial normal bundle.

Now assume that bM has no homology in dimensions k, $k-1$, so that the intersection pairing has determinant ± 1. Then one can choose a symplectic basis for H_kM: that is a basis $\{\lambda_1, \cdots, \lambda_r, \mu_1, \cdots, \mu_r\}$ such that

$$\lambda_i \cdot \lambda_j = 0 , \qquad \mu_i \cdot \mu_j = 0 , \qquad \lambda_i \cdot \mu_j = \delta_{ij} .$$

DEFINITION. The Arf *invariant* $c(M)$ is defined to be the residue class[12]

$$\psi_0(\lambda_1)\psi_0(\mu_1) + \cdots + \psi_0(\lambda_r)\psi_0(\mu_r) \in Z_2 .$$

(Compare [3].) This residue class modulo 2 does not depend on the choice of symplectic basis.

LEMMA 8.4. *If* $c(M)=0$ *then* H_kM *can be killed by a sequence of framed spherical modifications.*

The proof will depend on Lemma 7.1. Let $\{\lambda_1, \cdots, \lambda_r, \mu_1, \cdots, \mu_r\}$ be a symplectic basis for H_kM. By permuting the λ_i and μ_i we may assume that

$$\psi_0(\lambda_i) = \psi_0(\mu_i) = 1 \qquad \text{for } i \leqq s ,$$
$$\psi_0(\lambda_i) = 0 \qquad \text{for } i > s ,$$

where s is an integer between 0 and r. The hypothesis

$$c(M) = \sum \psi_0(\lambda_i)\psi_0(\mu_i) = 0$$

implies that $s \equiv 0 \pmod 2$.

Construct a new basis $\{\lambda'_1, \cdots, \mu'_r\}$ for H_kM by the substitutions

$$\lambda'_{2i-1} = \lambda_{2i-1} + \lambda_{2i} , \qquad \lambda'_{2i} = \mu_{2i-1} - \mu_{2i} ,$$
$$\mu'_{2i-1} = \mu_{2i-1} , \qquad \mu'_{2i} = \lambda_{2i} ,$$

for $2i \leqq s$, with

$$\lambda'_i = \lambda_i , \qquad \mu'_i = \mu_i$$

for $i > s$. This new basis is again symplectic, and satisfies the condition:

$$\psi_0(\lambda'_1) = \cdots = \psi_0(\lambda'_r) = 0 .$$

For any sphere imbedded in M with homology class $\lambda = n_1\lambda'_1 + \cdots + n_r\lambda'_r$ the invariant $\psi_0(\lambda)$ is zero, and hence the normal bundle is trivial. Thus the basis $\{\lambda'_1, \cdots, \mu'_r\}$ satisfies the hypothesis of Lemma 7.1. Thus H_kM can be killed by spherical modifications.

If M is a framed manifold then it is only necessary to use framed modifications for this construction. This follows from Lemma 6.2, since

[12] This coincides with the invariant $\Phi(M)$ as defined by Kervaire [12].

the homomorphism $\pi_k(\mathrm{SO}_k) \to \pi_k(\mathrm{SO}_{2k+1})$ is onto for $k \neq 1, 3, 7$. This completes the proof of Lemma 8.4.

THEOREM 8.5. *For k odd the group bP_{2k} is either zero or cyclic of order* 2.

According to Lemma 7.2 the groups bP_2, bP_6 and bP_{14} are zero. Thus we may assume that $k \neq 1, 3, 7$.

Let M_1 and M_2 be s-parallelizable and $(k-1)$-connected manifolds of dimension $2k$, bounded by homotopy spheres. If

$$c(M_1) = c(M_2)$$

we will prove that bM_1 is h-cobordant to bM_2. This will clearly prove 8.5.

Form the connected sum $(M, bM) = (M_1, bM_1) \# (M_2, bM_2)$ along the boundary. Clearly

$$c(M) = c(M_1) + c(M_2) = 0 .$$

Therefore, according to 8.4, it follows that the boundary

$$bM = bM_1 \# bM_2$$

bounds a contractible manifold. Hence, according to Theorem 1.1 the manifold bM_1 is h-cobordant to $-bM_2$. Since a similar argument shows that bM_2 is h-cobordant to $-bM_2$, this completes the proof.

REMARK. It seems plausible that $bP_{2k} \simeq Z_2$ for all odd k other than 1, 3, 7; but this is known to be true only for $k = 5$ (compare Kervaire [12]) and $k = 9$.

NEW YORK UNIVERSITY
PRINCETON UNIVERSITY

REFERENCES

1. ADAMS, J. F., *On the J homomorphism*, (to appear). *
2. ———, and WALKER, G., *On complex Stiefel manifolds*, (to appear). +
3. ARF, C., *Untersuchungen über quadratische Formen in Körpern der Charakteristik* 2, Crelles Math. Journal, 183 (1941), 148–167.
4. BOTT, R., *The stable homotopy of the classical groups*, Ann. of Math., 70 (1959), 313-337.
5. CERF, J., *Topologie de certains espaces de plongements*, Bull. Soc. Math. France, 89 (1961), 227-380.
6. HAEFLIGER, A., *Plongements différentiables de variétés dans variétés*, Comment. Math. Helv., 36 (1961), 47–82.

---

* Adams, J. Frank, *On the groups J(X)*, in "Differential and Combinatorial Topology (A Symposium in Honor of Marston Morse)," Princeton University Press, Princeton, NJ, 1965, pp. 121-143. (For more detailed presentations see Topology, **2** (1963) 181-195; **3** (1965) 137-171 and 193-222; **5** (1966) 21-71; and **7** (1968) 331.)

+ Adams, J.F.; Walker, G. "On complex Stiefel manifolds," Proc. Cambridge Philos. Soc. **61** (1965) 81-103.

7. HIRZEBRUCH, F., Neue topologische Methoden in der algebraischen Geometrie, Springer Verlag, 1956.
8. KERVAIRE, M., *Relative characteristic classes*, Amer. J. Math., 79 (1957), 517-558.
9. ———, *An interpretation of G. Whitehead's generalization of the Hopf invariant*, Ann. of Math., 69 (1959), 345-364.
10. ———, *A note on obstructions and characteristic classes*, Amer. J. Math., 81 (1959), 773-784.
11. ———, *Some non-stable homotopy groups of Lie groups*, Illinois J. Math., 4 (1960), 161-169.
12. ———, *A manifold which does not admit any differentiable structure*, Comment. Math. Helv., 34 (1960), 257-270.
13. ———, and MILNOR, J., *On 2-spheres in 4-manifolds*, Proc. Nat. Acad. Sci. U.S.A., 47 (1961), 1651-1657.
14. MILNOR, J., *On manifolds homeomorphic to the 7-sphere*, Ann. of Math., 64 (1956), 399-405.
15. ———, Differentiable manifolds which are homotopy spheres, Mimeographed Notes, Princeton, 1958. *
16. ———, *Sommes de variétés différentiables et structures différentiables des sphères*, Bull. Soc. Math. France, 87 (1959), 439-444.
17. ———, *A procedure for killing the homotopy groups of differentiable manifolds*, Symposia in Pure Math. A.M.S., vol. III (1961), 39-55.
18. ———, and KERVAIRE, M., *Bernoulli numbers, homotopy groups and a theorem of Rohlin*, Proc. Int. Congress of Math., Edinburgh, 1958.
19. MUNKRES, J., *Obstructions to the smoothing of piecewise-linear homeomorphisms*, Bull. Amer. Math. Soc., 65 (1959), 332-334.
20. PALAIS, R., *Extending diffeomorphisms*, Proc. Amer. Math. Soc., 11 (1960), 274-277.
21. PONTRJAGIN, L., *Smooth manifolds and their applications in homotopy theory*, Trudy Mat. Inst. im. Steklov, 45 (1955), and A.M.S. translations, Series 2, 11, 1-114.
22. SEIFERT, H., *Konstruktion dreidimensionaler geschlossener Räume*, Ber. Verh Sächs. Akad. Wiss. Leipzig, 83 (1931), 26-66.
23. ———, and THRELFALL, W., Lehrbuch der Topologie, Teubner Verlag, 1934.
24. SERRE, J.-P., *Homologie singulière des espaces fibrés. Applications*, Ann. of Math., 54 (1951), 425-505.
25. SMALE, S., *Generalized Poincaré conjecture in dimensions greater than four*, Ann of Math., 74 (1961), 391-406.
26. ———, *On the structure of manifolds*, Amer. J. Math., 84 (1962), 387-399.
27. STALLINGS, J., *Polyhedral homotopy-spheres*, Bull. Amer. Math. Soc., 66 (1960), 485-488.
28. THOM, R., *Quelques propriétés globales des variétés différentiables*, Comment. Math. Helv., 28 (1954), 17-86.
29. WALL, C. T. C., *Killing the middle homotopy groups of odd dimensional manifolds*, Trans. Amer. Math. Soc., 103 (1962), 421-433.
30. WALLACE, A. H., *Modifications and cobounding manifolds*, Canadian J. Math., 12 (1960), 503-528.
31. WHITEHEAD, J. H. C., *Manifolds with transverse fields in euclidean space*, Ann. of Math., 73 (1961), 154-212.
32. ———, *On the homotopy type of manifolds*, Ann. of Math., 41 (1940), 825-832.
33. ZEEMAN, C., *The generalized Poincaré conjecture*, Bull. Amer. Math. Soc., 67 (1961), 270.

---

* This volume pp. 65-87.

6

Differential Topology

J. Milnor

The basic objects studied in differential topology are smooth manifolds, sometimes with boundary, and smooth mappings between such manifolds. Here the word "smooth" is used to mean "differentiable of class C^∞." To give a rough idea of the flavor of this field, let us list a few of its central problems.

The Diffeomorphism Problem. Given two smooth manifolds M and M', how can we decide whether or not there is a diffeomorphism from M to M' (i.e., a smooth homeomorphism with smooth inverse)?

The Cobordism Problem. Given a smooth compact manifold M without boundary, does there exist a smooth compact manifold W whose boundary is equal to M? We may refine this problem by putting extra structures on both M and W. For example, we can require an orientation, or a weakly complex structure, or we may require that M and W should be k-connected (compare [10]).

The Imbedding Problem. Given M and M', does there exist a smooth imbedding $M \to M'$? If so, can we classify all such imbeddings? For example, the problem of classifying imbeddings of the circle in 3-space forms the field of "knot theory."

CHARACTERIZATIONS OF THE n-SPHERE

Let us single out one particular case of the diffeomorphism problem for consideration, namely the problem of characterizing the n-sphere. The many different tools which can be brought to bear on this one question will provide a survey of much of the field of differential topology.

First let us ask what conditions on a smooth manifold guarantee that it is *homeomorphic* to S^n.

The first such characterization is due to Reeb [16]. Let M be a smooth n-dimensional manifold. By a *Morse function* f on M will be meant a smooth real valued function whose critical points are all nondegenerate. Thus in a neighborhood of each critical point we can choose local coordinates $u_1, \ldots, u_n$ so that $f = \text{constant} \pm u_1^2 \pm \cdots \pm u_n^2$. (Morse [12, Lemma 4] or [11, §2.2].)

Theorem 1. If M is compact, without boundary, and possesses a Morse function with only two critical points, then M is homeomorphic to S^n.

OUTLINE OF PROOF. (See Figure 1.) Let m_0, m_1 denote the minimum and maximum of f. Thus in a neighborhood of its minimum point we have

$$f = m_0 + u_1^2 + \cdots + u_n^2$$

It follows that the set of points x where $f(x) \leq m_0 + \epsilon$ is diffeomorphic to the disk D^n. Similarly, the set of x with $f(x) \geq m_1 - \epsilon$

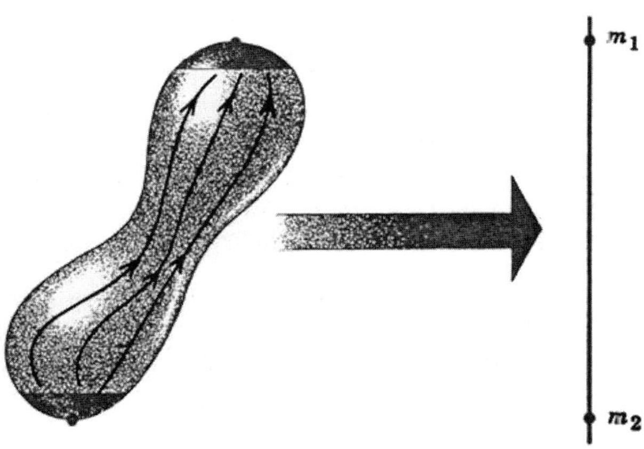

Figure 1

is diffeomorphic to D^n. But, deforming M into itself along the gradient lines of the function f [i.e., along the orthogonal trajectories of the surfaces f^{-1} (constant)] we can slide the disk, $f \leq m_0 + \epsilon$, up so that it covers precisely the disk $f \leq m_1 - \epsilon$. (Compare [7] or [11, §§3, 4].) Thus M can be considered as the union of two imbedded disks which intersect only along their common boundary. But this implies that M is homeomorphic to S^n. To see this, consider first the following.

Lemma 1. Any homeomorphism $h:S^{n-1} \to S^{n-1}$ extends to a homeomorphism $H:D^n \to D^n$.

PROOF. Set $H(tu) = th(u)$ for $0 \leq t \leq 1$, where u denotes an arbitrary unit vector.

Now if M is the union of two topologically imbedded disks $g_0(D^n)$, $g_1(D^n)$ which intersect precisely along their common boundary, we can first choose any homeomorphism from $g_0(S^{n-1}) = g_1(S^{n-1})$ to the equator of S^n, and then extend, using Lemma 1, to a homeomorphism which maps $g_0(D^n)$ to the southern hemisphere and $g_1(S^{n-1})$ to the northern hemisphere. This completes the proof.

Remark. Note that the differentiable structure is destroyed in the course of this proof. The reason is that there is no differentiable analogue of Lemma 1. (Even if $h:S^{n-1} \to S^{n-1}$ is a diffeomorphism, the extension we have constructed is highly nondifferentiable at the origin, unless h happens to be orthogonal.) We will return to this point later, in Section 3.

Following is another partial characterization of S^n (compare [17, 8]).

Theorem 2. Let M be a compact smooth manifold which is the union of two open sets, each diffeomorphic to a euclidean space. Then M is homeomorphic to a sphere.

In fact the proof will show that M with a single point deleted is diffeomorphic to the euclidean space R^n. The proof is based on two lemmas, both of which are interesting in their own right.

Lemma 2 (Palais and Cerf). Let f_1 and f_2 be smooth, orientation preserving imbeddings of the disk D^n into the interior of a connected manifold M^n. Then there exists a diffeomorphism h of M^n onto itself so that $h \circ f_1 = f_2$.

For the proof, which is quite elementary, the reader is referred to [3] or [14].

Lemma 3 (Brown, Stallings). Let M be a paracompact manifold such that every compact subset is contained in an open set diffeomorphic to euclidean space. Then M itself is diffeomorphic to euclidean space.

PROOF. (Compare [2, 21].) It is not difficult to show that M is a monotone union of disks. That is, we can find submanifolds with boundary

$$W_1 \subset W_2 \subset W_3 \cdots \subset M$$

with union M so that each W_i is diffeomorphic to D^n, and so that each W_i is contained in the interior of W_{i+1}. We wish to compare this sequence with the sequence

$$D_1^n \subset D_2^n \subset D_3^n \subset \cdots \subset R^n$$

where D_i^n denotes the disk of radius i in euclidean space. Start with any diffeomorphism $f_1 : D_1^n \to W_1$. Using Lemma 2, this can be extended to a diffeomorphism $f_2 : D_2^n \to W_2$, and so on. (To see this, consider the following diagram

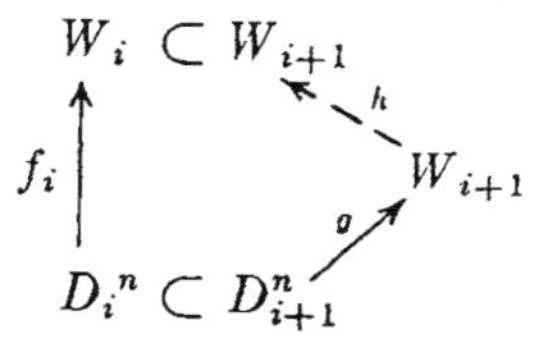

where g is an arbitrary orientation preserving diffeomorphism. Choosing a diffeomorphism h as in Lemma 2, we can now set f_{i+1} equal to $h \circ g$.) Finally, piecing together all these diffeomorphisms f_i, we obtain the required diffeomorphism $R^n \to M$.

Using these lemmas we can prove a sharpened form of Theorem 1. Let M be compact, without boundary, and let $f : M \to R$ be a smooth function with only two critical points.

Theorem 1′. Even if these critical points are allowed to be degenerate, it still follows that M is homeomorphic to S^n.

PROOF. (See Figure 2.) Let p and q be the critical points and let U be a neighborhood of p which is diffeomorphic to R^n, with $q \notin U$. By deforming M into itself along the gradient lines of f, we can stretch U so that it covers any compact subset of $M - q$.

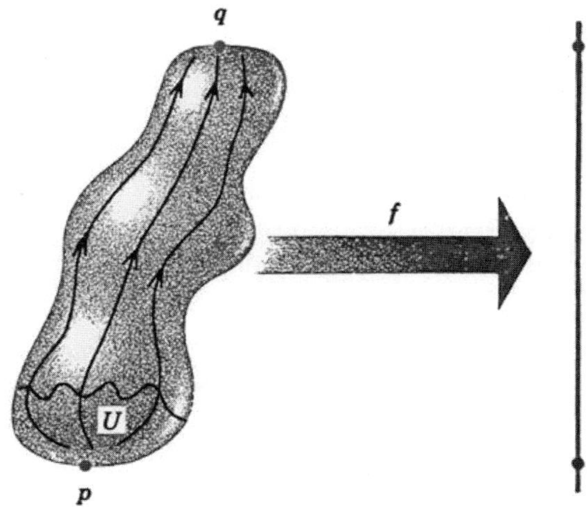

Figure 2

Hence it follows from Lemma 3 that $M - q$ is diffeomorphic to R^n, which clearly completes the proof.

The proof of Theorem 2 is similar. Let M be covered by open sets U and V which are diffeomorphic to R^n (Figure 3). Given $q \in M - U \subset V$ we will show that any compact subset of $M - q$ can be covered by an open set diffeomorphic to U.

Since V is diffeomorphic to R^n, it is easy to construct a diffeomorphism $h: V \to V$ which (1) satisfies $h(q) = q$, (2) shrinks the

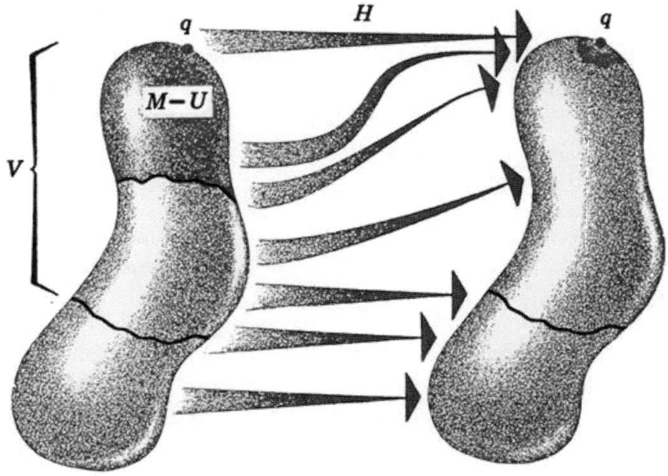

Figure 3

compact set $M - U$ down into an arbitrarily small neighborhood of q, and (3) coincides with the identity outside of a larger compact set. It follows from (3) that h extends to a diffeomorphism H of M. Clearly any compact subset of $M - q$ can be covered by $H(U)$ for suitable choice of the diffeomorphism H. Therefore it follows from Lemma 3 that $M - q$ is diffeomorphic to R^n. This completes the proof of Theorem 2.

One of the most striking properties of the sphere S^n is that the complement of each point is contractible.

Problem 1. Given a smooth manifold M *such that* M − p *is contractible, does it follow that* M *is homeomorphic to a sphere?*

It follows without too much difficulty from this hypothesis that M is compact and has the homotopy type of a sphere. Conversely, if M is compact, without boundary, and has the homotopy type of a sphere, it can be shown that $M - p$ is contractible.

If the dimension n *of* M *is* 0, 1, *or* 2, *then* M *must actually be diffeomorphic to* S^n. Here is a proof for $n = 2$. Choose a Riemannian metric and an orientation for M. By a classical theorem (the existence of "isothermal coordinates") each small neighborhood in M can be mapped conformally and diffeomorphically onto a region in the plane. Thus M becomes a Riemann surface. Since M is simply connected, the classical "uniformization theorem" asserts that M is conformally diffeomorphic to either the complex plane, the open unit disk, or the Gauss 2-sphere. But only the last possibility satisfies our hypothesis.

If the dimension n is 3, we have the classical Poincaré problem. So far, all attempts to solve this problem have foundered. For $n = 4$ the problem is also unsolved.

For $n \geq 5$ the problem has been solved affirmatively by Stallings and Zeeman [20, 24] and by Smale. In particular we have the following.

Theorem of Smale [19]. *If M is a smooth homotopy n-sphere of dimension $n \geq 5$, then M admits a Morse function with only two critical points.*

Hence Theorem 1 implies that M is homeomorphic to the n-sphere.

The proof of this theorem is much more difficult than anything we have encountered so far. The basic idea can be outlined very

Figure 4

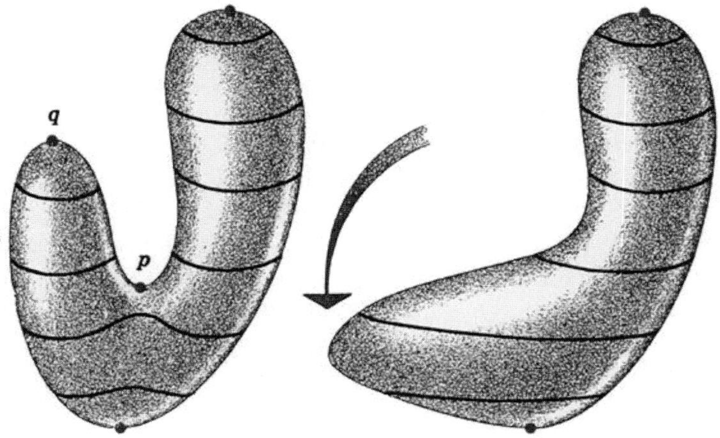

roughly as follows. It is not difficult to construct a Morse function $f: M \to R$ with many critical points. Furthermore f can be chosen so that the critical points occur in the proper order, in the following sense. If

$$f(u_1, \ldots, u_n) = \text{constant} - u_1{}^2 - \cdots - u_\lambda{}^2 + u_{\lambda+1}^2 + \cdots + u_n{}^2$$

in terms of local coordinates $u_1, \ldots, u_n$ near a critical point p, then the integer $\lambda = \lambda(p)$ is called the *index* of this critical point. Now f can be chosen so that $f(p) = \lambda(p)$ for each critical point p. Thus the minimum points, with $\lambda = 0$, occur in $f^{-1}(0)$, the maximum points occur in $f^{-1}(n)$, and the remaining critical points come in between.

The difficult part of the proof now consists in showing that a critical point of index λ and a critical point of index $\lambda + 1$ can sometimes be mutually cancelled. Thus in Figure 4 the critical point p of index 1 and the critical point q of index 2 can be mutually eliminated, by suitably changing the function. Repeating this procedure over and over, we eventually eliminate all critical points which are not essential in order to give the manifold M its proper homology groups. But a homotopy sphere has homology only in dimensions 0 and n. Hence we are left with only two critical points, with indices 0 and n respectively.

2. SOME EXOTIC SPHERES

This section will construct an example of a manifold which is homeomorphic, but not diffeomorphic, to a sphere. The proof will be based on the Hirzebruch signature theorem.

Let V be a compact oriented manifold of dimension $4k$. The signature (or index*) $\sigma(V)$ is defined as follows, following Hermann Weyl [22, p. 41]. Let $\alpha, \beta \in H_{2k}(V; Z)$ be homology classes with integer coefficients. Then the intersection number $\alpha \cdot \beta$ is a well-defined integer. This intersection number is symmetric in α and β since the dimension $2k$ is even. Passing to real coefficients, we can choose a basis $\alpha_1, \ldots, \alpha_r$ for the vector space $H_{2k}(V; R)$ so that the matrix of intersection numbers $(\alpha_i \cdot \alpha_j)$ is diagonal. Now the number of positive diagonal entries minus the number of negative diagonal entries is called the *signature* σ of V.

The following fundamental observation is due to Thom. *Suppose that* V *is the boundary of a compact oriented manifold* W^{4k+1}. *Then the signature* σ(V) *is zero.* Thom's proof is based on the Poincaré duality theorem.

We will also need to make use of the *Pontrjagin classes* of a smooth manifold M. Without attempting a definition, here are some basic properties.

1. To each smooth manifold M there are associated cohomology classes $p_i \in H^{4i}(M; Z)$ for $i = 1, 2, 3, \ldots$.
2. If U is an open subset of M, then $p_i(U)$ is equal to $p_i(M)$ restricted to U.
3. If M is parallelizable, then $p_i(M) = 0$.

Now let V be a closed oriented manifold of dimension $4k$. If $i_1 + i_2 + \cdots + i_r = k$, then the cohomology class $p_{i_1}p_{i_2} \cdots p_{i_r} \in H^{4k}(V; Z)$ gives rise to an integer which is denoted by $p_{i_1} \cdots p_{i_r}[V]$. These integers are called the *Pontrjagin numbers* of V.

If V is a boundary then these Pontrjagin numbers are all zero. Conversely, we have the following:

Thom Cobordism Theorem. If the Pontrjagin numbers $p_{i_1}p_{i_2} \cdots p_{i_r}[V]$ *are all zero, then the m-fold disjoint sum* $V + V + \cdots + V$ *is the boundary of a compact oriented manifold, for some* $m > 0$.

* The term "index" is preferred by Hirzebruch and others. I have substituted "signature" to avoid confusion with the Morse index, as used in Section 1.

An important consequence is the following.

Hirzebruch Signature Theorem. The signature of any closed, oriented 4k-manifold can be expressed as a linear combination of its Pontrjagin numbers, where the coefficients are rational numbers which depend only on the dimension. In particular:

$$\sigma(V^4) = \frac{p_1[V^4]}{3}$$

$$\sigma(V^8) = \frac{7p_2[V^8] - p_1^2[V^8]}{45}$$

$$\sigma(V^{12}) = \frac{62p_3[V^{12}]}{945} + \cdot \cdot \cdot$$

$$\sigma(V^{16}) = \frac{127p_4[V^{16}]}{4725} + \cdot \cdot \cdot$$

$$\sigma(V^{20}) = \frac{146p_5[V^{20}]}{13365} + \cdot \cdot \cdot$$

where the dots indicate terms in $p_1, \ldots, p_{k-1}$.

PROOF. Let V and V' be two manifolds with the same Pontrjagin numbers. Form $V - V'$ (the disjoint sum in which the orientation of V' has been reversed). Then all Pontrjagin numbers of $V - V'$ are zero. Hence some multiple

$$(V - V') + \cdot \cdot \cdot + (V - V')$$

is a boundary. This implies that the signature

$$\sigma((V - V') + \cdot \cdot \cdot + (V - V')) = m\sigma(V) - m\sigma(V')$$

is zero, and hence that $\sigma(V) = \sigma(V')$.

Thus $\sigma(V)$ is a function of the Pontrjagin numbers of V. Since both signature and Pontrjagin numbers are integers which behave additively when we form disjoint sums, it is clear that this function must be linear with rational coefficients.

The explicit formulas for $k = 1, 2, 3, 4, 5$ can be obtained by computing the signature and Pontrjagin numbers for suitable examples, and then solving the resulting linear equations. For further details the reader is referred to Hirzebruch [4].

For our purposes we will only need the following. A manifold will be called *almost parallelizable* if it becomes parallelizable when a single point is removed.

Corollary. The signature of a closed, almost parallelizable manifold of dimension 8 is always divisible by 7. Similarly, for a closed, almost parallelizable manifold of dimension 12, 16 or 20, the signature is divisible by 62, 127, or 146 respectively.

PROOF. We have $p_i(V^{4k} - x) = 0$, hence $p_i(V^{4k}) = 0$ for $i < k$. Thus the signature theorem reduces to

$$\sigma(V^8) = \frac{7p_2[V^8]}{45} \equiv 0 \pmod 7$$

$$\sigma(V^{12}) = \frac{62p_3[V^{12}]}{945} \equiv 0 \pmod{62}$$

and so on.

In higher dimensions we obtain analogous results. (However, sharper results in this direction can be obtained by a different method: see [5].)

Lemma 4. For $k \geq 2$ there exists a compact parallelizable $4k$-dimensional manifold W with signature $+8$, such that the boundary of W is a homotopy sphere.

PROOF. We will construct W in such a way that the matrix of intersection numbers is as follows:

$$\begin{bmatrix} 2 & 1 & 0 & 0 & 0 & 0 & 0 & 0 \\ 1 & 2 & 1 & 0 & 0 & 0 & 0 & 0 \\ 0 & 1 & 2 & 1 & 0 & 0 & 0 & 0 \\ 0 & 0 & 1 & 2 & 1 & 0 & 0 & 0 \\ 0 & 0 & 0 & 1 & 2 & 1 & 0 & 1 \\ 0 & 0 & 0 & 0 & 1 & 2 & 1 & 0 \\ 0 & 0 & 0 & 0 & 0 & 1 & 2 & 0 \\ 0 & 0 & 0 & 0 & 1 & 0 & 0 & 2 \end{bmatrix}$$

This remarkable matrix was suggested to the author by Hirzebruch. Note that it is positive definite, with determinant $+1$, and has only even entries on the diagonal.

As basic building block for the manifold W we take a tubular neighborhood T of the diagonal in $S^{2k} \times S^{2k}$. The homology group $H_{2k}(T; Z)$ is infinite cyclic, and the intersection number of a gener-

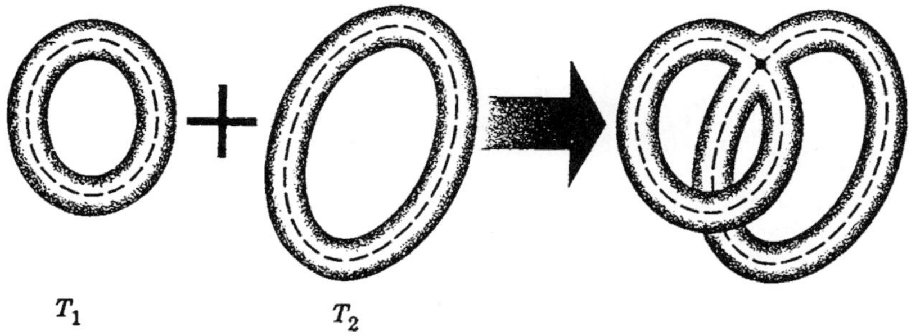

Figure 5

ator with itself is $+2$. [PROOF: if $\alpha, \beta \in H_{2k}(S^{2k} \times S^{2k})$ are the obvious generators, then $(\alpha + \beta) \cdot (\alpha + \beta) = 2\alpha \cdot \beta = 2$.]

Proof that T is parallelizable. Note that $S^{2k} \times S^{2k}$ can be imbedded in R^{4k+1} (as the boundary of a neighborhood of S^{2k}). Hence its tangent bundle is induced from that of the unit sphere by means of the Gauss map $g\colon S^{2k} \times S^{2k} \to S^{4k}$. Since $g|T$ is homotopic to a constant, it follows that T is parallelizable.

Next we introduce the operation of "plumbing together" two copies T_1 and T_2 of T. By this we mean the operation of matching together a region in T_1 and a region in T_2 in such a way that the central $2k$-sphere of T_1 will have intersection number $+1$ with the central $2k$-sphere of T_2. We must then "round off" the corners, so as to obtain a smooth manifold with boundary. For the 2-dimensional case this operation is illustrated in Figure 5.

In practice we need not two but eight copies of T. These are to be plumbed together according to the following schema:

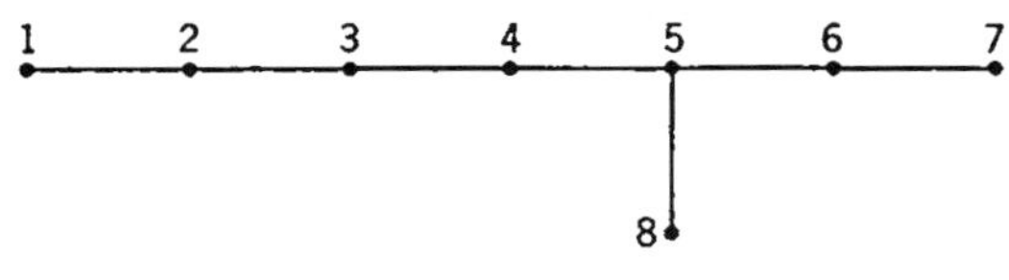

That is, T_2 is to be plumbed together with T_1 and T_3; T_5 is to be plumbed together with T_4, T_6, and T_8; and so on. In this way we obtain a smooth manifold W which clearly has the following properties:

1. The manifold W has the homotopy type of a bouquet consisting of eight copies of S^{2k} which intersect at a single point.

2. If $\alpha_1, \ldots, \alpha_8 \in H_{2k}(W; Z)$ is the corresponding homology basis, then the matrix $(\alpha_i \cdot \alpha_j)$ of intersection numbers is as indicated previously, positive definite with determinant $+1$.

Thus it follows that the signature of W is $+8$.

3. The boundary ∂W is a homology sphere.

For the Poincaré duality theorem implies that

$$H_i(W, \partial W; Z) = \begin{cases} Z + Z + Z + Z + Z + Z + Z + Z & \text{for } i = 2k \\ Z & \text{for } i = 4k \\ 0 & \text{otherwise} \end{cases}$$

Furthermore the homomorphism

$$H_{2k}(W; Z) \to H_{2k}(W, \partial W; Z)$$

corresponds to our intersection matrix, and hence is an isomorphism. Now the homology exact sequence of the pair $(W, \partial W)$ shows that $H_*(\partial W; Z) \cong H_*(S^{4k-1}; Z)$.

4. ∂W is simply connected.

PROOF. Any circle in ∂W bounds a disk in W. By a general position argument, since $2k + 2 < 4k$, this disk can be pushed off of the central $2k$-spheres of the tubes T_i, and hence can be pushed into the boundary of W.

It now follows, by standard arguments in homotopy theory, that ∂W is a homotopy sphere. Since W is clearly parallelizable, this completes the proof of Lemma 4.

Theorem 3. The manifold ∂W is homeomorphic to the sphere S^{4k-1} but is not diffeomorphic to S^{4k-1}.

PROOF. It follows immediately from Smale or Stallings that the homotopy sphere ∂W is homeomorphic to S^{4k-1}. (Recall that the dimension $4k - 1$ is ≥ 7.)

We will only prove the second statement for the special cases $k = 2, 3, 4, 5$. Suppose that ∂W were diffeomorphic to S^{4k-1}. Then by pasting a $4k$-disk onto the boundary of W we would obtain a smooth closed manifold $V = W \cup D^{4k}$. Clearly V is almost parallelizable. Thus, according to the corollary given earlier, the signature of V must be a multiple of 7 or of 62 or 127 or 146 respectively. But $\sigma(V) = \sigma(W) = 8$. This contradiction completes the proof for $2 \leq k \leq 5$. For higher values of k the reader is referred to [6].

3. THE GROUP Γ_n

Since we have seen that exotic spheres exist, it is natural to try to classify them. For this purpose we introduce an abelian group Γ_n which can be described in two different ways.

First Description. Let $\text{Diff}^+(D^n)$ be the group of all orientation preserving diffeomorphisms from a closed n-disk onto itself, and let $\text{Diff}^+(S^{n-1})$ be the corresponding group of diffeomorphisms of its boundary. Consider the restriction homomorphism

$$r\colon \text{Diff}^+(D^n) \to \text{Diff}^+(S^{n-1})$$

We assert that the image of this homomorphism is a normal subgroup. Hence the quotient group

$$\Gamma_n = \frac{\text{Diff}^+(S^{n-1})}{r\,\text{Diff}^+(D^n)}$$

is defined.

Second Description. A closed oriented manifold M will be called a *twisted sphere* if it admits a Morse function with two critical points. Let Γ'_n denote the set of all oriented-diffeomorphism classes of twisted n-spheres. To make this into a group we introduce the *connected sum* operation.

Given connected oriented n-manifolds M_1, M_2, choose imbeddings

$$f_1\colon D^n \to M_1, \qquad f_2\colon D^n \to M_2$$

To make the orientations come out right *it is important that one of these two imbeddings should preserve orientation and the other one should reverse orientation.* The connected sum $M_1 \# M_2$ is now formed from

$$[M_1 - f_1(\text{Int } D^n)] \cup [M_2 - f_2(\text{Int } D^n)]$$

by pasting together the two boundaries under the diffeomorphism

$$f_2 \circ f_1^{-1}\colon f_1(S^{n-1}) \to f_2(S^{n-1})$$

Better still, in order to make the differentiable structure clear, extend f_1 and f_2 to imbeddings $F_i\colon R^n \to M_i$. Then $M_1 \# M_2$ can be formed from

$$[M_1 - F_1(0)] \cup [M_2 - F_2(0)]$$

by identifying each $F_1(x)$ with $F_2(x/\|x\|^2)$. (Note that this correspondence preserves orientation.) Applying Lemma 2 to F_i restricted to some large disk, we can prove that $M_1 \# M_2$ is well defined up to orientation preserving diffeomorphism. (Compare [8, 6].)

To compare the group Γ_n with Γ'_n we will construct a homomorphism

$$\mathrm{Diff}^+(S^{n-1}) \to \Gamma'_n$$

Given $h \in \mathrm{Diff}^+(S^{n-1})$ let $M(h)$ be the twisted sphere which is obtained from two copies of R^n by identifying each $x \in R^n - 0$ in the first with $y = h(x/\|x\|)/\|x\|$ in the second. Then

$$f = \frac{\|x\|^2}{1 + \|x\|^2} = \frac{1}{1 + \|y\|^2}$$

is a Morse function with only two critical points. Taking the orientation of $M(h)$ from the first copy of R^n we obtain a well-defined twisted sphere.

It is easily verified that this construction defines a homomorphism from $\mathrm{Diff}^+(S^{n-1})$ onto Γ'_n. (Compare [7, p. 402].) Let us look at the kernel. Suppose that there is an orientation preserving diffeomorphism g from $M(h_1)$ to $M(h_2)$. According to Lemma 2 we may assume that g carries the point of $M(h_1)$ with coordinate x to the point of $M(h_2)$ with the same coordinate x for all $\|x\| \leq 1$. In terms of the y coordinates, this means that g carries the point with coordinate $y = th_1(u)$ to the point with coordinate $y = th_2(u)$ for all $t \geq 1$ and all $u \in S^{n-1}$. Thus we have a diffeomorphism $R^n \to R^n$ which takes S^{n-1} into itself by the diffeomorphism $h_2 \circ h_1{}^{-1}$. Hence $h_2 \circ h_1{}^{-1}$ belongs to the image

$$r\, \mathrm{Diff}^+(D^n) \subset \mathrm{Diff}^+(S^{n-1})$$

Conversely it can be shown that any element $h|S^{n-1}$ of the image gives rise to a manifold $M(h|S^{n-1})$ which is diffeomorphic to S^n. (The proof can be based on Munkres [13, §6.1].) *Henceforth we will drop the prime, and identify* Γ_n *with* Γ'_n.

The main properties of these groups can be described as follows.

Theorem 4. The group Γ_n is finite abelian for all n. For $n \leq 6$ we have $\Gamma_n = 0$, but for $n = 7$ the group $\Gamma_7 \cong Z_{28}$ is non-zero.

In fact it follows from Theorem 3 that $\Gamma_{4k-1} \neq 0$ for all $k \geq 2$.

The fact that $\Gamma_n = 0$ for $n \leq 3$ is due to Munkres. For $n = 4$ this assertion has recently been announced by J. Cerf. In these cases the proof is based on the first definition, in terms of diffeomorphisms of S^{n-1}. For $n \geq 5$ the proof is based rather on the second definition, in terms of twisted spheres (see [6]). Some indication of the method, for $n \geq 5$, is given in the following section.

4. HOW TO RECOGNIZE AN HONEST SPHERE

Let M be a twisted n-sphere. How can we decide whether or not M is diffeomorphic to the standard n-sphere?

First choose an imbedding $M \subset R^{n+k}$ for some large value of k. This is possible by a well-known theorem of Whitney [23].

Lemma 5. The normal bundle of M is trivial. That is, there exist k continuous linearly independent normal vector fields.

OUTLINE OF PROOF. Since $M - x$ is contractible, it is certainly possible to choose such vector fields in the complement of x. Now the "obstruction" to extending over x is described by an element of the homotopy group $\pi_{n-1}\mathrm{SO}(k)$. These groups (for k large) have been computed by Bott [1].

Case 1. If $n \equiv 3, 5, 6$, or $7 \pmod 8$, then $\pi_{n-1}\mathrm{SO}(k) = 0$, hence there is no obstruction.

Case 2. If $n = 4i$ then the group $\pi_{n-1}\mathrm{SO}(k)$ is infinite cyclic. In this case the obstruction class can be identified with a multiple of the Pontrjagin number $p_i[M]$. (See [5].) But this number is zero by the signature theorem.

Case 3. If $n \equiv 1$ or $2 \pmod 8$, then $\pi_{n-1}\mathrm{SO}(k) \cong Z_2$. The proof in this case is more delicate: the obstruction class $\mathfrak{o}$ satisfies $J(\mathfrak{o}) = 0$ according to [5], but J. F. Adams has shown that the J homomorphism has kernel zero for these values of n. This completes the proof.

Next we must introduce the concept of *framed cobordism*. By a *framing* φ of an imbedded manifold $M^n \subset R^{n+k}$, k large, will be meant a set of k linearly independent normal vector fields. Two framed manifolds (M_0, φ_0) and (M_1, φ_1) in R^{n+k} are called *framed cobordant* if there exists a compact framed manifold (W, ψ) of dimension $n + 1$ in $R^{n+k} \times [0, 1]$ with $\partial W = M_0 \times 0 \cup M_1 \times 1$, where the framing ψ restricts to φ_0 at one end and to φ_1 at the other. Using the disjoint sum as composition operation, the framed cobordism classes of n-manifolds form a group, which will be denoted by Π_n.

There are two fundamental theorems concerning these groups.

Theorem of Pontrjagin [15]. *The framed cobordism group* Π_n *is canonically isomorphic to the stable homotopy group* $\pi_{n+k}(S^k)$, $n < k - 1$.

Theorem of Serre [18]. *The stable homotopy groups* $\pi_{n+k}(S^k)$, $0 < n < k - 1$, *are all finite abelian groups.*

Thus Π_n is finite.

Consider the class of all framed twisted n-spheres in R^{n+k}. Using a suitably defined connected sum operation, and defining an appropriate concept of "isomorphism," we see that *the set of all isomorphism classes of framed twisted* n-*spheres forms an abelian group, which will be denoted by* $F\Gamma_n$. There is an exact sequence

$$\pi_n SO(k) \to F\Gamma_n \to \Gamma_n \to 0$$

PROOF. Every twisted sphere can be framed. The kernel of the homomorphism $F\Gamma_n \to \Gamma_n$ is obtained by looking at standard spheres with exotic framings. But a framing of the standard $S^n \subset R^{n+k}$ is clearly described by an element of $\pi_n SO(k)$.

On the other hand, there is clearly a homomorphism

$$j: F\Gamma_n \to \Pi_n \cong \pi_{n+k}(S^k)$$

Kervaire has shown that this homomorphism j is onto, except possibly when $n \equiv 2 \pmod 4$. Thus *every framed cobordism class contains a twisted sphere*, except in dimensions 2, 6, 10, ...

Consider the kernel of this homomorphism j. Suppose, for example, that $n \equiv 3 \pmod 4$. Then we claim that kernel (j) is infinite cyclic. Let (M, φ) be a framed twisted n-sphere belonging to the kernel of j. Then $(M, \varphi) = \partial(W, \psi)$ where $W \subset R^{n+k} \times [0, 1)$ is a compact framed manifold of dimension $n + 1 = 4i$. Hence the signature $\sigma(W)$ is defined.

This integer $\sigma(W)$ is an invariant of the framed manifold (M, φ).

For if (M, φ) is also the boundary of (W', ψ'), then by placing W in $R^{n+k} \times [0, 1)$ and W' in $R^{n+k} \times (-1, 0]$, we can construct a closed framed manifold

$$W \cup W' \subset R^{n+k} \times R$$

Giving $W \cup W'$ the orientation which is compatible with that of W, we see that

$$\sigma(W \cup W') = \sigma(W) - \sigma(W')$$

But this signature must be zero: the fact that $W \cup W'$ is framed implies that its Pontrjagin classes are all zero and hence, by the signature theorem, that its signature is zero.

Thus $\sigma(W)$ is an invariant $\bar{\sigma}$ of (M, φ). In this way we define a homomorphism

$$\bar{\sigma}\colon \text{kernel } (j) \to Z$$

The construction of Section 2 shows that this homomorphism $\bar{\sigma}$ is nontrivial.*

Finally we claim that the kernel of $\bar{\sigma}$ is zero. That is, if (M, φ) bounds (W, ψ) with $\sigma(W) = 0$, then M is diffeomorphic to S^n, and φ corresponds to the standard framing of S^n. The proof, which is fairly formidable, is divided into two parts. First, using the method of "surgery" [9], we show that all of the homotopy groups of W can be killed. In other words W can be replaced by a contractible manifold W'. Second, a theorem of Smale [19] asserts that such a manifold W' must be diffeomorphic to the $(n + 1)$-disk. No further details are given here.

Thus kernel (j) is infinite cyclic. In other words there is an exact sequence

$$0 \to Z \to F\Gamma_{4i-1} \xrightarrow{j} \Pi_{4i-1} \to 0$$

It follows that $F\Gamma_{4i-1}$ is an infinite abelian group of rank 1. But $\pi_{4i-1}SO(k) \cong Z$ is also a group of rank 1. Using the sequence

$$\pi_{4i-1}SO(k) \to F\Gamma_{4i-1} \to \Gamma_{4i-1} \to 0$$

we finally see that the group Γ_{4i-1} is finite.

* It is curious that $\bar{\sigma}$ behaves somewhat differently in dimension 3 than in higher dimensions. Thus the image of $\bar{\sigma}$ is generated by 8 if $n = 4i - 1 \geq 7$, but is generated by 16 if $n = 3$.

Similar constructions work for other values of n. Thus

$$0 \to F\Gamma_n \to \pi_{n+k}S^k \to 0 \qquad \text{for } n \equiv 0\ (4)$$

$$Z_2 \to F\Gamma_n \to \pi_{n+k}S^k \to 0 \qquad \text{for } n \equiv 1\ (4)$$

$$0 \to F\Gamma_n \to \pi_{n+k}S^k \to Z_2 \qquad \text{for } n \equiv 2\ (4)$$

$$0 \to Z \to F\Gamma_n \to \pi_{n+k}S^k \to 0 \qquad \text{for } n \equiv 3\ (4)^*$$

For further information on these groups, the reader is referred to [6].

REFERENCES

1. R. Bott, "The stable homotopy of the classical groups," *Ann. of Math.* **70** (1959) 313–337.
2. M. Brown, "The monotone union of open n-cells, is an open n-cell," *Proc. Amer. Math. Soc.* **12** (1961) 812–814.
3. J. Cerf, "Topologie de certains espaces de plongements," *Bull. Soc. Math. France* **89** (1961) 227–380.
4. F. Hirzebruch, *Neue topologische Methoden in der algebraischen Geometrie*, Berlin (1956) Springer.
5. M. Kervaire and J. Milnor, "Bernoulli numbers, homotopy groups and a theorem of Rohlin," *Proc. Intern. Congr. Math.*, Edinburgh (1958).
6. ———, "Groups of homotopy spheres: I," *Ann. of Math.* **77** (1963) 504–537.
7. J. Milnor, "On manifolds homeomorphic to the 7-sphere," *Ann. of Math.* **64** (1956) 399–405.
8. ———, "Sommes de variétés différentiables et structures différentiables des sphères," *Bull. Soc. Math. France* **87** (1959) 439–444.
9. ———, "A procedure for killing the homotopy groups of differentiable manifolds," Symposia in Pure Math. A.M.S. vol. III (1961) 39–55.
10. ———, "A survey of cobordism theory," *L'Enseignement Math.* **8** (1962) 16–23.
11. ———, *Morse theory*, Princeton, N.J. (1963) Princeton University Press.
12. M. Morse, "Relations between the numbers of critical points of a real function of n independent variables," *Trans. Amer. Math. Soc.* **27** (1925) 345–396.
13. J. Munkres, *Elementary differential topology*, Princeton, N.J. (1963) Princeton University Press.
14. R. Palais, "Extending diffeomorphisms," *Proc. Amer. Math. Soc.* **11** (1960) 274–277.
15. L. Pontrjagin, "Smooth manifolds and their applications in homotopy theory," *Trudy Mat. Inst. im. Steklov* **45** (1955), and A.M.S. translations, Series 2, 11, 1–114.
16. G. Reeb, *Sur certaines propriétés topologiques des variétés feuillétées*, Paris (1952) Hermann.
17. R. Rosen, "A weak form of the star conjecture for manifolds," Abstract 570-28, *Notices Amer. Math. Soc.* **7** (1960) 380.

*Corrected 2006.

18. J.-P. Serre, "Homologie singulière des espaces fibrés. Applications," *Ann. of Math.* **54** (1951) 425–505.
19. S. Smale, "On the structure of manifolds," *Amer. J. Math.* **84** (1962) 387–399.
20. J. Stallings, "Polyhedral homotopy spheres," *Bull. Amer. Math. Soc.* **66** (1960) 485–488.
21. ———, "The piecewise linear structure of Euclidean space," *Proc. Cambridge Philos. Soc.* **58** (1962) 481–488.
22. H. Weyl, "Análisis situs combinatorio," *Revista Mat. Hispano-Americana*, **5–6** (1923–1924), 209–218ff.
23. H. Whitney, "Differentiable manifolds," *Ann. of Math.* **37** (1936) 645–680.
24. C. Zeeman, "The generalized Poincaré conjecture," *Bull. Amer. Math. Soc.* **67** (1961) 270.

PART 2. EXPOSITORY LECTURES

This section will consist of the following:

> Lectures on Differential Topology: Unpublished notes by J. Munkres, circulated at Princeton University in 1958.
>
> Lectures on Differentiable Structures: Unpublished notes circulated at Princeton University in 1961.
>
> Smooth Manifolds with Boundary. An expanded version of *Variedades lisas con frontera*, Anales Instituto de Matématicas Universidad Nacional Autónoma de México **1** (1961) 82–116; notes by H. Cárdenas, E. Lluis, F. Recillas and R. Vázquez; as translated by I. E. Colon-Roldan and S. Simanca.

Introduction

The notes on **Differential Topology** by Jim Munkres are based on lectures which I gave at Princeton University. They describe some of the foundations of the field: differentiable manifolds, mappings, embeddings,[1] and immersions, as well as transversality, vector bundles, and Grassmann manifolds. They conclude with a brief introduction to cobordism theory. The Munkres notes provide a very useful introduction to these ideas and I am very happy that they are finally in print, but extremely chagrined that it has taken almost 50 years.

The **Differentiable Structures** notes cover further fundamental material, starting with a study of the extension problem for "diffeotopy" (= smooth isotopy). They discuss the monoid of smooth connected oriented manifolds under the connected sum operation and its group of invertible elements, and describes a bilinear pairing from $\pi_p(\mathrm{SO_q}) \otimes \pi_\mathrm{q}(\mathrm{SO_p})$ to the group Γ_{p+q+1} of twisted spheres. They close with a discussion of tubular neighborhoods.

I lectured (in English) on **Smooth Manifolds with Boundary** at the Universidad Nacional Autónoma de México in 1960. The lectures covered some of the same material, but also other topics such as multiplicative sequences and the

[1]The spelling "imbedding" was used earlier, but "embedding" seems preferable since it is less likely to be confused with immersion.

signature theorem. A central objective was to show that the bilinear pairing

$$\pi_m(\mathrm{SO_n}) \otimes \pi_{\mathrm{n}}(\mathrm{SO_m}) \;\to\; \pi_0\mathrm{Diff}^+(\mathrm{S^{m+n}}) \;\to\; \Gamma_{\mathrm{m+n+1}} \tag{1}$$

described above can be used to construct exotic spheres which can be detected by a diffeomorphism invariant

$$\lambda : \Gamma_{4k-1} \;\to\; \mathbb{Q}/\mathbb{Z}$$

in many cases with $m \equiv n \equiv 3 \pmod 4$. However, I ran out of time, so that the lectures didn't actually get that far. For that reason, I have extended these lectures, translated back into English, by adding a final section which completes the argument.

Further Developments

The bilinear pairing of Equation (1) and other similar pairings have been studied or applied by several authors. See for example P. KAHN [1965], A. KOSINSKI [1967], and T. LAWSON [1973]. Perhaps the most important result which has been obtained by such methods is the following statement which was proved by ANTONELLI, BURGHELEA AND KAHN in [1972]:

> **Theorem.** *If $n \geq 7$, then the group $\mathrm{Diff}^+(S^n)$ of all orientation preserving diffeomorphisms of the sphere does not have the homotopy type of a finite complex.*

By way of contrast, it is known that the inclusion $\mathrm{SO_{n+1}} \hookrightarrow \mathrm{Diff}^+(\mathrm{S^n})$ is a homotopy equivalence for $n \leq 3$. The case $n = 1$ is quite easy, and for $n = 2$ this was proved by SMALE [1959b]. The proof for $n = 3$ by HATCHER [1983] is much more difficult.

Remark. The statement that $\mathrm{Diff}^+(S^2)$ has the homotopy type of SO_3 can also be proved by using quasiconformal methods. The *Beltrami differential* of an orientation preserving diffeomorphism f of a Riemann surface is C^∞-smooth, being given, in terms of a local coordinate z, by the expression

$$\left(\frac{\partial f/\partial \overline{z}}{\partial f/\partial z}\right)\frac{d\overline{z}}{dz} \qquad \text{with} \qquad \left|\frac{\partial f/\partial \overline{z}}{\partial f/\partial z}\right| < 1\,. \tag{2}$$

This vanishes if and only if f is a conformal diffeomorphism. If our Riemann surface is the Riemann sphere $\mathbb{C}\cup\infty$, then there is a converse statement: Following AHLFORS AND BERS, any smooth Beltrami differential satisfying Equation (2) can be integrated to yield a diffeomorphism f which is unique up to composition with a Möbius automorphism. If we normalize by requiring that f must fix three points, say 0, 1 and ∞, then f is uniquely determined. Thus the subgroup of $\mathrm{Diff}^+(\mathbb{C}\cup\infty)$ consisting of diffeomorphisms which fix these three points is homeomorphic to the convex set consisting of all such Beltrami differentials, and hence is contractible. But any element of $\mathrm{Diff}^+(\mathbb{C}\cup\infty)$ can be written uniquely as the composition of a Möbius automorphism with an element of this contractible subgroup. This proves that there is a deformation retraction from the group $\mathrm{Diff}^+(\mathbb{C}\cup\infty)$ onto the subgroup $\mathrm{PSL}_2(\mathbb{C})$ of Möbius automorphisms, which in turn deformation retracts onto its maximal compact subgroup SO_3. $\square$

Lectures on Differential Topology

Notes by James Munkres

Princeton University, Fall Term 1958

Differential topology may be defined as the study of those properties of differentiable manifolds which are invariant under diffeomorphism (differentiable homeomorphism). Typical problems falling under this heading are the following:

(1) Given two differentiable manifolds, under what conditions are they diffeomorphic?

(2) Given a differentiable manifold, is it the boundary of some differentiable manifold with-boundary?

(3) Given a differentiable manifold, is it parallelizable?

All of these problems concern more than the topology of the manifold, yet they do not belong to differential geometry, which usually assumes additional structure (e.g., a connection or a metric).

The most powerful tools in this subject have been derived from the methods of algebraic topology. In particular, the theory of characteristic classes is crucial, whereby one passes from the manifold M to its tangent bundle, and thence to a cohomology class in M which depends on this bundle.

These notes are intended as an introduction to the subject; we will try to go as far as possible without bringing in algebraic topology. Our two main goals are Whitney's theorem that a differentiable n-manifold can be embedded as a closed subset of the euclidean space $\mathbb{R}^{2n+1}$ (see Corollary 1.32); and Thom's theorem that the non-orientable cobordism group $\mathcal{N}_n$ is isomorphic to a certain stable homotopy group (see Theorem 3.15).

Chapter 1 is mainly concerned with approximation theorems. First the basic definitions are given and the inverse function theorem is exploited (§1.1–1.12). Next two local approximation theorems are proved, showing that a given map can be approximated by one of maximal rank (§1.13–1.21). Finally locally finite coverings are used to derive the corresponding global theorems: namely Whitney's embedding theorem and Thom's transversality lemma (§1.28–1.36).

Chapter 2 is an introduction to the theory of vector space bundles, with emphasis on the tangent bundle of a manifold.

Chapter 3 makes use of the preceding material in order to study the cobordism groups $\mathcal{N}_n$.

1. Embeddings and Immersions of Manifolds.

Notation. If x is in the euclidean space $\mathbb{R}^n$, the coordinates of x are denoted by $(x^1, \ldots, x^n)$. Let $||x|| = \max|x^i|$; let $C(r) = C^n(r)$ denote the set of x such that $||x|| < r$; and $C^n(x_0, r)$ the set of x such that $||x - x_0|| < r$. The closure of a cube C is denoted by $\overline{C}$.

A real valued function $f(x^1, \ldots, x^n)$ is ***differentiable*** if the partials of f of all orders exist and are continuous (i.e. "differentiable" means C^∞). A map $f : U \to \mathbb{R}^p$ (where U is an open set in $\mathbb{R}^n$) is differentiable if each of the coordinate functions $(f^1, \ldots, f^p)$ is differentiable. Df denotes the $p \times n$ Jacobian matrix $(\partial f^i/\partial x^j)$ of f; one verifies that the Jacobian of a composition of two differentiable functions is given by the rule $D(fg) = Dg \cdot Df$. The notation $\partial(f^1, \ldots, f^p)/\partial(x^1, \ldots, x^n)$ is also used. If $n = p$, $\det(Df)$ denotes the determinant.

Definition 1.1. A ***topological n-manifold*** M^n is a Hausdorff space with a countable basis which is locally homeomorphic to $\mathbb{R}^n$.

A ***differentiable structure*** $\mathcal{D}$ on a topological manifold M^n is a collection of real-valued functions, each defined on an open subset of M, such that:

- **(1)** For every point p of M there is a neighborhood U of p and a homeomorphism h of U onto an open subset of $\mathbb{R}^n$ such that a function f, defined on the open subset W of U, is in $\mathcal{D}$ if and only if fh^{-1} is differentiable.
- **(2)** If U_i are open sets contained in the domain of f with union U, then $f|_U \in \mathcal{D}$ if and only if $f|_{U_i}$ is in $\mathcal{D}$, for each i.

A ***differentiable manifold*** M^n is a manifold provided with a differentiable structure $\mathcal{D}$; the elements of $\mathcal{D}$ are called the differentiable functions on M. Any open set U and homeomorphism h which satisfy the requirements of (1) above are called a ***coordinate system*** on M. *Notation.* A coordinate system is sometimes denoted by the coordinate functions: $h(p) = (u^1(p), \ldots, u^n(p))$.

Definition 1.2. (Alternate). Let a collection (U_i, h_i) be given, where h_i is a homeomorphism of the open subset U_i of M^n onto an open subset of $\mathbb{R}^n$, such that

- **(a)** the U_i cover M
- **(b)** $h_j h_i^{-1}$ is a differentiable map on the open set $h_i(U_i \cap U_j)$, for all i, j.

Define a ***coordinate system*** as an open set U and a homeomorphism h of U onto an open subset of $\mathbb{R}^n$ such that $h_i h^{-1}$ and hh_i^{-1} are differentiable on $h(U \cap U_i)$ and $h_i(U \cap U_i)$ respectively, for each i. Define the associated ***differentiable structure*** on M as the collection of all such coordinate systems. A function f, defined on the open set V, is ***differentiable*** if fh^{-1} is differentiable on $h(U \cap V)$, for all coordinate systems (U, h).

One shows readily that these two definitions are entirely equivalent.

Definition 1.3. Let M_1, M_2 be differentiable manifolds. If U is an open subset of M_1, $f : U \to M_2$ is ***differentiable*** if for every differentiable function g on an open set $V \subset M_2$, the composition gf is differentiable on $f^{-1}(V) \subset M_1$.

A function $f : M_1 \to M_2$ is a ***diffeomorphism*** if f and f^{-1} are defined and differentiable.

If $A \subset M_1$, a function $f : A \to M_2$ is ***differentiable*** if it can be extended to a differentiable function defined on a neighborhood U of A.

(A coordinate system (U, h) on M^n can then be defined as an open set U in M and a diffeomorphism h of U onto an open set in $\mathbb{R}^n$.)

If $A \subset M$, we have just defined the notion of differentiable function for subsets of A. Suppose that A is locally diffeomorphic to $\mathbb{R}^k$: this collection is easily shown to be a differentiable structure on A. In this case, A is said to be a ***differentiable submanifold*** of M.

The following lemma is familiar from elementary calculus.

LEMMA 1.4. *Let $f : C^n(r) \to \mathbb{R}^p$ satisfy the condition $|\partial f^i/\partial x^j| \leq b$, for all i, j and all $x \in C^n(r)$. Then $||f(x) - f(y)|| \leq bn||x - y||$, for all $x, y, \in C^n(r)$.*

THEOREM 1.5. (Inverse Function Theorem). *Let U be an open subset of $\mathbb{R}^n$, let $f : U \to \mathbb{R}^n$ be differentiable, and let Df be non-singular at x_0. Then f maps some neighborhood of x_0 diffeomorphically onto some neighborhood of $f(x_0)$.*

Proof. We may assume $x_0 = f(x_0) = 0$, and that $Df(x_0)$ is the identity matrix.

Let $g(x) = f(x) - x$, so that $Dg(0)$ is the zero matrix. Choose $r > 0$ so that $x \in U$ and $Df(x)$ is non-singular and $|\partial g^i/\partial x_j| \leq \frac{1}{2n}$, for all x with $||x|| < r$.

Assertion. *If $y \in C(r/2)$, there is exactly one $x \in C(r)$ such that $f(x) = y$.* In fact, by the previous lemma,

$$||g(x) - g(x_0)|| \leq \tfrac{1}{2}||x - x_0|| \tag{1}$$

on $C(r)$. Let us define a sequence $x_0, x_1, \ldots$ by $x_0 = 0$, $x_1 = y$, $x_{n+1} = y - g(x_n)$. This sequence is defined, since $x_n - x_{n-1} = g(x_{n-2}) - g(x_{n-1})$, so that

$$||x_n - x_{n-1}|| \leq \frac{1}{2}||x_{n-2} - x_{n-1}|| \leq \frac{1}{2^{n-1}}||y||;$$

and thus $||x_n|| \leq 2||y||$ for each n. Hence the sequence x_n converges to a point x with $||x|| \leq 2||y||$, so that $x \in C(r)$. Then $x = y - g(x)$, so that $f(x) = y$. This proves the existence of x. To show uniqueness, note that if $f(x) = f(x_1) = y$, then $g(x_1) - g(x) = x - x_1$, contradicting (1).

Hence $f^{-1} : C(r/2) \to C(r)$ exists. Note that

$$||f(x) - f(x_1)|| \geq ||x - x_1|| - ||g(x) - g(x_1)|| \geq \frac{1}{2}||x - x_1||$$

so that $||y - y_1|| \geq 1/2||f^{-1}(y) - f^{-1}(y_1)||$. Hence f^{-1} is continuous; the image of $C(r/2)$ under f^{-1} is open because it equals $C(r) \cap f^{-1}(C(r/2))$, the intersection of two open sets.

To show that f^{-1} is differentiable, note that

$$f(x) \;=\; f(x_1) \;+\; Df(x_1) \cdot (x - x_1) \;+\; h(x, x_1),$$

where $(x - x_1)$ is written as a column matrix and the dot stands for matrix multiplication. Here $h(x, x_1)/||x - x_1|| \to 0$ as $x \to x_1$. Let A be the inverse matrix of $DF(x_1)$. Then

$$A \cdot (f(x) - f(x_1)) = (x - x_1) + A \cdot h(x, x_1), \qquad \text{or}$$
$$A \cdot (y - y_1) + A \cdot h_1(y, y_1) = f^{-1}(y) - f^{-1}(y_1),$$

where $h_1(y, y_1) = -h(f^{-1}(y), f^{-1}(y_1))$. Now

$$\frac{h_1(y, y_1)}{||y - y_1||} = -\frac{h(x, x_1)}{||x - x_1||}\frac{||x - x_1||}{||y - y_1||}.$$

Since $||x - x_1||/||y - y_1|| \leq 2$, $h_1(y, y_1)/||y - y_1|| \to 0$ as $y \to y_1$. Hence

$$D(f^{-1}) = A = (D(f))^{-1}.$$

This means that $D(f^{-1})$ is obtained as the composition of the following maps:

$$C(r/2) \xrightarrow{f^{-1}} C(r) \xrightarrow{Df} \mathrm{GL}(n) \xrightarrow{\text{matrix inversion}} \mathrm{GL}(n);$$

where $\mathrm{GL}(n)$ denotes the set of non-singular $n \times n$ matrices, considered as a subspace of n^2-dimensional euclidean space. Since f^{-1} is continuous and Df and matrix inversion are C^∞-functions, $D(f^{-1})$ is continuous, i.e., f^{-1} is C^1. In general, if f^{-1} is C^k, then by this argument $D(f^{-1})$ is also, i.e., f^{-1} is of class C^{k+1}. This completes the proof. □

LEMMA 1.6. *Let U be an open subset of $\mathbb{R}^n$, let $f : U \to \mathbb{R}^p$ $(n \leq p)$, $f(0) = 0$, and let $Df(0)$ have rank n. Then there exists a diffeomorphism g of one neighborhood of the origin in $\mathbb{R}^p$ onto another so that $g(0) = 0$ and $gf(x^1, \ldots, x^n) = (x^1, \ldots, x^n, 0, \ldots, 0)$, in some neighborhood of the origin.*

Proof. Since $\partial(f^1, \ldots, f^p)/\partial(x^1, \ldots, x^n)$ has rank n, we may assume that

$$\partial(f^1, \ldots, f^n)/\partial(x^1, \ldots, x^n)$$

is the submatrix which is non-singular. Define $F : U \times \mathbb{R}^{p-n} \to \mathbb{R}^p$ by the equation

$$F(x^1, \ldots, x^p) = f(x^1, \ldots, x^n) + (0, \ldots, 0, x^{n+1}, \ldots, x^p).$$

F is an extension of f, since $F(x^1, \ldots, x^n, 0, \ldots, 0) = f(x^1, \ldots, x^n)$. The matrix DF is non-singular at the origin, since its determinant is equal to

$$\det\left(\partial(f^1, \ldots, f^n)/\partial(x^1, \ldots, x^n)\right),$$

which is non-zero. Hence F has a local inverse g. Thus g maps one neighborhood of the origin in $\mathbb{R}^p$ onto another with

$$gF(x^1, \ldots, x^p) = (x^1, \ldots, x^p),$$

and hence

$$gf(x^1, \ldots, x^n) = gF(x^1, \ldots, x^n, 0, \ldots, 0) = (x^1, \ldots, x^n, 0, \ldots, 0). \quad \square$$

COROLLARY 1.7. *Let $A = A^k$ be a differentiable submanifold of M^n. Given $x \in A$, there is a coordinate system (U, h) on M about x, such that $h(U \cap A) = h(U) \cap \mathbb{R}^k$ (where $\mathbb{R}^k$ is considered as the subspace $\mathbb{R}^k \times 0$ of $\mathbb{R}^k \times \mathbb{R}^{n-k} = \mathbb{R}^n$).*

Proof. Let (U_1, h_1) be a coordinate system on M about x; by hypothesis, there is a differentiable map f of a neighborhood V of x in M into $\mathbb{R}^k$ such that $f|_{V \cap A} = f_1$ is a diffeomorphism whose range is an open set W in $\mathbb{R}^k$. We may assume $U_1 = V$, and $h_1(x) = f(x) = 0$.

Now $fh_1^{-1}h_1f_1^{-1}$ is the identity on W, so that its Jacobian, which equals $D(fh_1^{-1}) \cdot D(h_1f_1^{-1})$ is non-singular. Hence $D(h_1f_1^{-1})$ has rank k, so that by the previous lemma, there is a diffeomorphism g of some neighborhood $V_1 \subset h_1(U_1)$ of the origin onto another such that $g(0) = 0$ and $gh_1f_1^{-1}(x^1, \ldots, x^k) = (x^1, \ldots, x^k, 0, \ldots, 0)$. Then $U = h_1^{-1}(V_1)$ and $h = gh_1$ will satisfy the requirements of the lemma. $\square$

LEMMA 1.8. *Let U be an open subset of $\mathbb{R}^n$, let $f : U \to \mathbb{R}^p$, $f(0) = 0$, $(n \geq p)$, and let $Df(0)$ have rank p. Then there is a diffeomorphism h of some neighborhood of the origin in $\mathbb{R}^n$ onto another such that $h(0) = 0$ and $fh(x^1, \ldots, x^n) = (x^1, \ldots, x^p)$.*

Proof. We may assume $\partial(f^1, \ldots, f^p)/\partial(x^1, \ldots, x^p)$ is non-singular at 0, since $Df(0)$ has rank p. Define $F : U \to \mathbb{R}^n$ by the equation

$$F(x^1, \ldots, x^n) = (f^1(x), \ldots, f^p(x), x^{p+1}, \ldots, x^n).$$

Then $DF(0)$ is non-singular; let h be the local inverse of F. Let g project $\mathbb{R}^n$ onto the subspace $\mathbb{R}^p$; $f = gF$. Then

$$fh(x^1, \ldots, x^n) = gFh(x^1, \ldots, x^n) = g(x^1, \ldots, x^n) = (x^1, \ldots, x^p). \qquad \square$$

Exercise 1.9. Let U be an open subset of $\mathbb{R}^n$, $f : U \to \mathbb{R}^p$, $f(0) = 0$; and let $Df(x)$ have rank k for all x in U. Then there are local diffeomorphisms h and g of $\mathbb{R}^n$ and $\mathbb{R}^p$ respectively such that

$$gfh(x^1, \ldots, x^n) = (x^1, \ldots, x^k, 0, \ldots, 0)$$

throughout some neighborhood of the origin.

Definition 1.10. If $f : M_1 \to M_2$, the ***rank*** of f at x is the rank of $D(h_2fh_1^{-1})$ at $h_1(x)$, where (U_1, h_1) and (U_2, h_2) are coordinate systems about x and $f(x)$, respectively. The differentiable map $f : M^n \to M^p$ is an ***immersion*** if rank $f = n$ everywhere $(n \leq p)$. It is an ***embedding*** if it is also a homeomorphism into.

If $f : M^n \to M^p$, then $y \in M^p$ is a ***regular value*** of f if rank $f = p$ on the entire set $f^{-1}(y)$. Otherwise, y is a ***critical value***. (If $y \notin f(M^n)$, then by definition y is a regular value of f.)

Exercise 1.11. If A is a differentiable submanifold of M, the inclusion $A \to M$ is an embedding; and conversely if $f : M_1 \to M$ is an embedding then $f(M_1)$ is a differentiable submanifold.

Exercise 1.12. If y is a regular value of $f : M^n \to M^p$, then $f^{-1}(y)$ is a differentiable submanifold of M^n of dimension $n - p$ (or is empty).

Definition 1.13. A subset A of $\mathbb{R}^n$ has ***measure zero*** if it may be covered by a countable collection of cubes $C(x, r)$ having arbitrarily small total volume. In such a case, $\mathbb{R}^n - A$ is everywhere dense (i.e., it intersects every non-empty open set).

LEMMA 1.14. *Let U be an open subset of $\mathbb{R}^n$; let $f : U \to \mathbb{R}^n$ be differentiable. If $A \subset U$ has measure 0, so does $f(A)$.*

Proof. Let C be any cube with $\overline{C} \subset U$. Let b denote the maximum of $|\partial f^i/\partial x^j|$ on $\overline{C}$ for all i, j. By Lemma 1.4, $||f(x) - f(y)|| \leq bn||x - y||$ for $x, y \in \overline{C}$.

Now $A \cap C$ has measure zero; let us cover $A \cap C$ by cubes $C(x_i, r_i)$, with closures contained in U, such that $\sum_{i=1}^{\infty} r_i^n < \epsilon$. Then $f(C(x_i, r_i)) \subset C(f(x_i), bnr_i)$, so that $f(A \cap C)$ is covered by cubes of total volume $b^n n^n \sum r_j^n < b^n n^n \epsilon$. Hence $f(A \cap C)$ has measure zero.

Since A can be covered by countably many such cubes C, $f(A)$ has measure zero. □

COROLLARY 1.15. *If $f : U \to \mathbb{R}^p$ is differentiable, where U is an open subset of $\mathbb{R}^n$ and $n < p$, then $f(U)$ has measure zero.*

Proof. Project $U \times \mathbb{R}^{p-n}$ onto U and apply f. Since $U \times 0$ has measure zero in $\mathbb{R}^p$, so does $f(U)$. □

Definition 1.16. If $A \subset M$, A has ***measure*** zero if $h(A \cap U)$ has measure zero for every coordinate system (U, h).

COROLLARY 1.17. *If $f : M^n \to M^p$ is differentiable and $n < p$, then $f(M^n)$ has measure zero.*

Definition 1.18. Let $\mathcal{M}(p, n)$ denote the space of $p \times n$ matrices, with the differentiable structure of the euclidean space $\mathbb{R}^{pn}$. Let $\mathcal{M}(p, n; k)$ denote the subspace consisting of matrices of rank k. Thus $\mathcal{M}(p, n; n)$ is an open subset of $\mathcal{M}(p, n)$ if $p \geq n$; the determinental criterion for rank proves this. More generally, we have:

LEMMA 1.19. *$\mathcal{M}(p, n; k)$ is a differentiable submanifold of $\mathcal{M}(p, n)$ of dimension $k(p + n - k)$, where $k \leq \min(p, n)$.*

Proof. Let $E_0 \in \mathcal{M}(p, n; k)$; we may assume that E_0 is of the form $\begin{pmatrix} A_0 & B_0 \\ C_0 & D_0 \end{pmatrix}$, where A_0 is a non-singular $k \times k$ matrix. There is an $\epsilon > 0$ such that if all the entries of $A - A_0$ are less than ϵ, A must also be nonsingular. Let U consist of all matrices in $\mathcal{M}(p, n)$ of the form $E = \begin{pmatrix} A & B \\ C & D \end{pmatrix}$, with all the entries of $A - A_0$ less than ϵ.

Then E is in $\mathcal{M}(p, n; k)$ if and only if $D = CA^{-1}B$; for the matrix

$$\begin{pmatrix} I_k & 0 \\ X & I_{p-k} \end{pmatrix} \begin{pmatrix} A & B \\ C & D \end{pmatrix} = \begin{pmatrix} A & B \\ XA + C & XB + D \end{pmatrix}$$

has the same rank as E. If $X = -CA^{-1}$, this matrix is

$$\begin{pmatrix} A & B \\ 0 & -CA^{-1}B + D \end{pmatrix}.$$

If $D = CA^{-1}B$, this matrix has rank k. The converse also holds, for if any element of $-CA^{-1}B + D$ is different from zero, this matrix has rank $> k$.

Let W be the open set in euclidean space of dimension

$$(pn - (p - k)(n - k)) = k(p + n - k)$$

consisting of matrices $\begin{pmatrix} A & B \\ C & 0 \end{pmatrix}$, with all the entries of $A - A_0$ less than ϵ. The map

$$\begin{pmatrix} A & B \\ C & 0 \end{pmatrix} \to \begin{pmatrix} A & B \\ C & -CA^{-1}B \end{pmatrix}$$

is then a diffeomorphism of W onto the neighborhood $U \cap \mathcal{M}(p, n; k)$ of E_0. □

THEOREM 1.20. *Let U be an open set in $\mathbb{R}^n$, and let $f : U \to \mathbb{R}^p$ be differentiable, where $p \geq 2n$. Given $\epsilon > 0$, there is a $p \times n$ matrix $A = (a^i_j)$ with each $|a^i_j| < \epsilon$, such that $g(x) = f(x) + A \cdot x$ is an immersion. (Here x is written as a column matrix.)*

Proof. $Dg(x) = Df(x) + A$; we would like to choose A in such a way that $Dg(x)$ has rank n for all x. I.e., A should be of the form $Q - Df$, where Q has rank n.

We define $F_k : \mathcal{M}(p, n; k) \times U \to \mathcal{M}(p, n)$ by the equation

$$F_k(Q, x) = Q - Df(x).$$

Now F_k is a differentiable map, and the domain of F_k has dimension $k(p+n-k)+n$. As long as $k < n$, this expression is monotonic in k (its partial with respect to k is $p + n - 2k$). Hence the domain of F_k has dimension not greater than

$$(n - 1)(p + n - (n - 1)) + n = (2n - p) + pn - 1$$

for $k < n$. Since $p \geq 2n$, this dimension is strictly less than $pn = \dim \mathcal{M}(p, n)$.

Hence the range of F_k has measure zero in $\mathcal{M}(p, n)$, so that there is an element A of $\mathcal{M}(p, n)$, arbitrarily close to the zero matrix, which is not in the range of F_k for $k = 0, \ldots, n - 1$. Then $A + Df(x) = Dg(x)$ has rank n, for each x. □

THEOREM 1.21. *Let U be an open subset of $\mathbb{R}^n$; and let $f : U \to \mathbb{R}^p$ be differentiable. Given $\epsilon > 0$, there is a $p \times n$ matrix A and a $p \times 1$ matrix B, with entries less than ϵ in absolute value, such that the map*

$$g(x) = f(x) + A \cdot x + B$$

has the origin as a regular value.

Remark. The following much more delicate result has been proved by [Sard, A.]: The set of critical values of any differentiable map has measure zero.

Proof of Theorem 1.21. Note that the theorem is trivial if $p > n$, since then $f(U)$ has measure zero, and we may choose $A = 0$ and B small in such a way that 0 is not in the image of g.

Assume $p \leq n$. We wish $Dg(x_0) = Df(x_0) + A$ to have rank p, where x_0 ranges over all points such that

$$g(x_0) \;=\; 0 \;=\; f(x_0) + A \cdot x_0 + B.$$

Hence A is of the form $Q - Df(x)$, and B is of the form $-f(x) - A \cdot x$, where Q is to have rank p.

We define $F_k : \mathcal{M}(p, n; k) \times U \to \mathcal{M}(p, n) \times \mathbb{R}^p$ by the equation

$$F_k(Q, x) \;=\; (Q - Df(x),\, -f(x) - (Q - Df(x)) \cdot x).$$

Then F_k is differentiable. If $k < p$, the dimension of its domain is not greater than $(p-1)(p+n-(p-1))+n = p+pn-1$. Hence the image of F_k, $k = 0, \ldots, p-1$ has measure zero; so that there is a point (A, B) arbitrarily close to the origin which is not in any such image set. This completes the proof. □

Definition 1.22. A covering of X is *locally finite* if every point has a neighborhood which intersects only finitely many elements of the covering. A *refinement* of a covering of X is a second covering each element of which is contained in an element of the first covering. A Hausdorff space is *paracompact* if every open covering has a locally finite open refinement.

If X is paracompact, and U_α is an open covering, there is a locally-finite indexed open covering V_α with $V_\alpha \subset U_\alpha$ for each α. (Each point has a neighborhood that intersects V_α for only finitely many α.) For let W_β be a locally-finite refinement of U_α such that distinct indices β correspond to distinct sets; choose $\alpha(\beta)$ so that $W_\beta \subset U_{\alpha(\beta)}$ for each β. Set $V_{\alpha_0} = \bigcup_{\alpha(\beta)=\alpha_0} W_\beta$. Given a neighborhood intersecting only finitely many W_β, it intersects only finitely many V_α as well.

THEOREM 1.23. *If X is locally compact and Hausdorff, having a countable basis, X is paracompact.*

Thus every "manifold", in the sense of Definition 1.1, is automatically paracompact.

Proof of Theorem 1.23. Let $U_1, U_2, \dots$ be a basis for X with $\overline{U}_i$ compact for each i. We first construct a sequence $A_1, A_2, \dots$ of compact sets whose union is X, such that $A_i \subset \mathrm{Int}A_{i+1}$. Start with $A_1 = \overline{U}_1$. Given A_i compact, let k be the smallest integer such that $k \geq i$, and such that A_i is contained in $U_1 \cup \cdots \cup U_k$; and let A_{i+1} be the closure of this union.

Let $\mathcal{O}$ be an open covering of X. Cover the compact set $A_{i+1} - \mathrm{Int}A_i$ by a finite number of open sets $V_1, \dots, V_n$ where each V_i is contained in some element of $\mathcal{O}$, and in the open set $\mathrm{Int}A_{i+2} - A_{i-1}$. Let P_i denote the collection $(V_1, \dots, V_n)$, and let $P = P_0 \cup P_1 \cup \cdots$. P refines $\mathcal{O}$, and since any compact closed neighborhood C is contained in some A_i, C can intersect only finitely many elements of P. □

Exercise 1.24. Prove that every paracompact space is normal (disjoint closed subsets have disjoint neighborhoods). First prove that it is regular (the special case where one of the two closed sets is a point).

THEOREM 1.25. *Let M^n be a differentiable manifold, $\{U_\alpha\}$ an open covering of M. There is a collection (V_j, h_j) of coordinate systems on M such that*

(1) *$\{V_j\}$ is a locally-finite refinement of $\{U_\alpha\}$.*
(2) *$h_j(V_j) = C^n(3)$.*
(3) *If $W_j = h_j^{-1}(C^n(1))$, then $\{W_j\}$ covers M.*

Proof. The proof proceeds along lines similar to the previous one. The only difference is that one chooses the V_j to satisfy (2), and makes sure that the sets $h_j^{-1}(C(1))$ also cover $A_{i+1} \setminus \mathrm{Int}A_i$. □

LEMMA 1.26. *There exists a C^∞ function $\varphi(x^1, \dots, x^n)$ such that $\varphi = 1$ on $\overline{C}(1)$, $0 < \varphi < 1$ on $C(2) \setminus \overline{C}(1)$, $\varphi = 0$ on $\mathbb{R}^n \setminus C(2)$.*

In fact, this function may be defined by the equation $\varphi(x^1, \dots, x^n) = \prod_1^n \psi(x_i)$, where

$$\psi(x) = \frac{\lambda(2+x)\cdot\lambda(2-x)}{\lambda(2+x)\cdot\lambda(2-x) + \lambda(x-1) + \lambda(-x-1)}$$

and

$$\lambda(x) = \begin{cases} e^{-1/x}, & \text{if} \quad x > 0 \\ 0, & \text{if} \quad x \leq 0. \end{cases}$$

Note that the denominator in the expression for ψ is always positive, and that

$$\begin{array}{rcrcl} & & \psi(x) & = & 1 \quad \text{for} \quad |x| \leq 1 \\ 0 & < & \psi(x) & < & 1 \quad \text{if} \quad 1 < |x| < 2 \\ & & \psi(x) & = & 0 \quad \text{if} \quad |x| \geq 2. \end{array} \qquad \square$$

Definition 1.27. Let $f, g : X \to Y$, where Y is a metric space, and let $\delta(x)$ be a positive continuous function defined on X. Then g is a δ-*approximation* to f if $d(f(x), g(x)) < \delta(x)$ for all x. [One can impose a topology on the space $F(X,Y)$ of all differentiable maps by taking the δ-approximations to f as basic neighborhoods of f. This topology is independent of the choice of metric on Y, provided that X, Y are paracompact.]

THEOREM 1.28. *Given a differentiable map $f : M^n \to \mathbb{R}^p$ where $p \geq 2n$, and a continuous positive function δ on M^n, there exists an immersion $g : M^n \to \mathbb{R}^p$ which is a δ-approximation to f. If* $\operatorname{rank} f = n$ *on the closed set N, we may choose $g|_N = f|_N$.*

Proof. Clearly the rank of f is equal to n on a neighborhood U of N. Cover M^n by U and $M^n - N$. Let (V_i, h_i) be a refinement of this covering, constructed as in Theorem 1.25. As before, $h_i(W_i) = C(1)$ and $h_i(V_i) = C(3)$. Let $h_i(V_i') = C(2)$, so that $V_i \supset V_i' \supset W_i$. Let the V_i be indexed with positive and negative integers so that those V_i with $i \leq 0$ are the ones contained in U. Let $\epsilon_i = \min \delta(x)$ on the compact set $\overline{V}'_i$.

Set $f_0 = f$. Given $f_{k-1} : M^n \to \mathbb{R}^p$, having rank n on $N_{k-1} = \bigcup_{j<k} \overline{W}_j$, consider $f_{k-1}h_k^{-1} : C(3) \to \mathbb{R}^p$. Let A be a $p \times n$ matrix; let $F_A : C(3) \to \mathbb{R}^p$ be defined by the equation

$$F_A(x) = f_{k-1}h_k^{-1}(x) + \varphi(x)\, A \cdot (x),$$

where (x) is written as usual as an $n \times 1$ column matrix; A is yet to be chosen; and $\varphi(x)$ is the function defined in Lemma 1.26.

First, we want $F_A(x)$ to have rank n on the set $K = h_j(N_{k-1} \cap \overline{U}_k)$; we are given that $f_{k-1}h_k^{-1}$ has rank n on K. Now

$$D(F_A(x)) = D(f_{k-1}h_k^{-1}(x)) + A \cdot (x) \cdot D\varphi(x) + \varphi(x)A.$$

(Here $D\varphi$ is a $1 \times n$ matrix.) The map of $K \times \mathcal{M}(p,n)$ into $\mathcal{M}(p,n)$ which carries (x, A) into $D(F_A(x))$ is continuous. It carries $K \times (0)$ into the open subset $\mathcal{M}(p,n;n)$ of $\mathcal{M}(p,n)$. Hence if A is sufficiently small, this map will carry $K \times A$ into $\mathcal{M}(p,n;n)$; our first requirement is that A be this small.

Secondly, we require A to be small enough that $||A \cdot (x)|| < \epsilon_k/2^k$ for all $x \in C(3)$.

Finally, by Theorem 1.20, A may be chosen arbitrarily small so that

$$f_{k-1}h_k^{-1}(x) + A \cdot (x)$$

has rank n on $C(2)$. Let A be chosen to satisfy this requirement. We then define $f_k : M^n \to \mathbb{R}^p$ by the equation:

$$f_k(y) = \begin{cases} f_{k-1}(y) + \varphi(h_k(y))\, A \cdot (h_k(y)), & \text{for} \quad y \in V_k \\ f_{k-1}(y), & \text{for} \quad y \in M - \overline{U}_k. \end{cases}$$

These definitions agree on the overlapping domains, so that f_k is differentiable. By the first condition on A, it has rank n on N_{k-1}; by the third condition it has rank n on $\overline{W}_k$. By the second condition, f_k is a $\delta/2^k$ approximation to f_{k-1}.

We define $g(x) = \lim_{k\to\infty} f_k(x)$. Since the covering V_i is locally finite, all the f_k agree on a given compact set for k sufficiently large; it follows that g is differentiable and has rank n everywhere. It is also a δ-approximation to f. □

LEMMA 1.29. *If $p > 2n$, any immersion $f : M^n \to \mathbb{R}^p$ can be δ-approximated by an injective immersion g. If f is injective in a neighborhood U of the closed set N, we may choose $g|_N = f|_N$.*

Proof. Choose a covering $\{U_\alpha\}$ of M such that each $f|_{U_\alpha}$ is an embedding. Let (V_i, h_i) be the locally finite refinement constructed in Theorem 1.25, and let $\varphi(x)$ be the function constructed in Lemma 1.26. Then we can define a differentiable function from M to $\mathbb{R}$ by the formula

$$\varphi_i(y) = \begin{cases} \varphi(h_i(y)), & \text{if} \quad y \in V_i \\ 0, & \text{otherwise.} \end{cases}$$

As before, we assume (V_i, h_i) refines the covering $\{U, M - N\}$ and that those V_i with $i \leq 0$ are the ones contained in U. Let $f_0 = f$. Given the immersion $f_{k-1} : M^n \to \mathbb{R}^p$, we define f_k inductively by the equation

$$f_k(y) = f_{k-1}(y) + \varphi_k(y) b_k\,,$$

where b_k is a point of $\mathbb{R}^p$ yet to be chosen. By the argument of the previous theorem, if b_k is sufficiently small, f_k will have rank n everywhere. The first requirement is that b_k be this small, and the second requirement is that b_k be small enough that f_k be a $\delta/2^k$ approximation to f_{k-1}.

Finally, let N^{2n} be the open subset of $M^n \times M^n$ consisting of pairs (y, y'), with $\varphi_k(y) \neq \varphi_k(y')$. Consider the differentiable map

$$(y, y') \mapsto -\frac{f_{k-1}(y) - f_{k-1}(y')}{\varphi_k(y) - \varphi_k(y')}$$

from N^{2n} into $\mathbb{R}^p$. Since $2n < p$, the image of N^{2n} has measure 0, so that b_k may be chosen arbitrarily small and ***not*** in this image. For every $k > 0$, it follows easily that $f_k(y) = f_k(y')$ if and only if both $\varphi_k(y) - \varphi_k(y') = 0$ and $f_{k-1}(y) - f_{k-1}(y') = 0$.

Define $g(y) = \lim_{k\to\infty} f_k(y)$. This limit exists and is differentiable since the covering $\{V_i\}$ is locally finite. If $g(y) = g(y')$ with $y \neq y'$, it would follow that $f_{k-1}(y) = f_{k-1}(y')$ and $\varphi_k(y) = \varphi_k(y')$ for all $k > 0$. The former condition implies that $f(y) = f(y')$, so that y and y' cannot belong to any one set V_i. Because of the latter condition, this means that neither is in any set V_i' for $i > 0$. Hence, they lie in U, contradicting the fact that f is injective on U. □

Definition 1.30. Let $f : M^n \to \mathbb{R}^p$. The *limit set* $L(f)$ is the set of $y \in \mathbb{R}^p$ such that $y = \lim f(x_n)$ for some sequence $\{x_1, x_2, \dots\}$ which has no subsequence converging to a point on M^n.

Exercise. Show the following:

(1) $f(M)$ is a closed subset of $\mathbb{R}^p$ if and only if $L(f) \subset f(M)$.

(2) f is a topological embedding if and only if f is injective and $L(f) \cap f(M)$ is vacuous.

LEMMA 1.31. *There exists a differentiable map $f : M^n \to \mathbb{R}$ with $L(f)$ empty.*

Proof. Let (V_i, h_i) and φ be chosen as in Theorem 1.25 and Lemma 1.26, with i ranging over positive integers; and again let

$$\varphi_i(y) = \begin{cases} \varphi\big(h_i(y)\big), & \text{if } \quad y \in V_i \\ 0, & \text{otherwise.} \end{cases}$$

Define $f(y) = \sum_j \big(j\, \varphi_j(y)\big)$. This sum is finite, since V_i is a locally finite covering. If $\{x_i\}$ is a set of points of M having no limit point, only finitely many lie in any compact subset of M. Given m, there is an integer i such that x_i is not in $\overline{W}_1 \cup \cdots \cup \overline{W}_m$. Hence $x_i \in \overline{W}_j$ for some $j > m$, whence $f(x_i) > m$. Thus the sequence $f(x_m)$ cannot converge. □

COROLLARY 1.32. *Every M^n can be differentiably embedded in $\mathbb{R}^{2n+1}$ as a closed subset.*

Proof. Let $f : M^n \to \mathbb{R} \subset \mathbb{R}^{2n+1}$ differentiably, with $L(f) = 0$. Set $\delta(x) \equiv 1$, and let g be an injective immersion which is a δ-approximation to f. Then $L(g)$ is empty, so that g is a homeomorphism. □

Definition 1.33. Consider a differentiable map $f : M^n \to N^p$, together with a codimension q differentiable submanifold $N_1 = N_1^{p-q} \subset N^p$. Given a point $x \in f^{-1}(N_1) \subset M$, let $(u, \dots, u^n)$ be a coordinate system about x; and let $(v^1, \dots, v^p)$ be a coordinate system about $f(x)$ chosen so that the intersection of N_1 with the associated coordinate neighborhood is defined by the equations $v^1 = \cdots = v^q = 0$. (Compare Lemma 1.6.) By definition, the ***transverse regularity condition*** for f and N_1 is satisfied at x if the $q \times n$ matrix

$$\big(\partial v^i / \partial u^j\big) \qquad \begin{matrix} i = 1, \dots, q \\ j = 1, \dots, n \end{matrix}$$

has rank q at x.

Remark 1.34. This condition is independent of the particular choice of local coordinates. In terms of the first derivative map from the tangent space $T_x(M) \to T_y(N)$ where $y = f(x)$, and the quotient map from $T_y(N)$ to $T_y(N)/T_y(N_1)$, it is just the condition that the composition maps $T_x(M)$ onto $T_y(N)/T_y(N_1)$. Compare the discussion in Definition 2.6 below.

Note that the set of points on which this transverse regularity condition is satisfied is open as a subset of $f^{-1}(N_1)$. The map f is said to be ***transverse regular on*** N_1 if the condition is satisfied for every x in $f^{-1}(N_1)$.

LEMMA 1.35. *If $f : M^n \to N^p$ is transverse regular on N_1^{p-q} then $f^{-1}(N_1)$ is a differentiable submanifold of dimension $n-q$ (or is empty).*

Proof. Let π project $\mathbb{R}^p$ onto its first q components; $\pi : \mathbb{R}^p \to \mathbb{R}^q$. If $(V, h) = (v^1, \ldots, v^p)$ is the coordinate system hypothesized in Definition 1.33, then

$$N_1 \cap V = h^{-1}\pi^{-1}(0),$$

where 0 denotes the origin in $\mathbb{R}^q$, and $f^{-1}(N_1 \cap V) = (\pi h f)^{-1}(0)$. Since $\pi h f$ has rank q at $x \in f^{-1}(N_1 \cap V)$, the origin is a regular value of $\pi h f$. Hence $(\pi h f)^{-1}(0)$ is a differentiable submanifold of M of dimension $n-q$ (see Exercise 1.12). □

THEOREM 1.36. *Let $f : M^n \to N^p$ be differentiable; let $N_1 = N_1^{p-q}$ be a closed differentiable submanifold of N. Let A be a closed subset of M such that the transverse regularity condition for f and N_1 holds at each x in $A \cap f^{-1}(N_1)$. Let δ be a positive continuous function on M. There exists a differentiable map $g : M^n \to N^p$ such that*

(1) *g is a δ-approximation to f,*
(2) *g is transverse regular on N_1, and*
(3) *$g|_A = f|_A$.*

Proof. There is a neighborhood U of A in M such that f satisfies the transverse regularity condition on $U \cap f^{-1}(N_1)$. Cover N by $Y_0 = N \setminus N_1$, together with coordinate systems (Y_i, η_i) for $i > 0$ each of which has coordinate functions $(v^1, \ldots, v^n)$ such that $v^1 = \cdots = v^q = 0$ on N_1. Now the open sets $f^{-1}(Y_i)$ cover M, as do the open sets U and $M \setminus A$. Let $\{(V_j, h_j)\}$ be a refinement of both coverings, constructed as in Theorem 1.25. Recall that $h_j(V_j) = C(3)$, $h_j(V_j') = C(2)$, $h_j(W_j) = C(1)$, and the W_j cover M. The V_j are to be indexed with positive and negative integers so that those V_j which are contained in U are the ones with $j \leq 0$.

Let φ be as in Lemma 1.26, and define $\varphi_i(x) = \varphi(h_i(x))$ for $x \in V_i$ and $\varphi_i(x) = 0$ elsewhere. For each j choose $i(j) \geq 0$ so that $f(V_j)$ is contained in $Y_{i(j)}$.

Set $f_0 = f$. Suppose $f_{k-1} : M \to N$ is defined and satisfies the transverse regularity condition for N_1 at each point of the intersection of $f_{k-1}^{-1}(N_1)$ with $\bigcup_{j<k} \overline{W}_j$. Furthermore suppose that $f_{k-1}(\overline{V}_j') \subset Y_{i(j)}$ for each j. Setting $i = i(k)$, it follows in particular that $f_{k-1}(\overline{V}_k') \subset Y_i$.

Consider the composition

$$\pi \eta_i f_{k-1} h_k^{-1} : C(2) \to \mathbb{R}^q,$$

where again π projects $\mathbb{R}^p$ onto $\mathbb{R}^q$. By Theorem 1.21, there is an arbitrarily small affine function $L(x) = A \cdot (x) + B$ from $\mathbb{R}^p$ to $\mathbb{R}^q$ such that when added to the previous function, the resulting map has the origin as a regular value. Consider $\mathbb{R}^q$ as the first q coordinates in $\mathbb{R}^p$, and define

$$f_k(x) = \begin{cases} \eta_i^{-1}\left(\eta_i f_{k-1}(x) + L(h_k(x))\,\varphi_k(x)\right), & \text{for } x \text{ in a neighborhood of } \overline{V}_k' \\ f_{k-1}(x), & \text{for } x \text{ in } \ M \setminus V_k'. \end{cases}$$

Here L is yet to be chosen. Of course, we must choose L small enough that

$$\eta_i f_{k-1} + L\varphi_k$$

lies in $C(1)$ for $x \in \overline{V}'_k$, in order that η_i^{-1} may be applied to it. This is the first requirement on L. Secondly, we choose L small enough that f_k is a $\delta/2^k$ approximation to f_{k-1}. Thirdly choose L small enough so that $f_k(\overline{V}'_j)$ is contained in $Y_{i(j)}$ for each j. This is possible since only a finite number of the sets $\overline{V}'_j$ can intersect $\overline{V}'_k$.

Now f_k by definition satisfies the transverse regularity condition for N_1 at each point of $f_k^{-1}(N_1) \cap \overline{W}_k$. We want to choose L small enough that the condition is satisfied at each point of the intersection of $f_k^{-1}(N_1)$ with $\bigcup_{j<k} \overline{W}_j$. It is sufficient to consider the intersection of the latter set with $\overline{V}'_k$; let this intersection be denoted by K. Consider the function which maps each pair (x, L) with $x \in K$ to the pair

$$\Big(f_k(x)\ ,\ D(\pi\eta_i f_k h_k^{-1}) \cdot (h_k(x)\Big) \ \in\ N \times \mathcal{M}(q,n)\,.$$

(Here the dot means that the derivative matrix is to be evaluated at the point $h_k(x)$.) This function is continuous and carries $K \times (0)$ into the set

$$\Big((N - N_1) \times \mathcal{M}(q,n)\Big) \cup \Big(N \times \mathcal{M}(q,n;q)\Big),$$

which is open in $N \times \mathcal{M}(q,n)$. Hence for L sufficiently small, (x, L) is carried into this set, so that f_k satisfies the transverse regularity condition for N_1 at each point of $f_k^{-1}(N_1) \cap \Big(\bigcup_{j\le k} \overline{W}_j\Big)$.

We define $g(x) = \lim_{k\to\infty} f_k(x)$, as usual. $\square$

2. Vector Space Bundles

Definition 2.1. An n-dimensional real *vector space bundle* ξ is a triple (π, a, s). Here π is a continuous map of E onto B, where E and B are Hausdorff spaces; and each set $F_b = \pi^{-1}(b)$ is called a *fibre*. The map $s : \mathbb{R} \times E \to E$ must carry each $\mathbb{R} \times F_b$ onto F_b; while the map a is defined on $\bigcup_b (F_b \times F_b) \subset E \times E$ and carries each $F_b \times F_b$ onto F_b.

The following must be satisfied:

(1) Each F_b is an n-dimensional real vector space with s and a as scalar product and vector addition, respectively.

(2) (Local triviality.) For each b in B, there is a neighborhood U of b and a homeomorphism $\varphi : U \times \mathbb{R}^n \to \pi^{-1}(U)$ such that φ is a vector space isomorphism of $b' \times \mathbb{R}^n$ onto $F_{b'}$ for each b' in U.

If in (2) the neighborhood U may be taken as all of B, the bundle is said to be the *trivial bundle*.

If ξ, η are n-dimensional and p-dimensional vector space bundles, respectively, we define the *product bundle* $\xi \times \eta$ as follows:

$$\begin{aligned} E(\xi \times \eta) &= E(\xi) \times E(\eta) \\ B(\xi \times \eta) &= B(\xi) \times B(\eta) \\ (\pi \times \lambda)(x,y) &= (\pi(x), \lambda(y)) \end{aligned}$$

where π, λ are the projections in ξ, η respectively and $F_b(\xi \times \eta)$ has the usual product structure for vector spaces.

If U is a subset of $B(\xi)$, then $\xi|_U$ denotes the bundle $\pi : \pi^{-1}(U) \to U$. It is called the ***restriction*** of the bundle ξ to U.

Definition 2.2. Let M^n be a differentiable manifold and let x_0 be in M. A ***tangent vector at*** x_0 is an operation X which assigns to each differentiable function f defined in a neighborhood of x_0, a real number. The following conditions must be satisfied:

(1) If g is the restriction of f, $X(g) = X(f)$.
(2) $X(cf + dg) = cX(f) + dX(g)$ (where c, d are real numbers).
(3) $X(f \cdot g) = X(f) \cdot g(x_0) + f(x_0) \cdot X(g)$, where the dot means ordinary real multiplication.

For the constant function 1, we have $X(1) = X(1 \cdot 1) = X(1) + X(1)$, by (3); hence $X(1) = 0$. It follows by (2) that $X(c) = 0$ for any constant function c.

If one thinks of a tangent vector as being the velocity vector of a parameterized curve lying in the manifold, then $X(f)$ is merely the derivative of f with respect to the parameter of the curve. This is made more precise below.

LEMMA 2.3. *Let $(u^1, \ldots, u^n)$ be a coordinate system about x. Let X be a tangent vector at x. Then X may be written uniquely as a linear combination of the operators $\partial/\partial u^i$ evaluated at x,*

$$X = \sum \alpha^i \frac{\partial}{\partial u^i}.$$

Proof. We assume $u(x)$ is the origin. Given any differentiable $f(u^1, \ldots, u^n)$ define

$$g(u^1, \ldots, u^n) = \begin{cases} (f(u^1, \ldots, u^n) - f(0, u^2, \ldots, u^n))/u^1, & \text{if } u^1 \neq 0 \\ \partial f(u^1, \ldots, u^n)/\partial u^1, & \text{if } u^1 = 0. \end{cases}$$

To see that g is differentiable, note that

$$g(s, u^2, \ldots, u^n) = \int_0^1 \frac{\partial f}{\partial u^1}(st, u^2, \ldots, u^n)\, dt.$$

(Then $f(u^1, \ldots, u^n) = u^1 g_1(u^1, \ldots, u^n) + f(0, u^2, \ldots, u^n)$.) Similarly,

$$f(0, u^2, \ldots, u^n) = u^2 g_2(u^2, \ldots, u^n) + f(0, 0, u^3, \ldots, u^n),$$

where $g_2(0) = \partial f/\partial u^2(0)$. Finally, we have

$$f(u^1, \ldots, u^n) = \sum u^i g_i + f(0), \quad \text{where} \quad g_i(0) = \frac{\partial f}{\partial u^i}(0).$$

Thus

$$\begin{aligned} X(f) &= \sum X(u^i) g_i(0) + 0 \cdot X(g_i) \\ &= \sum \alpha^i \frac{\partial f}{\partial u^i}(0), \quad \text{where} \quad \alpha^i = X(u^i). \end{aligned}$$ □

Remark. The α^i are called the ***components*** of the vector X with respect to the coordinate system $(u^1, \ldots, u^n)$. If $(v^1, \ldots, v^n)$ is another coordinate system about x, and $X = \sum \beta^j \partial/\partial v^j$, then $\alpha^i = X(u^i) = \sum \beta^j \partial u^i/\partial v^j$.

Definition 2.4. (Alternate). A tangent vector at x is an assignment to every coordinate system $(u^1, \ldots, u^n)$ about x of an element $(\alpha^1, \ldots, \alpha^n)$ of $\mathbb{R}^n$, with the requirement that if (β^j) is assigned to the system $(v^1, \ldots, v^n)$, then $\alpha^i = \sum \beta^j \partial u^i/\partial v^j$. The derivation operator X is then defined as $\sum \alpha^i \, \partial/\partial u^i$. One checks readily that

- **(a)** $X(f)$ is independent of the coordinate system used, and
- **(b)** $X(f)$ satisfies requirements (1), (2), and (3) for a tangent vector.

Definition 2.5. The *tangent bundle* τ of an n-dimensional differentiable manifold M is constructed as follows. For each x in M, the tangent vectors at x form an n-dimensional vector space. (The operations $\partial/\partial u^i$ form a basis, by Lemma 2.3.) Let the disjoint union of these be denoted $E(\tau)$; and define $\pi : E(\tau) \to M$ as mapping the tangent vector X at x_0 into x_0. The local product structure is given by $\varphi_u : U \times \mathbb{R}^n \to E$, where $(U, h) = (u^1, \ldots, u^n)$ is a coordinate system on M, and where $\varphi_u(x_0, a^1, \ldots a^n)$ is defined to be the tangent vector

$$X = \sum a^i \frac{\partial}{\partial u^i}$$

at the point x_0. One checks immediately that φ_u gives us a vector space isomorphism for each fibre. Since φ_u is to be a homeomorphism, this structure imposes a topology on E; and since each $\varphi_v^{-1}\varphi_u$ is a homeomorphism on $(U \cap V) \times \mathbb{R}^n$ this topology is unambiguously determined.

Indeed, $\varphi_v^{-1}\varphi_u$ is a C^∞ map on $(U \cap V) \times \mathbb{R}^n$, so that E is a differentiable manifold of dimension $2n$ (using Definition 1.2 of a differentiable manifold). The map π is differentiable of rank n.

Definition 2.6. If $f : M_1 \to M_2$, there is an induced map $df : E(\tau_1) \to E(\tau_2)$ defined as follows: $df(X) = Y$, where $Y(g) = X(gf)$. If X is a vector at x_0, Y is a vector at $f(x_0)$. Clearly df is linear on each fibre; it is called the *derivative map*.

If (U, h) and (V, k) are coordinate systems about x_0 and $f(x_0)$ respectively, and if (α^i), (β^j) are the respective components of X and $Y = df(X)$ with respect to these coordinate systems, then $(\beta^j) = D(kfh^{-1})\cdot(\alpha^i)$ where the vector components are written as column matrices, as usual.

Definition 2.7. Let ξ, η be two n-dimensional vector space bundles. A *bundle map* $f : \xi \to \eta$ is a continuous map of $E(\xi)$ into $E(\eta)$ which carries each fibre isomorphically onto a fibre. The induced map $f_B : B(\xi) \to B(\eta)$ is automatically continuous.

If $B(\xi) = B(\eta)$ and the induced map is the identity, f is said to be an *equivalence*. Note that if f is an equivalence, it is a homeomorphism: Locally f is just a map $U \times \mathbb{R}^n \to V \times \mathbb{R}^n$, and can be described by an expression of the form

$$\big(x, (\alpha^1, \ldots \alpha^n)\big) \mapsto \big(x, (\beta^1, \ldots \beta^n)\big) \qquad \text{where} \qquad \beta^i = \sum_j A^i_j(x)\, \alpha^j$$

and where $x \mapsto \big(A^i_j(x)\big)$ is a continuous function from $U \cap V$ to the group of non-singular $n \times n$ matrices. Since the operation of matrix inversion is continuous, it follows easily that f^{-1} is also continuous.

If there is an equivalence of ξ onto η, we write $\xi \simeq \eta$.

LEMMA 2.8. *Given a bundle η with projection map $\lambda : E(\eta) \to B(\eta)$, and a map $f : B_1 \to B(\eta)$, there is a bundle $\pi : E_1 \to B_1$ and a bundle map $g : E_1 \to E(\eta)$ such that $\lambda g = f\pi$. Furthermore, E_1 is unique up to an equivalence.*

This E_1 is called the *induced bundle* and is often denoted by $f^*\eta$.

Proof of Lemma 2.8. Let E_1 be the subset of $B_1 \times E(\eta)$ consisting of points (b, e) such that $f(b) = \lambda(e)$. Define $\pi(b, e) = b$ and $g(b, e) = e$. The map g is an isomorphism on each fibre. To show that E_1 is a vector space bundle, let $\varphi : V \times \mathbb{R}^n \to E(\eta)$ be a product neighborhood in $E(\eta)$, and let $f(U) \subset V$. Then define $\varphi_1 : U \times \mathbb{R}^n \to E_1$ by $\varphi_1(b, x) = (b, \varphi(f(b), x))$. This is continuous and injective; and its image equals $\pi^{-1}(U)$. Its inverse carries (b, e) into $(b, p\varphi^{-1}(e))$ (where p projects $V \times \mathbb{R}^n$ onto $\mathbb{R}^n$), so that it is continuous.

Now suppose $g' : E' \to E(\eta)$ is another bundle map, where $\pi' : E' \to B_1$ is a bundle and $\lambda g' = f\pi'$. We map $E' \to E_1$ by the formula

$$e' \mapsto \big(\pi'(e'), g'(e')\big) \in E_1 . \tag{2}$$

Because g' is an isomorphism on each fibre, this map (2) is also. It is an equivalence, since it induces the identity on the base space. □

Definition 2.9. Let ξ, η be two bundles over B. The *Whitney sum* $\xi \oplus \eta$ is a bundle defined as follows: Consider the product bundle $E(\xi) \times E(\eta) \to B \times B$; let d be the diagonal map $B \to B \times B$. The induced bundle $d^*(\xi \times \eta)$ is defined as the Whitney sum $\xi \oplus \eta$.

Note that the fibre over b in $\xi \oplus \eta$ is merely $F_b(\xi) \times F_b(\eta)$, so that

$$\dim(\xi \oplus \eta) = \dim \xi + \dim \eta.$$

Note also the commutativity and associativity of $\oplus$. I.e., $\xi \oplus \eta \simeq \eta \oplus \xi$ and $(\xi \oplus \eta) \oplus \zeta \simeq \xi \oplus (\eta \oplus \zeta)$. The proof is left as an exercise.

Definition 2.10. If ξ, η are bundles over B, then $g : E(\xi) \to E(\eta)$ is a *homomorphism* if

(1) it maps each fibre linearly into a fibre, and
(2) the induced map on B is the identity.

Note that an equivalence is both a bundle map and a homomorphism. An *embedding* of bundles is a injective homomorphism.

THEOREM 2.11. *If $f : E(\xi) \to E(\eta)$ maps each fibre linearly into a fibre, then f may be factored into a homomorphism followed by a bundle map.*

Proof. Let π_1, π_2 be the projections in ξ, η, respectively.

Let $f_B : B(\xi) \to B(\eta)$ be the map induced by f. Let $E_1 = f_B^*\eta$ be the bundle induced by f_B; let g be the bundle map $E_1 - E(\eta)$ and π the projection $E_1 \to B(\xi)$.

Define $h : E(\xi) \to B(\xi) \times E(\eta)$ by the equation $h(e) = (\pi_1(e), f(e))$. The image of h actually lies in that subset of $B(\xi) \times E(\eta)$ which is E_1; then h is a homomorphism. From the definition, $f = gh$.

$$\begin{array}{ccccc} E(\xi) & \xrightarrow{h} & E_1 & \xrightarrow{g} & E(\eta) \\ \downarrow{\scriptstyle \pi_1} & & \downarrow{\scriptstyle \pi} & & \downarrow{\scriptstyle \pi_2} \\ B(\xi) & \xrightarrow[i]{} & B(\xi) & \xrightarrow[f_B]{} & E(\eta) \end{array}$$

□

LEMMA 2.12. *Let ξ, η be bundles over B of dimensions n, p, respectively; let $g : \xi \to \eta$ be a homomorphism. If g is onto, then the* **kernel** *of g is a well defined vector bundle. If g is injective, then the* **cokernel** *of g, i.e., the quotient $\eta/\mathrm{image}\, g$ is a well defined vector bundle.*

Proof. Suppose g is injective (i.e., has rank n when restricted to each fibre). In $E(\eta)$, we define $e \sim e'$ if $e - e'$ exists and is in the image of g. We identify the elements of these equivalence classes.; the resulting identification space is defined to be $E(\eta/g(\xi))$. It is a bundle over B with projection naturally defined and each fibre is a vector space of dim $p - n$. We need only to show the existence of a local product structure.

Let U be an open set in B, with $\xi|_U$ equivalent to $U \times \mathbb{R}^n$ and $\eta|_U$ equivalent to $U \times \mathbb{R}^p$. Let g_0 denote the homomorphism of $U \times \mathbb{R}^n \to U \times \mathbb{R}^p$ induced by g. Now $(\eta/g(\xi))|_U$ is equivalent to the quotient $U \times \mathbb{R}^p/g_0(U \times \mathbb{R}^n)$, so that it suffices to show that this latter quotient is locally a product.

g_0 is given by a matrix $M(b) \in \mathcal{M}(p, n)$ which depends continuously on the point $b \in U$. Given b_0, we may assume that in a neighborhood U_0 of b_0, the first n rows are independent. We define $h : U_0 \times \mathbb{R}^n \times \mathbb{R}^{p-n} \to U_0 \times \mathbb{R}^p$ as the linear function on $\mathbb{R}^p$ whose matrix (non-singular) is

$$\left(M(b) \left| \begin{array}{c} 0 \\ \hline I_{p-n} \end{array} \right. \right)$$

The image of $U_0 \times \mathbb{R}^n \times 0$ under h is just $g_0(U_0 \times \mathbb{R}^n)$; since h is an equivalence, it induces an equivalence of

$$U_0 \times \mathbb{R}^{p-n} \simeq \frac{U_0 \times \mathbb{R}^n \times \mathbb{R}^{p-n}}{U_0 \times \mathbb{R}^n \times 0} \quad \text{onto} \quad \frac{U_0 \times \mathbb{R}^p}{g_0(U_0 \times \mathbb{R}^n)}.$$

Secondly, suppose g is onto (i.e., it has rank p on each fibre). $E(g^{-1}(0))$ is defined as that subset of $E(\xi)$ consisting of points e with $g(e) = 0$. Again, we need to show the existence of a local product structure. Let U, g_0, and $M(b)$ be as above. Given b_0, we may assume that the first p columns of $M(b)$ are independent in the neighborhood U_0 of b_0. We define $h : U_0 \times \mathbb{R}^n \to U_0 \times \mathbb{R}^p \times \mathbb{R}^{n-p}$ by the matrix function

$$\left(\begin{array}{c} M(b) \\ \hline 0 \;\; | \;\; I_{n-p} \end{array} \right)$$

Now h followed by the natural projection of $U_0 \times \mathbb{R}^p \times \mathbb{R}^{n-p}$ onto $U_0 \times \mathbb{R}^p$ equals $g_0|_{U_0}$. Hence h^{-1} maps $U_0 \times 0 \times \mathbb{R}^{n-p}$ onto $g_0^{-1}(U_0 \times 0)$; since h is an equivalence, the restriction of h^{-1} to $U_0 \times 0 \times \mathbb{R}^{n-p}$ is also. □

Remark. If g is onto, $\xi/g^{-1}(0)$ is a bundle, being the quotient of the inclusion homomorphism $g^{-1}(0) \to \xi$. If g is injective, $g(\xi)$ is a bundle, being the kernel of the projection homomorphism $\eta \to \eta/\,g(\xi)$.

Definition 2.13. If φ is a nonnegative function on B, the ***support*** of φ is the closure of the set of x with $\varphi(x) > 0$. A ***partition of unity*** is a collection φ_α of continuous non-negative functions on B, such that the sets C_α = support φ_α form a locally-finite covering of B, and $\sum \varphi_\alpha(x) = 1$ (this is a finite sum for each x).

LEMMA 2.14. *Let B be a normal space; U_α a locally-finite open covering of B. Then there is a partition of unity φ_α with support $\varphi_\alpha \subset U_\alpha$ for each α.*

Proof. First, we show that there is an open covering V_α of B with $\overline{V}_\alpha \subset U_\alpha$ for each α. Assume the U_α indexed by a set of ordinals (well-ordering theorem). Let V_α be defined for all $\alpha < \beta$ and assume that the sets V_α along with the sets U_α for $\alpha \geq \beta$ cover B. Consider the set $A(B) = B - \bigcup_{\alpha<\beta} V_\alpha - \bigcup_{\alpha>\beta} U_\alpha$. Then $A(\beta) \subset U_\beta$. Let V_β be an open set containing the closed set $A(\beta)$ with $\overline{V}_\beta \subset U_\beta$ (normality). This completes the construction of the V_α.

Now let g_α be a function which is positive on $\overline{V}_\alpha$ and 0 outside U_α (normality again). Define $\varphi_{\alpha_0}(x) = g_{\alpha_0}(x)/\sum g_{\alpha(x)}$. Since U_α is locally-finite, the sum in the denominator is finite and positive, so φ_α is well-defined. □

Remark. If B is a differentiable manifold, φ_α may be chosen to be differentiable: For fixed α, we can cover B with coordinate systems (V_i, h_i) as in Theorem 1.25 refining the covering $\{U_\alpha,\, B - \overline{V}_\alpha\}$. As in Lemma 1.31, let $\varphi_i(y) = \varphi(h_i(y))$ for $y \in V_i$, and $= 0$ otherwise (with φ as in Lemma 1.26). Let $g_\alpha(y) = \sum \varphi_i(y)$, where the sum extends over all i such that $V_i \subset U_\alpha$, and then proceed as above.

LEMMA 2.15. *Let B be paracompact and let $0 \to \xi \xrightarrow{i} \eta \xrightarrow{\varphi} \zeta \to 0$ be an exact sequence of homomorphisms of bundles. Then there is an equivalence $f : \eta \to \xi \oplus \zeta$, with fi the natural inclusion and φf^{-1} the natural projection.*

Proof. Let $\dim \xi = n$; $\dim \zeta = p$. We first construct a Riemannian metric on η (i.e., a continuous inner product in $E(\eta)$). Let U_α be a locally finite covering of B with $\eta|_{U_\alpha}$ trivial; let g_α be the corresponding projection of $\eta|_{U_\alpha}$ onto $\mathbb{R}^{n+p}$. Let φ_α be a partition of unity with support $\varphi_\alpha \subset U_\alpha$.

If e, e' are in $E(\eta)$ and $\pi(e) = \pi(e')$, define $e \cdot e' = \sum_\alpha \varphi_\alpha(\pi(e))\, g_\alpha(e) \cdot g_\alpha(e')$, where the dot on the right hand side is the ordinary scalar product in $\mathbb{R}^{n+p}$. This is a finite sum; it satisfies the axioms for a scalar product.

The way we use the Riemannian metric is to break η up into $iE(\xi)$ and its orthogonal complement. Let ξ' be the image of ξ in η and let $E(\zeta')$ be defined as that subset of $E(\eta)$ consisting of elements which are orthogonal to $i(E(\xi))$. In order to show that ζ' has a local product structure, consider the homomorphism

$$h : \eta \to \xi'$$

which sends each vector into its orthogonal projection in ξ'. [Verification that h is continuous. Over any coordinate neighborhood U we can choose a basis $a_1, \ldots, a_n$ for the fibre of ξ'. Then the function h carries $v \in E(\eta)$ into $\sum t_j a_j \in E(\xi') \subset E(\eta)$, where $t_j = \sum B_{jk}(v \cdot a_k)$ and where (B_{jk}) denotes the inverse matrix to $(a_j \cdot a_k)$.] Since h is onto, its kernel ζ' is again a vector space bundle.

Now the bundle $i(\xi) = \xi'$ is equivalent to ξ. It remains to show that ζ' is equivalent to ζ and that η is equivalent to $\xi' \oplus \zeta'$. The former follows immediately from the fact that $\varphi|_{\zeta'}$ is a homomorphism; from rank considerations it must be injective and onto as well. The latter follows by noting that $E(\xi' \oplus \zeta')$ is defined as the subset of $E(\xi') \times E(\zeta')$ consisting of points (e_1, e_2) such that $\pi(e_1) = \pi(e_2)$. Consider the map f of $E(\xi' \oplus \zeta')$ into $E(\eta)$ obtained by taking (e_1, e_2) into their sum in $E(\eta)$ (this sum exists because e_1 and e_2 lie in the same fibre). This is clearly a homomorphism; from rank considerations, it must be injective and onto. □

Definition 2.16. Let M_1, M_2 be differentiable manifolds, and let $f : M_1 \to M_2$ be an immersion. The ***normal bundle*** ν_f is defined as follows. Let τ_1, τ_2 be the tangent bundles of M_1, M_2 respectively. By Theorem 2.11, the map $df : E(\tau_1) \to E(\tau_2)$ may be factored into a homomorphism h of $E(\tau_1)$ into $E(f^*\tau_2)$ followed by a bundle map g. Now h is an injective homomorphism because f is an immersion; hence by Lemma 2.12, $f^*\tau_2/\text{image } h$ is a bundle over M_1. It is called the normal bundle ν_f.

Then $0 \to \tau_1 \to f^*\tau_2 \to \nu_f \to 0$ is an exact sequence of homomorphisms, so that by Lemma 2.15, $f^*\tau_2$ is equivalent to $\tau_1 \oplus \nu_f$. Indeed, given a Riemannian metric on $f^*\tau_2$, ν_f is equivalent to the orthogonal complement of the image of τ_1.

Let us consider the case $M_2 = \mathbb{R}^{n+p}$, where $\dim M_1 = n$. Then τ_2 is the trivial bundle, so that $f^*\tau_2$ is as well. (Proof: If $f : B \to B(\eta)$ and η is trivial, so is $f^*\eta$. We have the diagram

$$\begin{array}{ccc} & & B \times \mathbb{R}^n \\ & & \downarrow \pi \\ f : B_1 & \longrightarrow & B. \end{array}$$

$E(f^*\eta)$ is defined as that subset of $B_1 \times (B \times \mathbb{R}^n)$ consisting of points (b_1, b, x) such that $f(b_1) = \pi(b, x)$; i.e., of all points $(b_1, f(b_1), x)$. If we map this into (b_1, x), we obtain an equivalence of $f^*\eta$ with the bundle $B_1 \times \mathbb{R}^n \to B_1$.)

Thus $\tau_1 \oplus \nu_f$ is equivalent to a trivial bundle. In what follows, we investigate the following question: Given ξ, does there exist an η with $\xi \oplus \eta$ trivial? Using Theorem 1.28, this is always the case for ξ the tangent bundle of an n-manifold, and indeed η may be chosen also to have dimension n. A more general answer appears in Lemma 2.19.

Definition 2.17. Let $f : M_1 \to M_2$; let $\dim M_1 = n$, $\dim M_2 = p$. If f has rank p at every point of M_1, it is said to be ***regular***. If f is regular, the homomorphism $h : \tau_1 \to f^*\tau_2$ given by Theorem 2.11 is an onto map. By Lemma 2.12, the kernel of h is a bundle α_f. It is called the ***bundle along the fibre***.

Note that $f^{-1}(y)$ is a submanifold of M_1 of dimension $n-p$ (by Exercise 1.12 or Lemma 1.35). The inclusion i_y of $f^{-1}(y)$ into M_1 induces an inclusion di_y of its tangent bundle into τ_1. The kernel of h consists precisely of the vectors which are in the image of some di_y, i.e., the vectors tangent to the submanifolds $f^{-1}(y)$ are the ones carried into 0 by h.

One has the exact sequence $0 \to \alpha_f \to \tau_1 \xrightarrow{h} f^*\tau_2 \to 0$, so that by Lemma 2.15, τ_1 is equivalent to $\alpha_f \oplus f^*\tau_2$.

Definition 2.18. A bundle ξ is of *finite type* if B is normal and may be covered by a finite number of neighborhoods $U_1, \ldots, U_k$ such that $\xi|_{U_i}$ is trivial for each i.

LEMMA 2.19. *ξ is of finite type if B is compact, or paracompact finite dimensional.*

The former statement is clear; let us consider the latter. By definition, the dimension of B satisfies $\dim(B) \leq n$ if every open covering has an open refinement such that

> $(*)$ no point of B is contained in more than $n+1$ elements of the refinement.

It is a standard theorem of topology that an n-manifold has dimension n in this sense.

Cover B by open sets U, with $\xi|_U$ trivial; let $\{V_\alpha\}$ be an open refinement of this covering satisfying $(*)$. Since B is paracompact (Definition 1.22), we may assume that $\{V_\alpha\}$ is locally-finite as well. Let φ_α be a partition of unity with support $\varphi_\alpha \subset V_\alpha$ for each α (Lemma 2.14).

Let A_i be the set of unordered $(i+1)$-tuples of distinct elements of the index set $\{\varphi_\alpha\}$. Given a in A_i, where $a = \{\alpha_0, \ldots, \alpha_i\}$, let W_{ia} be the set of all x such that $\varphi_\alpha(x) < \min[\varphi_{\alpha_0}(x), \ldots, \varphi_{\alpha_i}(x)]$ for all $\alpha \neq \alpha_0, \ldots, \alpha_i$. Each set W_{ia} is open, and W_{ia} and W_{ib} are disjoint if $a \neq b$. Also W_{ia} is contained in the intersection of the supports of $\varphi_{\alpha_0}, \ldots, \varphi_{\alpha_i}$, and hence in some set V_α. If we set X_i equal to the union of all sets W_{ia}, for fixed i, the result is that $\xi|_{X_i}$ is trivial. (For $\xi|_{W_{ia}}$ is trivial, and the W_{ia} are disjoint.)

Finally, the sets $X_0, \ldots, X_n$ cover B. Given x in B, x is contained in at most $n+1$ of the sets V_α, so that at most $n+1$ of the functions φ_α are positive at x. Since some φ_α is positive at x, x is contained in one of the sets W_{ia} for $0 \leq i \leq n$.

[The intuitive idea of the proof is as follows: Consider an n-dimensional simplicial complex, with φ_α the barycentric coordinate of x with respect to the vertex α. The sets W_{0a} will be disjoint neighborhoods of the vertices, the sets W_{1a} disjoint neighborhoods of the open 1-simplices, and so on.]

THEOREM 2.20. *If ξ is of finite type, there is a bundle η such that $\xi \oplus \eta$ is trivial.*

Proof. We proceed by showing that ξ may be embedded in a trivial bundle $B \times \mathbb{R}^m$, so that we have the exact sequence $0 \to \xi \xrightarrow{i} B \times \mathbb{R}^m \to B \times \mathbb{R}^m / i(\xi) \to 0$ by Lemma 2.12. The theorem then follows from Lemma 2.15. (Paracompactness is not needed since the trivial bundle clearly has a Riemannian metric.)

Cover B by finitely many neighborhoods $U_1, \ldots, U_k$ with $\xi|_{U_i}$ trivial for each i. Let $\varphi_1, \ldots, \varphi_k$ be a partition of unity with support $\varphi_i \subset U_i$ for each i (Lemma 2.14). Let f_i denote the equivalence of $E(\xi|_{U_i})$ onto $U_i \times \mathbb{R}^n$; let $f_i^1, \ldots, f_i^n$ denote the coordinate functions of its projection into $\mathbb{R}^n$.

We define $h : E(\xi) \to B \times \mathbb{R}^{nk}$ as follows:

$$h(e) = \Big(\pi(e),\ \varphi_i(\pi(e)) \cdot f_i^j(e)\Big) \qquad \begin{matrix} i = 1, \ldots, k \\ j = 1, \ldots, n \end{matrix}$$

(no summation is indicated). This is well-defined, since $\varphi_i(\pi(e)) = 0$ unless $e \in E(\xi|_{U_i})$. It is clearly a homomorphism, since each f_i^j is linear on $E(\xi|_{U_i})$. To show that it is injective, let $e \neq 0$. Then for some i, $\varphi_i(\pi(e)) > 0$. Since f_i is an equivalence, $f_i^j(e) \neq 0$ for some j. Hence $h(e) \neq (\pi(e), 0)$, as desired. □

Definition 2.21. The bundle ξ is *s-equivalent* to η if there are trivial bundles o^p, o^n such that $\xi \oplus o^p \simeq \eta \oplus o^n$.

Here $o^p = B \times \mathbb{R}^p$. Symmetry and reflexivity are clear. To show transitivity, assume $\xi \oplus o^p \simeq \eta \oplus o^q$ and $\eta \oplus o^r \simeq \zeta \oplus o^s$. Then $\xi \oplus o^p \oplus o^r \simeq \zeta \oplus o^s \oplus o^q$.

Note that s-equivalence differs from equivalence. E.g., consider the two-sphere S^2 in $\mathbb{R}^3$. Then $\tau^2 \oplus \nu^1 = o^3$. The normal bundle ν^1 is easily seen to be trivial; but it is a classical theorem of topology that τ^2 is not (it does not admit a non-zero cross-section). Hence τ^2 is s-trivial, but not trivial.

THEOREM 2.22. *The set of s-equivalence classes of vector space bundles of finite type over B forms an abelian group*[1] *under* $\oplus$.

Proof. To avoid logical difficulties, we consider only subbundles of $B \times \mathbb{R}^m$, for all m. This suffices, since any bundle ξ of finite type may be embedded in some $B \times \mathbb{R}^m$, by Theorem 2.20. The class of trivial bundles o^p is the identity element, and the existence of inverses is the substance of Theorem 2.20. □

COROLLARY 2.23. *Given two immersions of the differentiable manifold M in euclidean space, their normal bundles are s-equivalent.*

Definition 2.24. M^n is a *π-manifold* if M may be immersed in some $\mathbb{R}^{n+p}$ so that its normal bundle is trivial.

This is equivalent to the requirement that τ^n be s-trivial: Let τ^n be s-trivial. If we take some immersion of M into $\mathbb{R}^{n+p}$, then $\tau^n \oplus \nu^p$ is trivial by Definition 2.16, so that ν^p is s-trivial, i.e., $\nu^p \oplus o^q = o^{p+q}$ for some q. Consider the composite immersion $M \to \mathbb{R}^{n+p} \subset \mathbb{R}^{n+p+q}$. The normal bundle of M in $\mathbb{R}^{n+p+q}$ is just $\nu^p \oplus o^q$, which is trivial.

Conversely, if ν^p is trivial for some immersion, then τ^n is s-trivial because $\tau^n \oplus \nu^p$ is trivial.

Definition 2.25. Let $G_{p,n}$ denote the set of all n-dimensional vector subspaces of $\mathbb{R}^{p+n}$ (i.e., all n-dim hyperplanes through the origin). It is called the *Grassmann manifold* of n-planes in $n + p$ space.

[1](Added 2006.) In the language of K-theory, which was introduced into topology shortly afterwards by Atiyah and Hirzebruch, the group of s-equivalence classes of vector bundles over B is denoted by $\widetilde{K}_{\mathbb{R}}(B)$ (or $\widetilde{KO}(B)$), where $K_{\mathbb{R}}(B) \cong \mathbb{Z} \oplus \widetilde{K}_{\mathbb{R}}(B)$ is the Grothendieck ring of virtual real vector bundles over the compact connected space B, and $\mathbb{Z}$ is the subring generated by the trivial line bundle. See for example D. Husemoller, "Fibre Bundles", McGraw-Hill 1966.

Its topology is obtained as follows: Consider $\mathcal{M}(n, n+p; n)$ (Definition 1.18). We identify two elements of this set if the hyperplanes spanned by their row vectors are the same. $G_{p,n}$ is in injective correspondence with this identification space, and is given the identification topology. If ρ is the projection

$$\rho : \mathcal{M}(n,\, n+p;\, n) \;\to\; G_{p,\,n}\,,$$

then $\rho(A) = \rho(B)$ if and only if $A = CB$ for some non-singular $n \times n$ matrix C. The hyperplane $\rho(A)$ consists of all points $(x^1, \ldots, x^{n+p}) \in \mathbb{R}^{n+p}$ which equal $(c^1, \ldots, c^n) \cdot A$ for some choice of constants c^i. If $\rho(A) = \rho(B)$, then

$$\begin{aligned} (1,\, 0,\, \ldots,\, 0) \cdot A &= (c_1^1,\, \ldots,\, c_1^n) \cdot B \\ (0,\, 1,\, \ldots,\, 0) \cdot A &= (c_2^1,\, \ldots,\, c_2^n) \cdot B, \text{ etc., for some choice of } c_i^j. \end{aligned}$$

Then $IA = CB$, where C has rank n because A does. The converse is clear.

(a) $G_{p,n}$ is locally euclidean. Let $A \in \mathcal{M}(n,\, n+p;\, n)$; after permuting the columns, we may assume $A = (P, Q)$ where P is $n \times n$ and non-singular. Let U be the set of all such A; it is an open set in $\mathcal{M}(n,\, n+p;\, n)$, being the inverse image of the non-zero reals under the continuous map $(P,Q) \to \det P$. If $\rho(P,Q) = \rho(R,S)$, where P is non-singular, then $(P,Q) = (CR, CS)$ for some non-singular C. Hence R is necessarily non-singular; it follows that $\rho^{-1}(\rho(U)) = U$, so that $\rho(U)$ is open in $G_{p,n}$ (by definition of the identification topology).

We show $\rho(U)$ homeomorphic with $\mathbb{R}^{pn}$. Define $\varphi : U \to \mathbb{R}^{pn}$ by $\varphi(P,Q) = P^{-1}Q$. If $\rho(P,Q) = \rho(R,S)$ then $(P,Q) = (CR, CS)$, so that

$$P^{-1}Q = (CR)^{-1}(CS) = R^{-1}S.$$

Hence φ induces a continuous map $\varphi_0 : \rho(U) \to \mathbb{R}^{pn}$. Define $\psi : \mathbb{R}^{pn} \to \rho(U)$ by $\psi(Q) = \rho(I, Q)$ where Q is an $n \times p$ matrix. One checks immediately that ψ and φ_0 are inverses of each other.

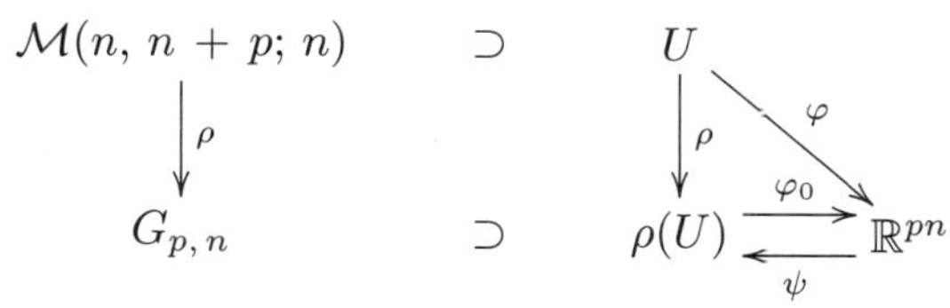

(b) To show that $G_{p,n}$ is Hausdorff, we show that ψ maps every compact set into a closed set (this will clearly suffice). Let K be a compact subset of $\mathbb{R}^{pn}$; we show $\varphi^{-1}(K)$ is closed in $\mathcal{M}(n, p+n; n)$. $\varphi^{-1}(K)$ consists of all matrices (P,Q) with P non-singular and $P^{-1}Q \in K$. Let $(P,Q) \in \mathcal{M}(n, n+p; n)$ be the limit of the sequence (P_i, Q_i) of elements of $\varphi^{-1}(K)$. Since K is compact some subsequence of the sequence $\varphi(P_i, Q_i) = P_i^{-1}Q_i$ converges to a point R of K. Then the corresponding subsequence of the sequence Q_i converges to PR, so that $(P,Q) = P(I,R)$. Since (P,Q) has rank n it follows that P is non-singular, so that $(P,Q) \in \varphi^{-1}(K)$, as desired.

Hence $G_{p,n}$ is a manifold of dimension pn.

(c) $G_{p,n}$ is a differentiable manifold and ρ is a differentiable map. A function f on the open set V in $G_{p,n}$ belongs to the differentiable structure $\mathcal{D}$ if $f\rho$ is differentiable. To show that this satisfies the conditions for a differentiable structure, we show that $(\rho(U), \varphi_0)$, as defined in (a), is a coordinate system. Let f be defined on $V \subset \rho(U)$. Given $Q \in \mathbb{R}^{pn}$, $f\varphi_0^{-1}(q) = f\rho(I, Q)$ so that $f\varphi_0^{-1}$ is differentiable if $f\rho$ is. Conversely, given $(P, Q) \in V$, $f\rho(P, Q) = f\varphi_0^{-1}\varphi_0\rho(P, Q) = f\varphi_0^{-1}(P^{-1}Q)$, so that $f\rho$ is differentiable if $f\varphi_0^{-1}$ is.

(d) $G_{p,n}$ is compact. Let L be the subset of $\mathcal{M}(n, n+p; n)$ consisting of matrices whose rows are orthonormal vectors. L is a closed and bounded subset of $\mathbb{R}^{n(n+p)}$. Since $\rho(L) = G_{p,n}$ (the Gram-Schmidt orthogonalization process proves this), $G_{p,n}$ is compact.

(e) $G_{p,n}$ is diffeomorphic to $G_{n\,p}$. Geometrically, the homeomorphism h is defined as carrying each hyperplane into its orthogonal complement. It is clearly injective; to show it is differentiable we use the coordinate system $(\rho(U), \varphi_0)$ defined in (a). Let g map U into $\mathcal{M}(n, n+p; p)$ by carrying (P, Q) into $(-(P^{-1}Q)^{\tau}, I_p)$; it is differentiable (τ denotes transpose). The row space of (P, Q) is the same as that of $(I_n, P^{-1}Q)$, while the row vectors of this matrix are orthogonal to those of $(-(P^{-1}Q)^{\tau}, I_p)$ (multiply the one by the transpose of the other). Hence g induces $h|_{\rho(U)}$, so that the latter is differentiable.

Definition 2.26. Let $E(\gamma_p^n)$ be defined as that subset of $G_{p,n} \times \mathbb{R}^{n+p}$ consisting of pairs (H, x) where x is a vector lying in the hyperplane H. It is called the ***universal bundle*** (for reasons we shall see). The projection π maps (H, x) into H; the fibre is thus an n-dimensional subspace of $\mathbb{R}^{n+p}$.

To show that γ_p^n is an n-dimensional vector space bundle over $G_{p,n}$, we need to show the existence of a local product structure. Let $(\rho(U), \varphi_0)$ be a coordinate neighborhood on $G_{p,n}$, as in (a) above. We define $h : \rho(U) \times \mathbb{R}^n \to \pi^{-1}\rho(U)$ as carrying $(H, (x^1, \ldots, x^n))$ into $(H, (x^1, \ldots, x^n) \cdot (I_n, Q))$ where $Q = \varphi_0(H)$. This is a vector in the hyperplane H; h is clearly an isomorphism on each fibre. Its inverse is continuous, since it sends $(H, (y^1, \ldots, y^n))$ in $G_{p,n} \times \mathbb{R}^{p+n}$ into $(H, (y^1, \ldots, y^n))$ in $\rho(U) \times \mathbb{R}^n$.

Definition 2.27. ξ is a ***differentiable vector space bundle*** if $E(\xi)$ and $B(\xi)$ are differentiable manifolds, and if the homeomorphisms

$$U \times \mathbb{R}^n \to \pi^{-1}(U)$$

which specify the local product structure can be chosen as diffeomorphisms.

It follows that $\pi : E \to B$ is differentiable of maximum rank. Note that B can be differentiably embedded in E by mapping b into the 0-vector of F_b. The normal bundle of this embedding is just ξ.

Examples of differentiable bundles include the tangent bundle of a manifold, the normal bundle of an immersed manifold, and the universal bundle γ_p^n above. In the latter case, $E(\gamma_p^n)$ is embedded differentiably in $G_{p,n} \times \mathbb{R}^{p+n}$.

THEOREM 2.28. *Let ξ^n be an n-dimensional vector space bundle over a normal base space. The following conditions are equivalent:*

(a) *ξ is of finite type.*

(b) *There is a bundle η^p such that $\xi^n \oplus \eta^p$ is trivial.*

(c) *There is a bundle map $\xi^n \to \gamma_p^n$ for some p. (Thus the terminology " universal bundle" for γ_p^n.)*

Proof. We have already shown that (a) implies (b) Theorem 2.20 ; the bundle η^p there constructed has dimension $n(k-1)$, where k is the number of elements in the covering $U_1, \ldots, U_k$ of $B(\xi) = B$ such that $\xi|_{U_i}$ is trivial.

(b) implies (c): Condition (b) means that ξ^n may be embedded in the trivial bundle $B(\xi) \times \mathbb{R}^{p+n}$; let f be this embedding. We wish to define g and g_B in the following diagram:

$$\begin{array}{ccc} E(\xi) & \xrightarrow{g} & E(\gamma_p^n) \\ \downarrow{\scriptstyle\pi} & & \downarrow \\ B(\xi) & \xrightarrow[g_B]{} & G_{p,n} \end{array}$$

Since f is an injective homomorphism, $f(F_b)$ is the cartesian product of b and an n-dimensional hyperplane H^n in $\mathbb{R}^{p+n}$; let $g_B(b)$ equal this hyperplane H^n. If $e \in F_b$, then $f(e) = (b, x)$, where x is a vector in the hyperplane H^n; let $g(e) = (H^n, x)$ in $G_{p,n} \times \mathbb{R}^{p+n}$. Then $g(e)$ actually lies in the subset of $G_{p,n} \times \mathbb{R}^{p+n}$ which constitutes $E(\gamma_p^n)$. From rank considerations, g is automatically an isomorphism on each fibre.

It remains to show that g is continuous. Locally, g just looks like a map

$$U \times \mathbb{R}^n \to G_{p,n} \times \mathbb{R}^{p+n}.$$

We factor it into a continuous map $h : U \times \mathbb{R}^n \to \mathcal{M}(n, n+p; n) \times \mathbb{R}^{p+n}$ followed by the projection $\rho \times 1$ into $G_{p,n} \times \mathbb{R}^{p+n}$. Locally, f looks like a map $U \times \mathbb{R}^n \to B \times \mathbb{R}^{n+p}$. Let $e_1, \ldots, e_n$ be a basis for $\mathbb{R}^n$; we define $h(b, x)$ as $(A, p_2 f(b, x))$. Here p_2 projects $B \times \mathbb{R}^{n+p}$ onto its second factor and A is the matrix having $p_2 f(b, e_1), \ldots, p_2 f(b, e_n)$ as its rows. Then h is continuous, and $(\rho \times 1)h$ equals g.

(Note: the converse assertion, (c) implies (b), can be proved by the same argument.)

(c) implies (a): Being compact, $G_{p,n}$ is covered by finitely many neighborhoods U_i with $\gamma_p^n|_{U_i}$ trivial. (In fact $(n+p)!/n!p!$ neighborhoods will suffice.) If f is a bundle map $\xi^n \to \gamma_p^n$ then the sets $f_B^{-1}(U_i) = V_i$ cover B, and $\xi|_{V_i}$ is equivalent to the bundle induced by $f_B : V_i \to G_{p,n}$ (the uniqueness part of Lemma 2.8). Then $\xi|_{V_i}$ is trivial (since it is induced from a trivial bundle). □

3. The Cobordism Theory of Thom.

Definition 3.1. An *n-manifold-with-boundary* Q is a Hausdorff space with a countable basis which is locally homeomorphic with $\mathbb{H}^n$ (the subset of $\mathbb{R}^n$ such that $x^1 \geq 0$). The *boundary* ∂Q is that subset of Q corresponding to $\mathbb{R}^{n-1}$ under the local homeomorphism ($\mathbb{R}^{n-1}$ being the subset of $\mathbb{R}^n$ with $x^1 = 0$). ∂Q is well-defined, since the image of an open set in $\mathbb{R}^n$ under a homeomorphism of it into

$\mathbb{R}^n$ must be open (Brouwer theorem on invariance of domain). It is clear that ∂Q is an $(n-1)$-manifold.

A ***differentiable structure*** $\mathcal{D}$ on Q is a collection of real-valued functions f defined on open subsets of Q such that

(1) every point of Q has an open neighborhood U and a homeomorphism h of U into an open subset of $\mathbb{H}^n$, such that f is in $\mathcal{D}$ if and only if fh^{-1} is differentiable. (f is defined on an open subset of U; fh^{-1} differentiable means that it may be extended to a neighborhood of $h(U)$ *in* $\mathbb{R}^n$ so as to be differentiable).

(2) If U_i are open sets contained in the domain of f and $U = \bigcup U_i$, then $f|_U \in \mathcal{D}$ if and only if $f|_{U_i} \in \mathcal{D}$ for each i.

As before, (U, h) is called a ***coordinate system*** on Q, and one can define differentiable structures alternatively by means of coordinate systems.

We impose an additional condition on $\mathcal{D}$ in Definition 3.2.

Definition 3.2. Let M_1, M_2 be compact differentiable n-manifolds. They are said to lie in the same ***cobordism class*** ($M_1 \sim M_2$) if there is a compact differentiable $n+1$ manifold-with-boundary Q such that ∂Q is diffeomorphic with the disjoint union of M_1 and M_2 (denoted by $M_1 + M_2$) .

Symmetry and reflexivity of this relation are clear. To show transitivity, we impose the additional condition on $\mathcal{D}$ that there is a neighborhood U of ∂Q in Q which is diffeomorphic with $\partial Q \times [0,1)$, the diffeomorphism being the identity on $\partial Q \times 0$. This is redundant, but we assume it to avoid proving it. Transitivity follows:

Let $M_1 + M_2$ be diffeomorphic with ∂Q_1 and $M_2 + M_3$ diffeomorphic with ∂Q_2; let h_1, h_2 be the diffeomorphisms. We form a new space Q_3 from $Q_1 \cup Q_2$ by identifying each point of $h_1(M_2)$ with its image under $h_2 h_1^{-1}$. There is then a homeomorphism of $M_2 \times (-1,1)$ into this space which equals h_1 when restricted to $M_2 \times 0$, and is a diffeomorphism of $M_2 \times [0,(-1)^i)$ into Q_i for $i = 1,2$. (It is derived from the postulated "collar neighborhoods" $\partial Q_i \times [0,1)$.) If this is taken to be a coordinate system on Q_3, then Q_3 becomes a differentiable manifold-with-boundary, and $M_1 + M_3$ is diffeomorphic with ∂Q_3, while Q_1 and Q_2 are diffeomorphic with subsets of Q_3.

Definition 3.3. As usual, there are logical difficulties involved in considering the collection of *all* manifolds. One way of avoiding them is to consider only manifolds-with-boundary embedded in some euclidean space $\mathbb{R}^p$: If Q_1 is a differentiable manifold-with-boundary and $Q_2 = \partial Q_1 \times [0,1)$, then the space Q_3 constructed in the preceding paragraph is a differentiable manifold, so that it may be embedded in some euclidean space. Hence Q_1 may so be embedded.

With these restrictions, the set of cobordism classes of n-manifolds forms an abelian group (denoted by $\mathcal{N}_n$) under the operation + (disjoint union). If $M_1 \sim M_1'$ and $M_2 \sim M_2'$, this means that $M_i + M_i'$ is diffeomorphic with ∂Q_i. Then $(M_1 + M_2) + (M_1' + M_2')$ is diffeomorphic with $\partial(Q_1 \cup Q_2)$, so that $M_1 + M_2 \sim M_1' + M_2'$ and the operation + is well-defined on cobordism classes. The zero element is the vacuous manifold or the n-sphere (or ∂Q, where Q is any compact differentiable $(n + 1)$-manifold-with-boundary). The remaining axioms are clear. Note that $M + M$ is diffeomorphic with $\partial(M \times [0,1])$, so that every element is of order 2.

The groups $\mathcal{N}_n$ are called the (non-orientable) ***cobordism groups***. Let $\mathcal{N}$ denote the direct sum $\mathcal{N}_0 \oplus \mathcal{N}_1 \oplus \mathcal{N}_2 \oplus \cdots$. There is a bilinear symmetric pairing of $\mathcal{N}_i$, $\mathcal{N}_j$ into $\mathcal{N}_{i+j}$, i.e., a homomorphism of $\mathcal{N}_i \oplus \mathcal{N}_j$ intc $\mathcal{N}_{i+j}$ induced by the operation of cartesian product.

First, $(M_1 + M_2) \times M_3 = (M_1 \times M_3) + (M_2 \times M_3)$ by definition of cartesian product. Second, if $M_1 \sim 0$, i.e., $M_1 = \partial Q$, then $M_1 \times M_2$ is diffeomorphic with $\partial(Q \times M_2)$, so that $M_1 \times M_2 \sim 0$.

Since $M_1 \times M_2 \sim M_2 \times M_1$, and since $M_1 \times p \sim M_1$ (where p is a point-manifold), this pairing makes $\mathcal{N}$ into a (graded) commutative ring with unit. Indeed, it is a graded algebra over the field $\mathbb{Z}/2$.

Remark 3.4. The general result of Thom is the following

Theorem. *$\mathcal{N}$ is a polynomial algebra over $\mathbb{Z}/2$ with one generator in each positive dimension except those of the form $2^m - 1$. If n is even, the real projective n-space is a generator.*

This theorem means that there are compact manifolds M^2, M^4, M^5, ... such that every compact manifold is in the cobordism class of a disjoint union of products of these manifolds, and that there are no relations among the generators (except commutativity and associativity of products).

Thom's procedure is to show that $\mathcal{N}_n$ is isomorphic with the $(n+k)^{\text{th}}$ homotopy group of a certain space T_k, and then to compute these homotopy groups. We shall consider only the first of these two problems in the present notes.

Definition 3.5. Let h be an embedding of the differentiable manifold M^n in $\mathbb{R}^{n+k}$; consider the normal bundle of this embedding. Using the standard Riemannian metric for the tangent bundle to $\mathbb{R}^{n+k}$, this normal bundle is equivalent to the orthogonal complement of the image in the tangent bundle of $\mathbb{R}^{n+k}$ of the tangent bundle of M^n (Definition 2.16); this complement we denote by ν^k. Define e as the canonical map of $E(\nu^k)$ into $\mathbb{R}^{n+k}$ which maps the vector v normal to M^n at x into its end point. (Described differently, one maps the tangent bundle to $\mathbb{R}^{n+k}$ into itself canonically by mapping the vector v, based at x, into the point $x + v$ of $\mathbb{R}^{n+k}$. This map is differentiable; its restriction to $E(\nu^k)$ is the map e.)

Consider M^n as the set of zero vectors of $E(\nu^k)$.

THEOREM 3.6. *There is a neighborhood of M^n in $E(\nu^k)$ which is mapped diffeomorphically by e onto a neighborhood of M^n in $\mathbb{R}^{n+k}$.*

Proof. Note that e is differentiable, and that it has rank $n + k$ at points of $M^n \subset E(\nu^k)$. (This is easily checked by computing the derivative matrix of e with respect to a local coordinate system.) Hence e has rank $n+k$ in some neighborhood of M^n in $E(\nu^k)$, so that it is a local homeomorphism at points of M^n, that is, it maps a neighborhood of each $x \in M^n$ homeomorphically onto a neighborhood of $f(x)$. We then appeal to the topological lemma:

If $f : X \to Y$ is a local homeomorphism-onto and the restriction of f to a closed subset A is a homeomorphism, then f is a homeomorphism on some neighborhood V of A.

(X, Y are Hausdorff spaces with countable bases; X is locally compact.) This lemma is proved as follows:

(1) If A is compact, the lemma holds. For otherwise, there would be points x, y arbitrarily close to A such that $f(x) = f(y)$. Since A has a compact neighborhood, we may choose sequences x_n, y_n converging to x, y, respectively, in A such that $x_n \neq y_n$ and $f(x_n) = f(y_n)$. Hence $f(x) = f(y)$ so that $x = y$, f being a homeomorphism on A. But then f is not a local homeomorphism at x.

(2) Let A_0 be a compact subset of A. Then there is a neighborhood U_0 of A_0 such that $\overline{U}_0$ is compact and f is a homeomorphism on $\overline{U}_0 \cup A$: It will suffice for f to be injective, since f is a local homeomorphism-onto. By (1), let V_0 be a neighborhood of A_0 so that $f|_{\overline{V}_0}$ is injective. If no neighborhood of A_0 in V_0 satisfies the requirements for U_0, there is a sequence of points x_n of $X - A$ converging to $x \in A_0$ with $f(x_n) \in f(A)$. Choose $y_n \in A$ with $f(x_n) = f(y_n)$. Since f is continuous, $f(y_n)$ converges to $f(x)$; since f is a homeomorphism on A, y_n converges to x. Since $x_n \neq y_n$, this contradicts the fact that f is a local homeomorphism at x.

(3) Express A as the union of an ascending sequence of compact sets $A_1 \subset A_2 \subset \cdots$. Let V_1 be a neighborhood of A_1 such that $\overline{V}_1$ is compact and f is a homeomorphism on $\overline{V}_1 \cup A$ (by (2)). Given V_i a neighborhood of A_i satisfying these conditions, consider the set $\overline{V}_i \cup A_{i+1}$. It is a compact subset of $\overline{V}_i \cup A$, and f is a homeomorphism on $\overline{V}_i \cup A$. Hence by (2) there is a neighborhood V_{i+1} of $\overline{V}_i \cup A_{i+1}$ with $\overline{V}_{i+1}$ compact, such that f is a homeomorphism on $\overline{V}_{i+1} \cup A$. We proceed by induction. f is injective on $V = \bigcup V_{i+1}$, so that it is a homeomorphism of V (being a local homeomorphism-onto). □

COROLLARY 3.7. *Any differentiable submanifold (without boundary) of $\mathbb{R}^{n+k}$ is a differentiable neighborhood retract.*

The projection of $E(\nu^k) \to M^n$ induces (under e) a differentiable map of a neighborhood of M^n in $\mathbb{R}^{n+k}$ onto M^n which is the identity on M^n.

Definition 3.8. Let ξ be a vector space bundle with compact base space $B(\xi)$. The one-point compactification $T(\xi)$ of the total space $E(\xi)$ is called the *Thom space* of ξ. The added point will be denoted by ∞.

Let ξ have a Riemannian metric. Let $T_\epsilon(\xi)$ be obtained from $E(\xi)$ by identifying all vectors of length greater than or equal to ϵ to a point. Let $\alpha(x)$ be a C^∞ function with $\alpha'(x) \geq 0$ which equals 1 in a neighborhood of $x = 0$ and $\to \infty$ as $x \to 1$. The map of $E(\xi)$ into $T(\xi)$ which carries the vector e into the vector $e\,\alpha(||e||/\epsilon)$ induces a homeomorphism of $T_\epsilon(\xi)$ onto $T(\xi)$ which is a diffeomorphism on the set $E_\epsilon(\xi)$, consisting of vectors of length less than ϵ. The fact that B is compact is used here.

Definition 3.9. Let the compact manifold M^n be embedded in $\mathbb{R}^{n+k}$. ν^k is given the Riemannian metric of $\mathbb{R}^{n+k}$; by Theorem 3.6 there is a neighborhood of M^n in $\mathbb{R}^{n+k}$ which is diffeomorphic to the subset $E_{2\epsilon}(\nu^k)$ of $E(\nu^k)$. Such a neighborhood is called a ***tubular neighborhood*** of M^n.

By Definition 3.8, we see that $T(\nu^k)$ is homeomorphic with the space obtained from $\mathbb{R}^{n+k}$ by collapsing the exterior of the tubular ϵ-neighborhood of M to a point.

We will need three lemmas concerning approximation by differentiable functions.

LEMMA 3.10. *Let A be a closed subset of the differentiable manifold M, let $f : M \to \mathbb{R}^m$ be differentiable on A. Let δ be a positive continuous function on M. There exists $g : M \to \mathbb{R}^m$ such that*

(1) *g is differentiable*

(2) *g is a δ-approximation to f*

(3) *$g|_A = f|_A$.*

Proof. It suffices to prove this lemma in the case $m = 1$.

Given $x \in A$, $f|_A$ may be extended to a differentiable function f_x in a neighborhood N_x of x. Let N_x be chosen small enough that $|f_x(y) - f(y)| < \delta(y)$ for all $y \in N_x$.

Given $x \in M \setminus A$, choose a neighborhood N_x of x small enough that $|f(y) - f(x)| < \delta(y)$ for all $y \in N_x$. Define $f_x(y) \equiv f(x)$ for $y \in N_x$.

Now let φ_α be a differentiable partition of unity with support φ_α contained in some N_x, say $N_{x(\alpha)}$, for each α. Define $g(y) = \sum \varphi_\alpha(y)\, f_{x(\alpha)}(y)$. One checks the conditions of the lemma easily. □

More generally:

LEMMA 3.11. *Let $f : M_1 \to M_2$ be a continuous map of differentiable manifolds which is differentiable on the closed subset A of M_1. Let $\epsilon(x) > 0$ be given; and give M_2 the metric determined by some embedding $M_2 \subset \mathbb{R}^p$. Then there exists $g : M_1 \to M_2$ such that*

(1) *g is differentiable*

(2) *g is an ϵ-approximation to f*

(3) *$g|_A = f|_A$.*

Proof. There is a neighborhood U of M_2 in $\mathbb{R}^p$ of which M_2 is a differentiable retract (Corollary 3.7). Let ρ be the differentiable retraction of U onto M_2. Let $\delta(x)$ be a positive function on M_2 so chosen that the cubical neighborhood of $f(x)$ of radius $\delta(x)$ lies in U, and so that its image under ρ has radius less than $\epsilon(x)$. Let $f_1 : M_1 \to \mathbb{R}^p$ be a differentiable map which is a δ-approximation to f, such that $f_1|_A = f|_A$ (by Lemma 3.10). Define $g(x) = \rho(f_1(x))$. □

LEMMA 3.12. *Let $f : M_1 \to M_2$ be a continuous map of differentiable manifolds; let the metric on M_2 be obtained by embedding it in some euclidean space. Given $\epsilon(x)$, there is a $\delta(x)$ such that if $g : M_1 \to M_2$ is a δ-approximation to f, g is homotopic to f under a homotopy $F(x,t)$ with*

(1) *$F(x,t) = f(x)$ for any x such that $g(x) = f(x)$ and*

(2) *$F(x,t)$ is an ϵ-approximation to f for any t.*

Proof. Let U, ρ and $\delta(x)$ be chosen as in Lemma 3.11. Let $g : M_1 \to M_2$ be a δ-approximation to f.

Then the line segment from $g(x)$ to $f(x)$ lies in U, so that

$$F(x,t) = \rho(tg(x) + (1-t)f(x))$$

is well-defined. Furthermore $F(x,t)$ is an ϵ-approximation to $f(x)$ for any t. □

Definition 3.13. Let ξ^k be a differentiable vector space bundle with $B(\xi)$ compact and n-dimensional; let $E(\xi^k)$ be given a metric by embedding it as a closed differentiable submanifold in some euclidean space (it is an $(m+k)$-manifold).

Given an element of $\pi_{n+k}(T(\xi^k), \infty)$, let it be represented by the map

$$f : (\overline{C}_{n+k},\ \partial\overline{C}_{n+k}) \to (T(\xi^k),\ \infty), \tag{3}$$

where $\overline{C}_{n+k}$ is the closed cube $[0,1]^{n+k}$ and $\partial\overline{C}_{n+k}$ is its boundary. Let U denote the open subset $f^{-1}(E(\xi^k))$ of C_{n+k}. Let $g : U \to E(\xi^k)$ be a differentiable δ-approximation to $f|_U$, where δ is so chosen that $\delta < 1$ and g is homotopic to f, the homotopy F also being a 1-approximation to f. (This ensures that f will be continuous if we define $F(x,t) \equiv \infty$ for $x \in \overline{C}_{n+k} - U$.)

Now g may be approximated in turn by a differentiable map $h : U \to E(\xi)$ which is transverse regular on the submanifold $B(\xi)$ of $E(\xi)$. We choose the approximation close enough that g is homotopic to h, the homotopy H being a 1-approximation to g for each t. Extend h to $\overline{C}_{n+k}$ by defining $h(x) = \infty$ for $x \in \overline{C}_{n+k} - U$. Then h is in the homotopy class of f; and $h^{-1}(B(\xi))$ is a differentiable submanifold $M^n \subset U$ which is closed in $\overline{C}_{n+k}$, and thus compact.

THEOREM 3.14. *There is a well-defined homomorphism*

$$\lambda : \pi_{n+k}(T(\xi^k), \infty) \to \mathcal{N}_n$$

which assigns to each homotopy class of maps f the cobordism class of the manifold M^n constructed above.

Proof. Let $H : (\overline{C}_{n+k} \times I,\ \partial\overline{C}_{n+k} \times I) \to (T(\xi^k), \infty)$ be a homotopy between $h_0 = H(x,0)$ and $h_1 = H(x,1)$. Let h_0, h_1 satisfy the conditions

(1) h_i is differentiable on $h_i^{-1}(E(\xi))$

(2) h_i is transverse regular on $B(\xi)$, $(i = 1, 2.)$

We wish to show that $h_0^{-1}(B)$ and $h_1^{-1}(B)$ belong to the same cobordism class.

We may assume that $H(x,t) = H(x,0)$ for $t \leq 1/3$, and $H(x,t) = H(x,1)$ for $t \geq 2/3$. Let $U = h^{-1}(E(\xi)) \cap [\overline{C}_{n+k} \times (0,1)]$; then U is an open subset of $\mathbb{R}^{n+k+1}$. Let $G : U \to E(\xi)$ be a differentiable 1-approximation to H which equals H on the closed subset A, where $A = U \cap [\overline{C}_{n+k} \times ((0,1/4] \cup [3/4,1))]$. (See Lemma 3.11. H is differentiable on A.)

Now G satisfies the transverse regularity condition for $B(\xi)$ at points in A (since h_0 and h_1 are transverse regular on B) so that by Theorem 1.36 there is a differentiable map $F : U \to E(\xi)$ which equals G on A, is transverse regular on $B(\xi)$, and is a 1-approximation to G. Because F is a 2-approximation to H, it remains continuous if we define $F(x,t) = \infty$ for $(x,t) \in (\overline{C}_{n+k} \times (0,1)) - U$. Because F equals H on A, it remains continuous if we define $F(x,t) = H(x,t)$

for $t = 0, 1$. Hence $F^{-1}(B)$ is a compact subset of $\overline{C}_{n+k} \times I$, being closed and bounded.

Because $F|_U$ is transverse regular on B, $(F|_U)^{-1}(B)$ is a differentiable $(n+1)$-submanifold of $\overline{C}_{n+k} \times (0,1)$. Its intersection with $\overline{C}_{n+k} \times t$ equals $h_0^{-1}(B) \times t$ for $t \leq 1/4$ and $h_1^{-1}(B) \times t$ for $t \geq 3/4$. Hence $F^{-1}(B)$ is a differentiable manifold-with-boundary whose boundary is $h_0^{-1}(B) + h_1^{-1}(B)$. Thus λ is well-defined.

It is trivial to show that λ is a homomorphism, because the sum in $\mathcal{N}_n$ is derived from disjoint unions of representative manifolds. □

THEOREM 3.15. *If ξ^k is the universal bundle γ_m^k where $k \geq n+1$, $m \geq n$, then $\lambda : \pi_{n+k}(T(\gamma_m^k), \infty) \to \mathcal{N}_n$ is onto.*

Proof. Let M^n be a compact n-manifold; let $k \geq n+1$. Let M^n be embedded in C_{n+k} (Corollary 1.32); let ν^k be the normal bundle of this embedding. The Riemannian metric on $E(\nu^k)$ is that derived from the natural scalar product on the tangent bundle to $\mathbb{R}^{n+k}$, in which ν^k is contained.

By Theorem 3.6, for small ϵ the subset of $E_{2\epsilon}(\nu^k)$ of $E(\nu^k)$ is diffeomorphic with a tubular neighborhood of M^n in C_{n+k}; let U be the image of $E_\epsilon(\nu^k)$.

Let p_1 project $\overline{C}_{n+k}$ on the space obtained from $\overline{C}_{n+k}$ by identifying $\overline{C}_{n+k} - U$ to a point (denoted by $\overline{C}_{n+k}/\overline{C}_{n+k} - U$).

Let p_2 be the diffeomorphism of U onto $E_\epsilon(\nu^k)$, followed by the map of $E(\nu^k)$ into $T_\epsilon(\nu^k)$ which identifies all vectors of length $\geq \epsilon$ (Definition 3.8). p_2 is then extended by mapping $\overline{C}_{n+k} - U$ into ∞.

Let p_3 be the homeomorphism of $T_\epsilon(\nu^k)$ onto $T(\nu^k)$ constructed in Definition 3.8. The composite map $p_3\, p_2\, p_1$ is a diffeomorphism of U onto $E(\nu^k)$.

Finally, let p_4 be the bundle map of ν^k into γ_m^k induced from the embedding of M^n in $\mathbb{R}^{n+k} \subset \mathbb{R}^{m+k}$. Because both fibres have dimension k, this map satisfies the transverse regularity condition for $G_{k\,m}$ at each point of M^n. Extend p_4 in the obvious way to map $T(\nu^k)$ into $T(\gamma_m^k)$.

Let $g = p_4\, p_3\, p_2\, p_1$. Then $g : \partial\overline{C} \to \infty$. Let $\mu(M^n)$ denote the homotopy class of g in $\pi_{n+k}(T(\gamma_m^k), \infty)$. Now g is transverse regular on $G_{k\,m}$ and $M^n = g^{-1}(G_{k\,m})$. By definition, the cobordism class of M^n is the image of $\mu(M^n)$ under λ, so that $\lambda\mu(M^n) = [M^n]$. □

THEOREM 3.16. *If ξ^k is the universal bundle γ_m^k with $k \geq n+2$, $m > n$, then λ is one-to-one.*

Proof. Given an element of $\pi_{n+k}(T(\gamma_m^k), \infty)$, we may suppose it represented by a map

$$f : (\overline{C}_{n+k}, \partial\overline{C}_{n+k}) \to (T(\gamma_m^k), \infty)$$

which is differentiable on $f^{-1}(E)$ and transverse regular on $G_{m\,k}$ (by Definition 3.13). Let

$$M^n = f^{-1}(G_{m\,k});$$

we wish to show that if M^n is the boundary of an $(n-1)$-manifold-with boundary Q, then f is homotopic to the constant map.

M^n is a submanifold of C_{n+k}; let its normal bundle be ν^k. Let ϵ be chosen so that $E_{2\epsilon}(\nu^k)$ is diffeomorphic with the 2ϵ-neighborhood of M^n; let U_ϵ be the image of the vectors $E_\epsilon(\nu^k)$. Impose a Riemannian metric on γ^k_m; let δ be chosen so that $||x|| \geq \epsilon$ implies $||f(x)|| \geq \delta$ for $x \in E(\nu^k)$.

Step 1. f is homotopic to a map f_1 such that

(1) f_1 is differentiable on $f_1^{-1}(E)$ and transverse regular on $G_{m\,k}$.
(2) $f = f_1$ on $M^n = f_1^{-1}(G_{m\,k})$.
(3) f_1 carries everything outside U_ϵ into ∞.

Define $F : E(\gamma^k_m) \times I \to T(\gamma^k_m)$ by the equation $F(e,t) = e\,\alpha(t||e||/\delta)$, where α is the function defined in Definition 3.8. Let $f_1(x) = F(f(x), 1)$.

Step 2. By the diffeomorphism of $U_{2\epsilon}$ with $E_{2\epsilon}$, f_1 induces a map $\overline{f}_1$ of $\overline{E}_\epsilon(\nu^k)$ into $T(\gamma^k_m)$ which carries $\partial(E_\epsilon)$ into ∞. Any homotopy of $\overline{f}_1$ which leaves $\partial(E_\epsilon)$ at ∞ induces a homotopy of f_1.

Now $\overline{f}_1$ is homotopic to a map $\overline{f}_2$ such that

(1) $\overline{f}_2$ is differentiable on $\overline{f}_2^{\,-1}(E)$ and transverse regular on $G_{m\,k}$.
(2) $\overline{f}_2 = \overline{f}_1$ on $M^n = \overline{f}_2^{\,-1}(G_{m\,k})$.
(3) $\overline{f}_2$ is locally a bundle map in some neighborhood of M^n.

The homotopy leaves $\partial(E_\epsilon)$ at ∞.

Consider $G : \overline{E}_\epsilon(\nu^k) \times I \to T(\gamma^k_m)$ defined by the equation $G(e,t) = \overline{f}_1(te)/t$. As $t \to 0$, $G(e,t)$ approaches a limit which is non-zero if $e \neq 0$ (since $\overline{f}_1$ is differentiable and t-regular). It is easily seen to be a bundle map. It will not suffice for our purposes, since it does not carry $\partial(E_\epsilon)\times I$ into ∞. Choose $\delta > 0$ so that $||x|| \geq \epsilon$ implies $||G(x,t)|| \geq \delta$ for $x \in E(\nu^k)$, $t \in I$, and define $H(e,t) = [G(e,t)]\,\alpha\,(||G(e,t)||/\delta)$. If we set $\overline{f}_2 = H(e,0)$, then $\overline{f}_2$ is a bundle map for $||e||$ small (since $\alpha(x) \equiv 1$ for x small). The map $E(e,1) = \overline{f}_1(e)\,\alpha(||\overline{f}_1(e)||/\delta)$ does not equal $\overline{f}_1$, but it is homotopic to $\overline{f}_1$, the homotopy leaving $\partial(E_\epsilon)$ at ∞. The homotopy is defined by the equation

$$K(e,t) = \overline{f}_1(e)\,\alpha(t||\overline{f}_1(e)||/\delta), \quad \text{as in Step 1.}$$

Step 3. Let Q be the $n+1$ manifold-with-boundary such that $M^n = \partial Q$. Let h be a diffeomorphism of $M^n \times [0,1)$ into Q which carries $M^n \times 0$ onto ∂Q. Define $h_1 : Q \to C_{n+k} \times I$ as follows:

If $x = h(y,t)$ where $y \in M^n$ and $0 \leq t \leq 1/2$, let $h_1(x) = (y,t)$.
If $x \notin \operatorname{image} h$, let $h_1(x) = p$, where p is some fixed point interior to $C_{n+k}\times I$.
If $x = h(y,t)$ where $y \in M^n$ and $1/2 \leq t \leq 1$, let

$$h_1(x) = (1-\beta(t))\,h_1(y,1/2) + \beta(t)p, \quad \text{where}$$

$\beta(t)$ is a C^∞ function with $\beta'(t) \geq 0$, $\beta(t) \equiv 0$ in a neighborhood of $t = 1/2$ and $\beta(t) \equiv 1$ in a neighborhood of $t = 1$.

h_1 is a differentiable map of $\operatorname{Int} Q$ into $\operatorname{Int}(C_{n+k} \times I)$; and h_1 is an injective immersion in a neighborhood of ∂Q. Since $\dim(C_{n+k} \times I) > 2(n+1)$, h_1 may be approximated by an injective immersion h_2 which equals h_1 in a neighborhood of ∂Q (by Lemma 1.29). It may be extended to an embedding of Q into $C_{n+k} \times I$. (Since Q is compact, an injective immersion is automatically an embedding.) Let Q now be considered as this subset of $C_{n+k} \times I$.

Step 4. We have a map f_2 of $\overline{C}_{n+k} \times 0$ into $T(\gamma_m^k)$ which is a bundle map when restricted to a small tubular neighborhood of $M^n \times 0$ in $C_{n+k} \times 0$. We extend it to $\overline{C}_{n+k} \times [0, b)$ for b small in the trivial way. Suppose there exists a map g of the ϵ'-neighborhood N of Q in $C_{n+k} \times I$ into $T(\gamma_m^k)$ which equals f_2 in some neighborhood of ∂Q in $C_{n+k} \times I$ and maps each point of $N - Q$ into a non-zero vector in $E(\gamma_m^k)$. Our theorem then follows: Let δ be so chosen that, if the distance of x from Q is $\geq \epsilon\, 1/2$, then $||g(x)|| \geq \delta$.
Define $g_1 : C_{n+k} \times I \to T(\gamma_m^k)$ by the equation

$$g_1(x, s) = \begin{cases} g(x, s)\, \alpha(||g(x, s)||/\delta), & \text{for } (x, s) \in N, \quad \text{and} \\ \infty, & \text{otherwise.} \end{cases}$$

The restriction of g_1 to $C_{n+k} \times 0$ does not equal the map f_2, but it is homotopic to f_2, by the same technique as used at the end of Step 2. Thus g_1 provides the homotopy required for our theorem.

To show that the extension g exists, we refer to Steenrod, "Fibre Bundles" (Princeton University Press, 1951). According to §19.4 and 19.7 of this book, the principal bundle associated with γ_m^k is an m-universal bundle. That is: given a vector space bundle ξ^k over a complex of dimension $\leq m$, any bundle map of ξ^k, restricted to a subcomplex, into γ_m^k can be extended throughout ξ^k. We will assume the well known result that Q can be triangulated. The dimension $n+1$ of Q is $\leq m$. Hence any bundle map of the normal bundle ν^k of Q, restricted to a polyhedral neighborhood of ∂Q, into γ_m^k can be extended throughout ν^k.

Applying this result to the map f_2, this completes the proof of Theorem 3.16.
□

Letting T_k stand for the union of the Thom spaces $T(\gamma_m^k) \subset T(\gamma_{m+1}^k) \subset \cdots$, in the fine (direct limit) topology, Theorems 3.15 and Theorem 3.16 imply the following.

THEOREM 3.17. *The cobordism group $\mathcal{N}_n$ is canonically isomorphic to the stable homotopy group $\pi_{n+k}(T_k)$, for $k \geq n+2$.*

References

Morse, A.P.: *The behavior of a function on its critical set,* Annals of Math. **40** (1939), 62–70.

Sard, A.: *The measure of the critical values of differentiable maps,* Bull. Amer. Math. Soc. **48** (1942), 883–890.

Thom, R.: *Quelques propriétés globales des variétés différentiables,* Comment. Math. Helv. **28** (1954), 17–86.

Whitney, H.: *A function not constant on a connected critical set of points,* Duke Math. J., **1** (1935), 514–517.

———: *Differentiable manifolds,* Annals of Math. **37** (1936), 645–680.

———: *The self-intersections of a smooth n-manifold in 2n-space,* Annals of Math. **45** (1944), 220–246.

———: *The singularities of a smooth n-manifold in (2n − 1) space,* Annals of Math. **45** (1944), 247–293.

Lectures on Differentiable Structures

John Milnor

Princeton University, Spring 1961

1. Strong and Weak Diffeotopies: the diffeotopy extension problem.

We shall be concerned solely with *smooth*, i.e. C^∞, manifolds. Without making precise definitions for the moment, we can state our basic problem as follows: can every motion of a submanifold of a given manifold V be extended to all of V? Figure 1 illustrates a motion of a submanifold of $\mathbb{R}^3$ which cannot be extended to all of $\mathbb{R}^3$. The trefoil knot is moved so that the knotted portion approaches a

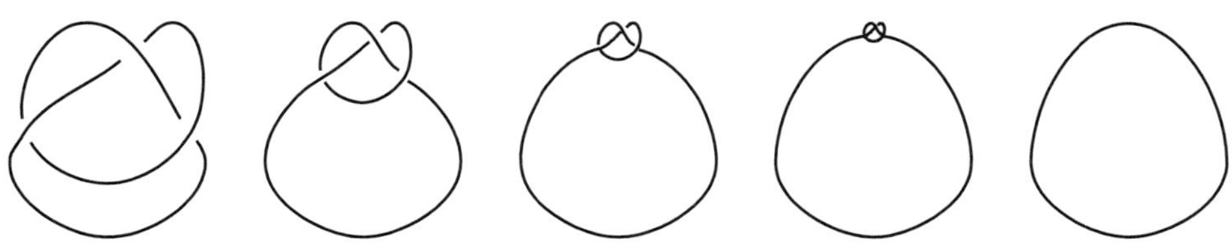

FIGURE 1.

point; at the final stage the knot is replaced by a circle. We will illustrate, after a few definitions, other motions of submanifolds which cannot be extended.

Definition. Let M and V be two smooth manifolds. A smooth map $f : M \to V$ is an *embedding* iff.

(1) f is a homeomorphism of M into V

(2) the rank of f_* at $p \in M$ is the dimension of M at p [where f_* is the induced map on the tangent space of M].

If $f(M) = V$ then f^{-1} will be smooth and f is called a *diffeomorphism*.

Definition. A *diffeotopy from* M *into* V is a one-parameter family of embeddings which is smooth as a function of both variables, i.e. a smooth map $f : M \times I \to V$ (I denotes the unit interval [0, 1]) such that each $f_t : M \to V$, defined by $f_t(x) = f(x, t)$ is an embedding. The map f is called a diffeotopy between f_0 and f_1, which are called diffeotopic maps.

The family of embeddings of S^1 into $\mathbb{R}^3$ indicated in Figure 1 is not a diffeotopy because it is not smooth at $(x, 1)$, where x is the point towards whose image the knotted portion of the circle converges.

Although the unit interval I is strongly suggested as a parameter space in the definition of homotopic maps, it has several disadvantages for defining diffeotopic maps. For one thing, it is not immediately clear that diffeotopy is an equivalence relation. Moreover $M \times I$ is a manifold with boundary if M has no boundary, and $M \times I$ has corners if M has a boundary. These objections present no real difficulties because, without loss of generality, we may assume that

$$f_t = \begin{cases} f_0, & \text{for} \quad t \leq 1/3 \\ f_1, & \text{for} \quad t \geq 2/3\,. \end{cases}$$

To see this simply define $\overline{f}_t$ as $f_{\lambda(t)}$ where λ is smooth and

(1) $\lambda = 0 \quad \text{for} \quad t \leq 1/3$
(2) $\lambda = 1 \quad \text{for} \quad t \geq 2/3$
(3) $0 \leq \lambda \leq 1\,.$

It now becomes clear that the relation of diffeotopy is transitive. Furthermore we see that the open unit interval (or the full real line) can be used in place of I in the definition.

Definition. A *strong diffeotopy of V to itself* is a diffeotopy f from V into V such that each f_t is a diffeomorphism. If $f_0, f_1 : M \to V$ are two embeddings, f_0 is *strongly diffeotopic* to f_1 iff there is a strong diffeotopy F of V to itself carrying f_0 to f_1, i.e. such that F_0 is the identity and $F_1 \circ f_0 = f_1$.

Two maps f and g may be diffeotopic without being strongly diffeotopic. Figure 2 indicates a family of embeddings of $\mathbb{R}$ into $\mathbb{R}^2$; at the last stage the image of $\mathbb{R}$ is a circle minus a point.

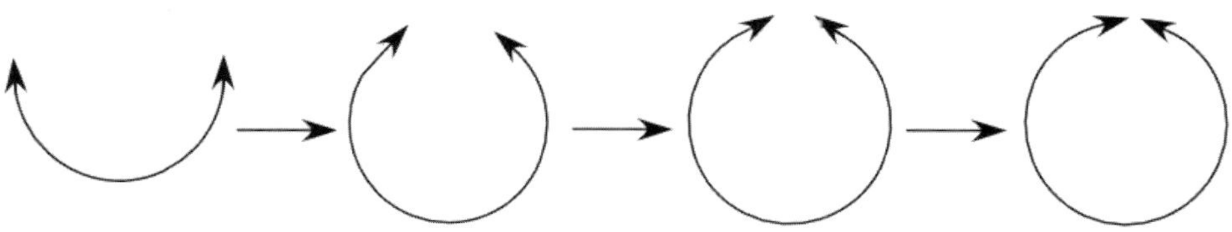

FIGURE 2.

Clearly there is not even a homeomorphism of $\mathbb{R}^2$ onto itself which will take the first embedding into the last. We might hope to avoid such situations by requiring that each $f_t(M)$ be a closed subset of V. However there is an embedding of $\mathbb{R}$ into $\mathbb{R}^3$ which shows that this requirement is not sufficient; it is indicated in Figure 3.

For $|u| \geq 1$ the embedding f_0 satisfies $f_0(u) = (u, 0, 0)$. The embedding f_0 is diffeotopic to the standard embedding of $\mathbb{R}$ in $\mathbb{R}^3$. This can be seen by sliding the knotted portion to the left to ∞, as follows: let

$$\begin{aligned} f_t(u) &= f_0(u + t/(1-t)) - (t/(1-t), 0, 0), \qquad 0 \leq t < 1 \\ f_1(u) &= (u, 0, 0). \end{aligned}$$

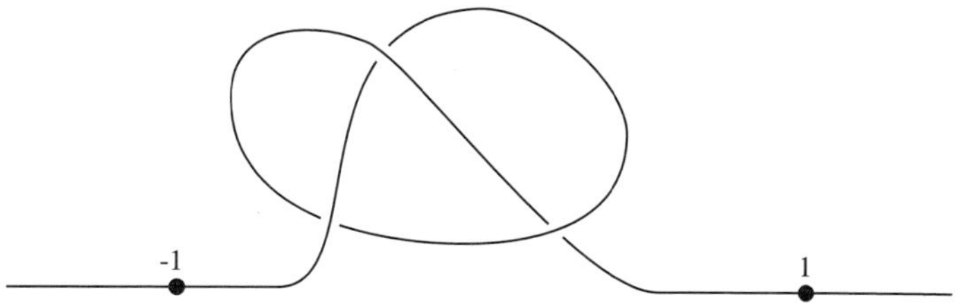

FIGURE 3.

It is again clear that f_0 and f_1 are not strongly diffeotopic. This example is really just that of Figure 1, with the bad point at ∞.

A somewhat more spectacular example of an embedding of $\mathbb{R}$ in $\mathbb{R}^3$ that is diffeotopic, but not strongly diffeotopic, to the usual embedding is illustrated in Figure 4.

FIGURE 4.

There are loops infinitely far to the left but there is a rightmost loop after which we have the usual embedding of a line. Clearly, by pulling out the last loop between $t = 0$ and $t = 1/2$, the second between $t = 1/2$ and $t = 3/4$, etc., we can get a diffeotopy with the standard embedding, which can certainly be extended to $\mathbb{R}^3$ between $t = 0$ and $t = t_0$ for any $t_0 < 1$. Nevertheless, the fundamental group of the complement of $\mathbb{R}^3$ under this embedding is non-abelian, so that the diffeotopy cannot be extended to $\mathbb{R}^3$ between $t = 0$ and $t = 1$, (see R. Fox, *A remarkable simple closed curve,* Annals of Math. **50** (1949), 264–265.)

These examples should justify all the hypothesis of the following theorem.

THEOREM 1.1. *Let M and V be smooth manifolds. Let $f_0, f_1 : M \to V$ be two smooth embeddings which are diffeotopic under a diffeotopy which leaves all points fixed outside a compact set $M_0 \subset M$. Then f_0 and f_1 are strongly diffeotopic, under a strong diffeotopy which leaves all points fixed outside a compact set $V_0 \subset V$.*

This theorem is due to R. Thom, *La classification des immersions,* Séminaire Bourbaki **157** (Dec. 1957). See also R. Palais, *Local triviality of the restriction map for embeddings,* Commentarii **34** (1960), 305–312.

Proof. We may assume that f_t is defined for all real t, and that

$$f_t = \begin{cases} f_0, & \text{for} \quad t \leq 1/3 \\ f_1, & \text{for} \quad t \geq 2/3\,. \end{cases}$$

Define a map $F : M \times \mathbb{R} \to V \times \mathbb{R}$ by $F(x, t) = (f_t(x), t)$. It is easy to see that F is an embedding. The map F is indicated in Figure 5 where V is a line and M a point.

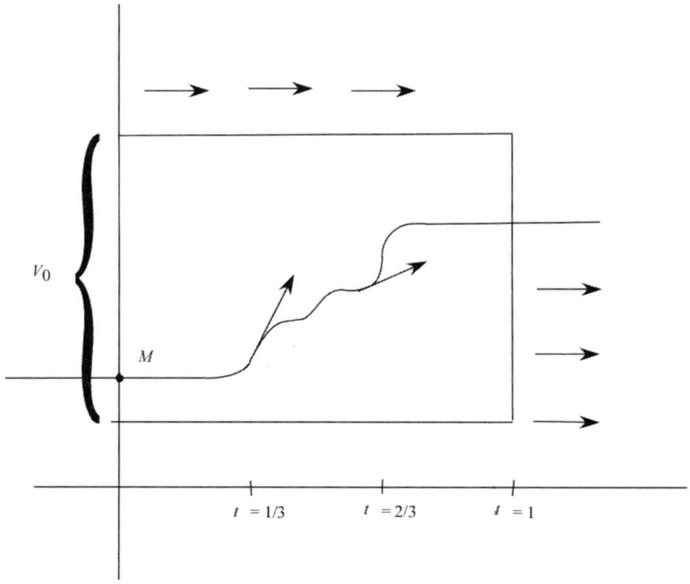

FIGURE 5.

Let V_0 be a compact neighborhood of $f(M_0 \times I)$.

We wish to define a vector field X on $V \times \mathbb{R}$ such that

(1) outside of $V_0 \times I$ the field X is $\frac{\partial}{\partial t}$.

(2) in $F(M \times \mathbb{R})$ the field X is $F_*(\frac{\partial}{\partial t})$.

(X is indicated by arrows in Figure 5).

If $y \notin (V_0 \times I)$ and $y \in F(M \times \mathbb{R})$ these requirements agree: if (2nd. coord y) $\notin [0, 1]$ they agree because f_t is constant for $t \notin [0, 1]$; if (1st. coord y) $\notin V_0$ then $y \notin f(M_0 \times I)$ so they agree since f_t is constant outside M_0. Therefore X has been defined on $F(M \times \mathbb{R})$ and outside $V_0 \times I$.

In order to extend X to all of $V \times \mathbb{R}$ it is sufficient to extend it locally (we then use partitions of unity). At points on the boundary of $V_0 \times I$ we can clearly extend X locally by using the field $\frac{\partial}{\partial t}$. If $y \in F(M \times I)$ there is, by the implicit function theorem, a local coordinate system $(u_1, \ldots, u_{n+1})$ for $V \times \mathbb{R}$ in a neighborhood U of y such that $U \cap F(M \times I) = \{p \in V \times I : u_1(p) = \ldots = u_{n-k}(p) = 0\}$ where $n = \dim V$, and $k = \dim M$. It is clear how to extend a vector field on a $(k+1)$-plane of $\mathbb{R}^{n+1}$ to all of $\mathbb{R}^{n+1}$, which gives an extension of X to the neighborhood of U. If y is in the interior of $V_0 \times I$ and $y \notin F(M \times I)$ then we can extend the vector field arbitrarily.

Thus we can get a vector field X on $V \times \mathbb{R}$ satisfying **(1)** and **(2)**. We may also assume that

(3) the $\mathbb{R}$ component of X is always $\frac{\partial}{\partial t}$.

In fact we may replace X by X' where the V-component of X' is equal to the V-component of X but the $\mathbb{R}$ component is $\frac{\partial}{\partial t}$.

Conditions **(1)** and **(2)** remain valid.

Let $\phi_t : V \to V$ be given by $\phi_t(x)$ = first component of solution at time t of the vector field X, passing through $(x, 0)$. (The second component is clearly t, by condition **(3)**). The solutions of the vector field X are known to exist locally; their global existence, from time $t = 0$ to $t = 1$, follows from conditions **(1)** and **(3)** and the compactness of $V_0 \times I$. By condition **(2)**, if $x \in M$ we have $\phi_1(x) = f_1(x)$. The family ϕ_t thus gives us the desired strong diffeotopy of V onto V. (Clearly ϕ leaves all points fixed outside V_0.) $\square$

Remark. The above proof works just as well if M is a manifold with boundary. However, the theorem is false if V has a boundary, as is seen by pushing a small circle to the boundary of a large disk, (Figure 6).

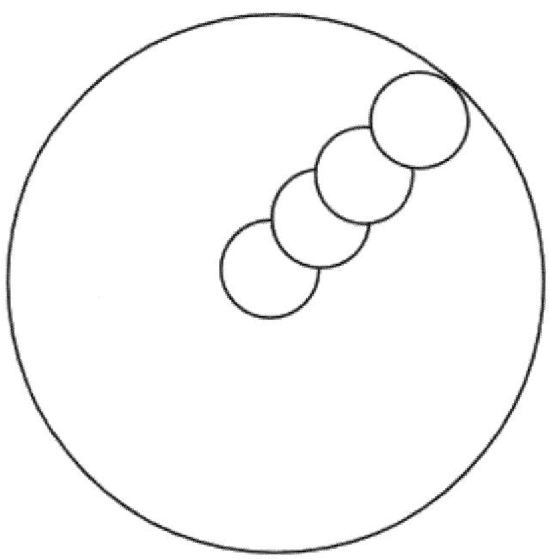

FIGURE 6.

2. Embedding a Cell in a Manifold.

Denote a point $(x_1, \dots, x_n) \in \mathbb{R}^n$ by $\vec{x}$, for convenience. $\vec{0}$ denotes $(0 \dots, 0)$.

LEMMA 2.1. *Let $\vec{f}$ be a smooth map of $\mathbb{R}^n$ into $\mathbb{R}^m$ such that $\vec{f}(\vec{0}) = \vec{0}$. Then there are smooth functions $\vec{g}_i$ such that*

$$\vec{f}(\vec{x}) = \sum_{i=1}^{n} x_i \vec{g}_i(\vec{x})$$

and

$$\vec{g}_i(\vec{0}) = \frac{\partial \vec{f}}{\partial x_i}(\vec{0}).$$

(This second condition actually follows from the first.)

Proof.

$$\vec{f}(\vec{x}) = \int_0^1 \frac{d\vec{f}(t\vec{x})}{dt} dt = \int_0^1 \sum_{i=1}^{n} x_i \frac{\partial \vec{f}(t\vec{x})}{\partial x_i} dt.$$

Therefore we can let

$$\vec{g}_i = \int_0^1 \frac{\partial \vec{f}(t\vec{x})}{\partial x_i} dt.$$

□

THEOREM 2.2. *Let M be a smooth, connected, orientable manifold of dimension n. Let $f, g : \mathbb{R}^n \to M$ be smooth, orientation preserving embeddings. Then f is (weakly) diffeotopic to g.*

Proof.

Case 1: $M = \mathbb{R}^n$. It is sufficient to show that f is diffeotopic to the standard embedding. We can assume that $\vec{f}(\vec{0}) = \vec{0}$, where we are now using the notation $\vec{f} = (f_1, \ldots, f_n)$, with $f_i : \mathbb{R}^n \to \mathbb{R}$.
Let

$$\vec{h}(\vec{x}, t) = t^{-1}\vec{f}(t\vec{x}) \quad \text{for} \quad t \neq 0.$$

To examine the behavior of $\vec{h} = (h_1, \ldots, h_n)$ as $t \to 0$ we write, using Lemma 2.1,

$$\begin{aligned} h_j(\vec{x}, t) &= t^{-1} f_j(t\vec{x}) \\ &= t^{-1} \sum_{i=1}^{n} t x_i g_{ij}(t\vec{x}) \\ &= \sum_{i=1}^{n} x_i g_{ij}(t\vec{x}). \end{aligned}$$

Therefore $\vec{h}$ is smooth if we define h_j by this formula on $\mathbb{R}^n \times \mathbb{R}$.

For $t \neq 0$ the map $\vec{x} \to h(\vec{x}, t)$ is clearly an embedding. At $t = 0$ we have

$$h_i(\vec{x}, 0) = \sum_{i=1}^{n} x_i g_{ij}(0) = \sum_{i=1}^{n} x_i \frac{\partial f_j}{\partial x_i}(\vec{0})$$

which is linear, non-singular, and of positive determinant, since the matrix

$$\left(\frac{\partial f_i}{\partial x_i}(\vec{0}) \right)$$

is non-singular and f is orientation preserving. Thus f is diffeotopic to a linear map which is diffeotopic to the identity.

Case 2: M unrestricted but $f(\mathbb{R}^n) \subset g(\mathbb{R}^n)$. Case 2 is clear from Case 1.

Case 3: $f(\mathbb{R}^n) \cap g(\mathbb{R}^n) \neq \emptyset$. We can choose an orientation preserving embedding h so that $h(\mathbb{R}^n)$ is a cell in $f(\mathbb{R}^n) \cap g(\mathbb{R}^n)$. By Case 2, we have $h \approx f$ and $h \approx g$, so that $f \approx g$.

Case 4: General case. Since M is connected there is a sequence of embeddings

$$f = h_0, h_1, \ldots, h_n = g.$$

such that $h_i(\mathbb{R}^n) \cap h_{i+1}(\mathbb{R}^n) \neq 0$, $(i = 0, \ldots, n-1)$. By Case 3

$$f = h_0 \approx h_1 \approx \ldots \approx h_n = g.$$

The following theorem is due to R. Palais and J. Cerf. (R. Palais, *Extending diffeomorphisms,* Proceedings Amer. Math. Soc. **11** (1960), 274–277.)

THEOREM 2.3. *Let D^n be the closed unit disk in $\mathbb{R}^n$. Let M be a smooth, connected, orientable manifold of dimension n. Let $f, g : D^n \to M$ be orientation preserving embeddings such that f and g can be extended to embeddings $F, G : \mathbb{R}^n \to M$. Then f and g are strongly diffeotopic.*

Actually, F and G will always exist if M has no boundary, so that this hypothesis is redundant.

Proof. This follows from Theorems 1.1 and 2.2. □

Definition. Let M be a smooth, compact, oriented manifold. The diffeotopy classes of all orientation preserving diffeomorphisms is a group under composition, denoted by $\pi_0 \operatorname{Aut} M$.

As an application of Theorem 2.3 we will prove the following.

THEOREM 2.4. *$\pi_0\operatorname{Aut} S^n$ is abelian.*

Proof. Let $f : S^n \to S^n$ be a smooth map and let $i : D^n \hookrightarrow S^n$ be the "standard" embedding of D^n into the northern hemisphere on S^n, given by

$$i(t\vec{u}) = ((\sin t\pi/2)\vec{u}, \cos t\pi/2).$$

Assume that S^n is so oriented that i is orientation preserving. We have two maps i, $f \circ i : S^n \to S^n$. By Theorem 2.3 there is a family of diffeomorphisms $h_t : S^n \to S^n$ with $h_0 = I$ and $h_1 \circ f \circ i = i$, which means that $h_1 \circ f$ is the identity on $i(D^n)$. Therefore $f = h_0 \circ f$ is diffeotopic to $h_1 \circ f$, a map which is the identity on $i(D^n)$.

Now given two maps $f, g : S^n \to S^n$ let

$$\begin{aligned} f &\approx f' \\ g &\approx g' \end{aligned}$$

where f' leaves the northern hemisphere fixed and g' leaves the southern hemisphere fixed. Then

$$f \circ g \circ f^{-1} \circ g^{-1} \approx f' \circ g' \circ (f')^{-1} \circ (g')^{-1}.$$

This last map clearly leaves all points of S^n fixed.

3. The Connected Sum of Two Manifolds.

Definition. Let M_1 and M_2 be smooth, connected, oriented manifolds, both of dimension n. Let D^n be the unit n-disk in $\mathbb{R}^n$ and let rD^n be the n-disk of radius r. Let $i_q : 2D^n \to M_q$ be embeddings $(q = 1, 2)$, one preserving orientation, one reversing it. Form the disjoint sum

$$[M_1 - i_1(\frac{1}{2}D^n)] + [M_2 - i_2(\frac{1}{2}D^n)]$$

and identify $i_1(t\vec{u})$ with $i_2(t^{-1}\vec{u})$ for $\vec{u} \in S^{n-1}$ and $1/2 < t < 2$. The result is denoted $M_1\#M_2$, and called the ***connected sum*** of M_1 and M_2. It has an obvious orientable smooth structure.

By Theorem 2.3, the manifold $M_1\#M_2$ is well defined (up to orientation preserving diffeomorphism). Moreover, it is clear that $\#$ is commutative, associative, and has S^n as a zero element (up to orientation preserving diffeomorphism).

For convenience we will consider only compact connected oriented n-manifolds without boundary. Let $\mathcal{M}_n$ be the set of classes of these manifolds under orientation preserving diffeomorphism. Then $\mathcal{M}_n$, with the composition operation $\#$, is a semi-group.

$\mathcal{M}_1 = 0$.
$\mathcal{M}_2$ is a free commutative semi-group on one generator, the torus.
$\mathcal{M}_3$ is a free commutative semi-group on $\mathcal{N}_0$ generators.

(This result is essentially due to H. Kneser. See J. Milnor, *A unique decomposition theorem for 3-manifolds,* to appear.[1])

The structure of $\mathcal{M}_4$ is not known. However, $\mathcal{M}_4$ does not satisfy the cancellation law, as shown by an example of Hirzebruch, and hence cannot be free commutative. Let $M^4 = \mathbb{CP}^2$ be complex projective 2-space. Let $\overline{M}^4 = M^4$ with the opposite orientation. Then it can be shown that

$$M^4 \# M^4 \# \overline{M}^4 \approx M^4 \# (S^2 \times S^2)$$

but

$$M^4 \# \overline{M}^4 \not\approx S^2 \times S^2.$$

For higher values of n it may happen that $M\#N \approx S^n$ but $M \not\approx S^n$. This definitely occurs for $n = 7$.

THEOREM 3.1. *If $M\#N \approx S^n$ then M is homeomorphic to S^n.*

Proof. We form the infinite sum $M \# N \# M \# N \# \cdots$. Since $M\#N \approx N\#M \approx S^n$ we have

$$\begin{aligned} M \# N \# M \# N \# \cdots &\approx (M \# N) \# (M \# N) \# \cdots \\ &\approx S^n \# S^n \# S^n \# \cdots \\ &\approx \mathbb{R}^n . \\ M \# N \# M \# N \# \cdots &\approx M \# (N \# M) \# (N \# M) \# \cdots \\ &\approx M \# S^n \# S^n \# \cdots \\ &\approx M \# \mathbb{R}^n . \end{aligned}$$

So $\mathbb{R}^n \approx M\#\mathbb{R}^n$. But since $M \approx M\#S^n$ we have M-point $\approx M\#\mathbb{R}^n \approx \mathbb{R}^n$. Therefore M is homeomorphic to S^n.

[1] *A unique decomposition theorem for 3-manifolds,* Amer. J. Math. (1962), 1–7; or Collected Papers II: The Fundamental Group. Publish or Perish (1995).

4. Differentiable Structures on Spheres.

This section will describe constructions by which one can obtain exotic differentiable structures on spheres. However the proof that some of the resulting manifolds are not diffeomorphic to the usual sphere will only be given later.

Definition. Let $f : S^{n-1} \to S^{n-1}$ be a diffeomorphism. We wish to define a manifold by taking the disjoint union of two hemispheres and identifying their boundary $(n-1)$-spheres by f. As a matter of convenience, for defining the differentiable structure on this new manifold, we adopt the following procedure: Take $\mathbb{R}^n + \mathbb{R}^n$ and identify $t\vec{u}$ in the first $\mathbb{R}^n$ with $t^{-1}f(\vec{u})$ in the second, for $\vec{u} \in S^{n-1}$ and $0 < t$. The resulting smooth manifold is denoted $M(f)$ and is called a ***twisted sphere with twist*** f. Clearly $M(f)$ is topologically a sphere.

As an example, if i denotes the identity map of S^{n-1}, we have $M(i) \approx S^n$.

Remark. It can be shown that a closed manifold M is a twisted sphere if and only if there exists a smooth real valued function on M with exactly two critical points, both being non-degenerate. Smale has recently proved that, at least for $n \neq 3, 4$, M is a twisted sphere if and only if M has the homotopy type of a sphere.

LEMMA 4.1. *The correspondence $f \to M(f)$ defines a homomorphism from the group $\pi_0 \operatorname{Aut} S^{n-1}$ into the semi-group $\mathcal{M}_n$. That is,*

(1) *if $f \approx g$ then $M(f)$ is orientation preserving diffeomorphic to $M(g)$,*

(2) *$M(g \circ f)$ is orientation preserving diffeomorphic to $M(f) \# M(g)$.*

Proof. If $f \approx g$ we have the diagram of figure 7, where $f : S_1^{n-1} \to S_3^{n-1}$ and $g : S_2^{n-1} \to S_4^{n-1}$; and there is an extension H from the annular region A between S_1^{n-1} and S_2^{n-1} to the annular region B between S_3^{n-1} and S_4^{n-1} (since $f \approx g$). Let $M = (S_+^n \cup A) + (S_-^n \cup B)$ with x identified with $H(x)$ for $x \in A$.

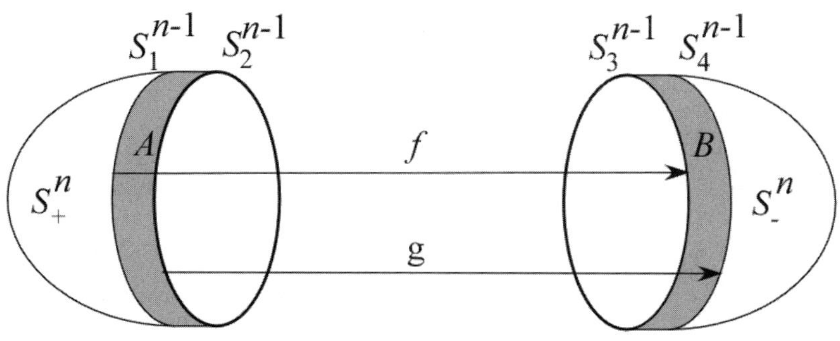

FIGURE 7.

It is clear that $M = M(f)$ and also $M = M(g)$.

To show $M(g \circ f) = M(f) \# M(g)$ consider the diagram shown in figure 8: To form $M(f) \# M(g)$ we may remove cell A from $M(f)$ and cell B from $M(g)$ and identify their boundary $n-1$ spheres by the identity map. This is clearly the same

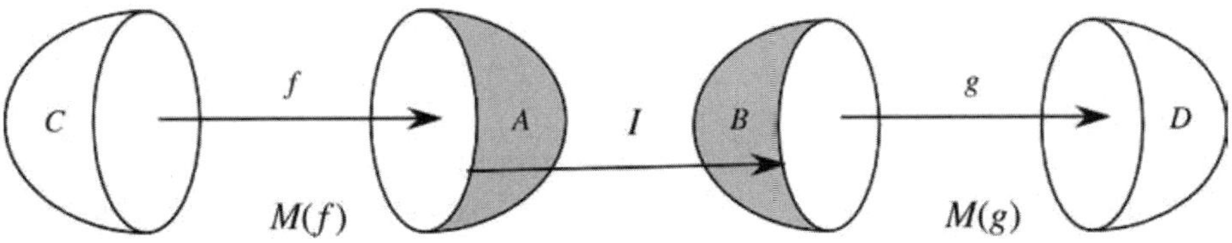

FIGURE 8.

as identifying the hemispheres C and D along their boundary $n-1$ spheres by $g \circ f$. □

Definition. The image group of $\pi_0 \operatorname{Aut} S^{n-1}$ under this homomorphism is denoted by Γ_n. We may describe Γ_n as the *group of twisted n-spheres.*

Next we describe a procedure for constructing non-trivial diffeomorphisms of S^{n-1}. Let $n-1 = p+q$. Start with two homotopy classes

$$(f) \in \pi_p(\mathrm{SO_q}) \text{ and } (\mathrm{g}) \in \pi_\mathrm{q}(\mathrm{SO_p})$$

where $\mathrm{SO_q}$ denotes the group of rotations of $\mathbb{R}^q$. These homotopy classes are represented by smooth maps

$$f : \mathbb{R}^p \to \mathrm{SO_q}, \text{ and } \mathrm{g} : \mathbb{R}^\mathrm{q} \to \mathrm{SO_p}$$

where $f(x) = e$ for $||x|| \geq 1$, $g(y) = e$ for $||y|| \geq 1$. Now define automorphisms F and G of $\mathbb{R}^{p+q} = \mathbb{R}^p \times \mathbb{R}^q$ by

$$\begin{aligned} F(x, y) &= (x, f(x) \cdot y) \\ G(x, y) &= (g(y) \cdot x, y). \end{aligned}$$

Then $FGF^{-1}G^{-1}$ is an automorphism of $\mathbb{R}^{p+q}$ which leaves everything outside of $D^p \times D^q$ fixed. Hence $FGF^{-1}G^{-1}$ extends to an automorphism of the smooth manifold S^{p+q}.

LEMMA 4.2. *This construction gives rise to a bilinear pairing*

$$\pi_p(\mathrm{SO_q}) \otimes \pi_\mathrm{q}(\mathrm{SO_p}) \to \pi_0 \operatorname{Aut} \mathrm{S^{p+q}}.$$

Proof. If f_0 and f_1 represent the same element of $\pi_p(\mathrm{SO_q})$ then one can choose a smooth homotopy f_t between them, where $f_t(x) = e$ for $||x|| \geq 1$. Now the formula $F_t G F_t^{-1} G^{-1}$ describes a diffeotopy between the corresponding automorphisms of S^{p+q}.

Next consider the sum of two homotopy classes (f_1) and (f_2) in $\pi_p(\mathrm{SO_q})$. We may assume that

$$f_1(x) = e \quad \text{for } ||x|| \geq \frac{1}{2}$$

and that $f_2(x) = e$ except for x lying in a ball contained in the region $1/2 \leq ||x|| \leq 1$. Then

$$f_3(x) = f_1(x) \cdot f_2(x)$$

represents the homotopy class $(f_1) + (f_2)$. The identity

$$(F_1 G F_1^{-1} G^{-1})(F_2 G F_2^{-1} G^{-1}) = F_3 G F_3^{-1} G^{-1}$$

now shows that our construction is bilinear.

As an example, suppose that $p = q = 3$. Then the rotation group SO_3 has the sphere S^3 as 2-fold covering space. Hence

$$\pi_3(\mathrm{SO}_3) = \pi_3(\mathrm{S}^3) = \mathbb{Z};$$

and we have homomorphisms

$$\mathbb{Z} = \mathbb{Z} \otimes \mathbb{Z} \xrightarrow{\alpha} \pi_0 \text{ Aut } S^6 \xrightarrow{\beta} \Gamma_7.$$

It will be shown later that $\beta \circ \alpha$ is non trivial. In fact it is known that Γ_7 is a cyclic group of order 28 generated by $\beta\alpha(1)$. Almost nothing is known about π_0 Aut S^6 or image α or kernel β.

5. Tubular Neighborhoods

Definition. An *ℓ-disk bundle* is a fiber bundle with the ℓ-disk D^ℓ as fiber and the orthogonal group $O(\ell)$ as a structure group. An *ℓ-dimensional vector bundle* is a fiber bundle with $\mathbb{R}^\ell$ as fiber and the general linear group $GL(\ell, \mathbb{R})$ as structure group. (For the definition of a fiber bundle see N. Steenrod, *The Topology of Fibre Bundles.*) A fiber bundle is *smooth* if all spaces and maps appearing in the definition are smooth.

Definition. Let M and V be smooth manifolds without boundary, of dimensions k and $n = k+\ell$, respectively. Suppose that M is compact and that $i : M \hookrightarrow V$ is an embedding. A *tubular neighborhood of M in V* is a smooth ℓ-disk bundle with base space M and total space a closed neighborhood N of $i(M)$, such that the embedding $i : M \hookrightarrow N$ agrees with the canonical cross-section of the bundle obtained by selecting the center of D^ℓ at each point. The projection map π of the bundle is therefore a retraction of N onto M.

We will first show that tubular neighborhoods always exist. For the case $V = \mathbb{R}^n$ we will give a detailed proof; we will then indicate how the proof may be modified for the general case.

If M is a smooth compact k-dimensional manifold without boundary and $i : M \hookrightarrow \mathbb{R}^n$ is an embedding, where $n = k + \ell$, define

$$E = \{(x, v) \in M \times \mathbb{R}^n : v \text{ is perpendicular to } i_*(M_x)\}.$$

If $(x, v) \in E$ we can choose a coordinate system $u^1, \ldots, u^n$ for a neighborhood U of $i(x)$ in $\mathbb{R}^n$ so that

$$i(M) \cap U = \{y \in \mathbb{R}^n \cap U : u^{k+1}(y) = \ldots = u^n(y) = 0\}.$$

Let $g_{ij}(y) =$ inner product of $\frac{\partial}{\partial u^i}\big|_y$ and $\frac{\partial}{\partial u^j}\big|_y$. Then a point $(y, a^1, \ldots, a^n)$ of $M \times \mathbb{R}^n$ is in E if and only if

$$\sum_{j=1}^{n} g_{ij}(y)a^j = 0 \quad \text{for } i = 1, \ldots, k.$$

Since the $k \times n$ matrix $g_{ij}(y)$ has rank k this shows that E is an n-dimensional smooth submanifold of $M \times \mathbb{R}^n$.

Let $\exp : E \to \mathbb{R}^n$ be the map (the exponential map) defined by $\exp(x, v) = i(x) + v$; this map is easily seen to be smooth. Let

$$E_\epsilon = \{(x, v) \in E : ||v|| < \epsilon\}$$

and

$$N_\epsilon = \{y \in \mathbb{R}^n : d(y, i(M)) < \epsilon\}.$$

THEOREM 5.1. *For sufficiently small ϵ, the map* exp *is a diffeomorphism from E_ϵ onto N_ϵ.*

It is then clear that $\overline{N}_{\epsilon/2}$ is a tubular neighborhood of M.

Proof. Let $U \subset E$ be the set of non-critical points of exp. It is easy to see that any pair $(x, 0) \in E$ belongs to U; and that $\exp|_U$ is a local diffeomorphism. Choose ϵ_1 so that $U \supset U_1 = \{(x, v) \in E : ||v|| \le \epsilon_1\}$.

Then U_1 is compact and we may use the following lemma.

LEMMA 5.2. *Let A_0 be a closed subset of a compact metric space A. Let $f : A \to B$ be a local homeomorphism such that $f|_{A_0}$ is one-to-one. Then there is a neighborhood U of A_0 such that $f|_U$ is one-to-one.*

Proof. Let $C \subset A \times A$ be the set of all pairs (x, y) with $x \neq y$ but $f(x) = f(y)$. Then C is closed: let (x_n, y_n) be a sequence in C with $x_n \to x$ and $y_n \to y$; since $f(x_n) = f(y_n)$ for every n we have $f(x) = f(y)$; moreover $x \neq y$ since f is locally one-to-one.

Let $g : C \to \mathbb{R}$ be defined by $g(x, y) = d(x, A_0) + d(y, A_0)$. Then g is everywhere positive. Since C is compact there is an $\epsilon > 0$ such that $g \ge 2\epsilon$ on C. Then f is one-to-one on the ϵ-neighborhood of A_0.

We may apply the lemma to the map exp which is a local homeomorphism on the compact metric space U_1 and one-to-one on $\{(x, 0) \in E\}$. It follows that for sufficiently small ϵ the map exp is a diffeomorphism of E_ϵ into N_ϵ.

If $p \in N_\epsilon$ there is a point x of M which is nearest to p. Then the vector $\overrightarrow{i(x), p}$ is perpendicular to $i_*(M_x)$ and has length $< \epsilon$. Therefore $p = \exp(x, \overrightarrow{i(x), p})$ so that exp takes E_ϵ diffeomorphically onto N_ϵ.

In the general case (when V is not $\mathbb{R}^n$) choose some Riemannian metric g for V. Define

$$\begin{aligned} E &= \{(x, v) : x \in M \text{ and } v \text{ is a tangent vector of } V_{i(x)} \text{ with } g(v, i_*(M_x)) = 0\} \\ E_\epsilon &= \{(x, v) \in E : \sqrt{g(v, v)} < \epsilon\} \\ N_\epsilon &= \{y \in V : \rho(y, i(M)) < \epsilon\} \end{aligned}$$

where ρ is the (topological) metric on V induced by g.

For small enough ϵ the map exp may be defined as follows: given a tangent vector v at $i(x)$ let w be the geodesic with $w(0) = i(x)$ and $w'(0) = v$. If ϵ is sufficiently small w may be defined on $[0, 1]$. The point $w(1)$ is defined as $\exp(x, v)$.

Theorem 5.1 remains true; the proof is the same.

Next we will prove a theorem which asserts that tubular neighborhoods are essentially unique. Let $i : M \to V$ be an embedding of the smooth compact manifold M into the smooth manifold V. Suppose that we are given two tubular neighborhoods of M with projection maps.

$$\pi : N \to M, \quad \pi' : N' \to M$$

respectively.

THEOREM 5.3. *There exists a strong diffeotopy F_t of V leaving $i(M)$ pointwise fixed so that*

(1) F_0 = *identity map of V, and*

(2) $F_1|_N$ *is a bundle map of N onto N'.*

Thus F_1 is an automorphism of V which throws each ℓ-disk $\pi^{-1}(x)$ linearly onto the disk $\pi'^{-1}(x)$.

The proof begins as follows. We may clearly assume that $N \subset E \subset V$ [or $N' \subset E' \subset V$] where E [or E'] has the structure of a smooth ℓ-dimensional vector bundle over M which extends the structure of the given k-disk bundle.

LEMMA 5.4. *There is a bundle map $f : E \to E'$ such that $f : E \to V$ is weakly diffeotopic to the inclusion $E \subset V$.*

Proof. Case 1: Suppose that $E \subset E'$, and let $j : E \to E'$ be the inclusion map. Define

$$j_t(e) = t^{-1}j(te) \text{ for } t \neq 0,\ e \in E;$$

where multiplication by t [or t^{-1}] is the multiplication within the fiber over $\pi(e)$ [or $\pi' j(te)$].

Choosing local coordinates (x, y) for E where $x = (x^1, \ldots, x^k)$ gives the point in the base space M and y the point in the fiber $\mathbb{R}^\ell$; and choosing similar coordinates (x', y') for E' we have

$$j_t(x, y) = (\alpha(x, ty), t^{-1}\beta(x, ty))$$

for certain smooth maps α and β, where

$$\alpha(x, 0) = x \qquad \beta(x, 0) = 0$$

(since j leaves M fixed). The first coordinate $\alpha(x, ty)$ is clearly a smooth function of x, y and t even when $t \to 0$.

We can write $\beta(x, y) = \sum y^i g_i(x, y)$ by Lemma 2.1. (The extra variables x do not affect the proof of Lemma 2.1.) Thus

$$j_t(x, y) = (\alpha(x, ty), \sum y^i g_i(x, ty))$$

which makes sense, and is smooth, even for $t = 0$. In fact

$$j_0(x, y) = (x, \sum y^i g_i(x, 0)).$$

Thus the second coordinate of j_0 is a linear function L of y, for fixed x. In order to complete the proof of Case 1 it is sufficient to show that L in non-singular.

Since $g_i(x, 0) = \frac{\partial \beta^j}{\partial y^i}(x, 0)$ the matrix of L is

$$\left(\frac{\partial \beta^j}{\partial y^i}(x, 0) \right).$$

The determinant of this matrix is the same as the determinant of the matrix of j_1, so it is non-zero. Thus j_0 is a bundle isomorphism from E to E'.

Case 2 (the General Case): The set $E \cap E'$ is a manifold containing M. There is an ℓ-dimensional vector bundle $E" \subset E \cap E'$ (since there is a tubular neighborhood of M in $E \cap E'$). Then Case 1 may be applied to E and $E"$ and to E' and $E"$. This completes the proof of Lemma 5.4. □

We now wish to prove that F_1 can be made a bundle map when the group is reduced to the orthogonal group $O(\ell)$. Let $L : W \to W'$ be a non-singular linear transformation between two real vector spaces with a Euclidean inner product. We will need the following facts:

(1) Any non-singular matrix A can be expressed uniquely as $A = OP$ where O is orthogonal and P is symmetric positive definite. (See C. Chevalley, *Theory of Lie Groups I,* p. 14.) Furthermore O and P depend differentiability on A.

(2) The set of symmetric positive definite matrices is convex.

Fact **(1)** implies that L can be factored uniquely $L = OP$ (where $P : W \to W$ and $O : W \to W'$) with P positive definite and O an isometry. We may apply this factorization to the bundle map f constructed in Lemma 5.4 fiber by fiber to obtain a factorization of f,

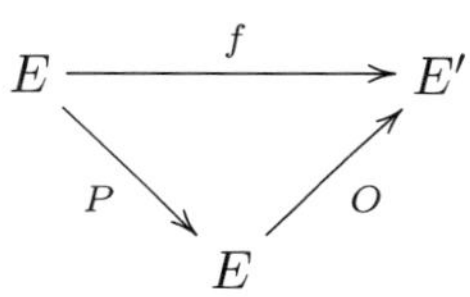

$f = OP$ where O and P are smooth fiber maps, P is symmetric positive definite on each fiber, and O is an isometry on each fiber. Since the symmetric positive definite matrices are convex the family of bundle maps

$$f_t = 0 \cdot (tI + (1 - t)P)$$

is a diffeotopy between f and a bundle map $E \to E'$ which is an isometry on each fiber. Therefore we have proved:

LEMMA 5.5. *There is a bundle map $g : E \to E'$ which is an isometry on each fiber such that $g : E \to V$ is weakly diffeotopic to the inclusion $E \subset V$.*

Proof of Theorem 5.3. Clearly $g|_N$ gives a bundle map from N to N', which is weakly diffeotopic, in V, to the inclusion $N \subset V$. Since N is compact we can apply Theorem 1.1 and conclude that $g|_N$ is strongly diffeotopic to the inclusion. This completes the proof. □

Smooth Manifolds with Boundary

John Milnor

Based on lectures at the University of Mexico in 1960.
notes by H. Cárdenas, E. Lluis, F. Recillas and R. Vázquez
as translated by I. E. Colon-Roldan and S. Simanca

Contents:

1. Basic Constructions: Smooth Manifolds and Smooth Maps.

Let $\mathbb{H}^n$ be the Euclidean half-space consisting of all points $x = (x_1, \ldots, x_n)$ of the real Euclidean space $\mathbb{R}^n$ such that $x_n \geq 0$. Define the *boundary* $\partial\mathbb{H}^n \cong \mathbb{R}^{n-1}$ of $\mathbb{H}^n$ as the set of all points in $\mathbb{H}^n$ with $x_n = 0$. (For the case $n = 0$, we define $\mathbb{H}^0 = \mathbb{R}^0$ to be a single point with vacuous boundary.)

The concept of smoothness (or more precisely C^∞-smoothness[1]) will first be defined for maps between open subsets of Euclidean spaces or half-spaces and then for more general smooth manifolds. If V is an open subset of $\mathbb{R}^n$ or of $\mathbb{H}^n$, and if V' is an open subset of $\mathbb{R}^m$ or $\mathbb{H}^m$, then a continuous map $f : V \to V'$ will be called *smooth* if the partial derivatives of all orders $\frac{\partial^r f_i}{\partial x_{j_1} \cdots \partial x_{j_r}}$ are defined and continuous as mappings from V to the real numbers. Note that the identity map

[1]For the purposes of differential topology, one could equally well work with C^r-smooth manifolds and C^r-smooth maps for any integer $r \geq 1$. In fact any C^r-smoothness structure can be upgraded to a C^∞-smoothness structure and any C^r-diffeomorphism can be approximated by a C^∞-diffeomorphism, provided that $r \geq 1$. However, C^∞ structures are particularly convenient to work with.

of V is smooth, and that the composition of smooth maps between open subsets of Euclidean spaces or half-spaces is again smooth.

Smooth functions to the real numbers will play a special role. We will make use of the fact that a smooth real valued function on an open subset of $\mathbb{H}^n$ can always be extended locally, near any point, to a smooth real valued function on an open subset of $\mathbb{R}^n$. (Compare the Appendix.[2])

Let V be an open subset of $\mathbb{H}^n$ or $\mathbb{R}^n$. By a *germ* of a function at a point $x \in V$ is meant an equivalence class of functions, each defined on some neighborhood of x in V, where two functions f_1, f_2 defined in neighborhoods V_1, V_2 of x are said to be equivalent if there is a neighborhood V_3 of x, with $V_3 \subset V_1 \cap V_2$, such that f_1 and f_2 coincide on V_3.

Fixing some point $x \in V$, let $\mathcal{S}_x(V)$ be the ring consisting of all germs of smooth real valued functions defined on open neighborhoods of the point x in V. (Note that $\mathcal{S}_x(V) \cong \mathcal{S}_x(\mathbb{H}^n)$ if V is open in $\mathbb{H}^n$, and that $\mathcal{S}_x(V) \cong \mathcal{S}_x(\mathbb{R}^n)$ if V is open in $\mathbb{R}^n$.) The union $\mathcal{S}(V) = \bigcup_{x \in V} \mathcal{S}_x(V)$ consists of all germs of smooth real valued functions at the points of V. Note that a function $f : V \to \mathbb{R}$ is smooth if and only if its germ at each point $x \in V$ belongs to this ring $\mathcal{S}_x(V)$. Furthermore, a function $x \mapsto f(x) = \big(f_1(x), \ldots, f_m(x)\big)$ from V to $V' \subset \mathbb{R}^m$ or $V' \subset \mathbb{H}^m$ is smooth if and only if each component function $f_i : V \to \mathbb{R}$ satisfies this condition. Still another equivalent condition can be given as follows.

LEMMA 1.1. A smoothness criterion in $\mathbb{R}^n$. *Again let V and V' be open subsets of Euclidean spaces or half-spaces. A function $f : V \to V'$ is smooth (that is, has continuous partial derivatives of all orders) if and only if, for each $x \in V$ and for each germ $g \in \mathcal{S}_{f(x)}(V')$ of a smooth real valued function at the point $y = f(x)$, the germ of the composition $g \circ f$ at x belongs to $\mathcal{S}_x(V)$.*

The proof is completely straightforward. These constructions should help to motivate the following.

Definition 1.2. A smooth manifold is a pair $(M, \mathcal{S}(M))$ consisting of a Hausdorff space M together with a set $\mathcal{S}(M)$ of germs of real valued functions on open subsets of M, satisfying the following condition:

For each point $p \in M$ there is a neighborhood U of p, an open subset V of $\mathbb{H}^n$ or of $\mathbb{R}^n$ for some $n \geq 0$, and a homeomorphism $h : U \to V$, such that a germ of real valued functions in V belongs to $\mathcal{S}(V)$ if and only if the corresponding germ in U belongs to $\mathcal{S}(M)$.

The set $\mathcal{S}(M)$ will be called a *smoothness structure* on M. The triple (U, V, h) is called a *coordinate chart* for $(M, \mathcal{S}(M))$, and U is called a *coordinate neighborhood* of the point p.

(Occasionally, when the smoothness structure $\mathcal{S}(M)$ is completely clear, we may simply refer to M as a smooth manifold.)

[2]The proof requires some work. For an easier presentation, one could just define a smooth function on an open subset of $\mathbb{H}^n$ to be one which extends locally to a smooth function on $\mathbb{R}^n$.

As examples, the half-space $(\mathbb{H}^n, \mathcal{S}(\mathbb{H}^n))$ and the full Euclidean space $(\mathbb{R}^n, \mathcal{S}(\mathbb{R}^n))$ are certainly examples of smooth manifolds. Similarly, any open subset $V \subset \mathbb{H}^n$ or $V \subset \mathbb{R}^n$ has a natural smoothness structure $\mathcal{S}(V)$, defined as above. Other examples will be described presently.

Remarks.

(1) The dimension n is a continuous integer valued function on M, and is therefore constant on each connected component of M. In most applications, this dimension function takes the same value everywhere, and one speaks of an n-dimensional manifold, or briefly an n-manifold. However, occasionally it can be convenient to allow the more general case, where different connected components may have different dimensions.

(2) If the charts (U, V, h) all have the property that each V is an open subset of $\mathbb{R}^n$ (or of the open half-space $x_n > 0$ in $\mathbb{H}^n$), then we speak of a manifold *without boundary*.

(3) It is often convenient to describe a smooth manifold structure by means of its collections of coordinate charts. We will see in Lemma 1.4 that this form of the definition is completely equivalent to the one given above.

The concept of a smooth map between arbitrary smooth manifolds can now be defined as follows.

Definition 1.3. Let $(M, \mathcal{S}(M))$, $(N, \mathcal{S}(N))$ be smooth manifolds. A continuous map $f : M \to N$ is *smooth* if for each $x \in M$ and each $\varphi \in \mathcal{S}_{f(x)}(N)$, the germ of $\varphi \circ f$ at x belongs to $\mathcal{S}_x(M)$. If the inverse map $f^{-1} : N \to M$ is also defined and smooth, then f is called a *diffeomorphism* (or more precisely a C^∞-*diffeomorphism*) between $(M, \mathcal{S}(M))$ and $(N, \mathcal{S}(N))$.

Evidently the identity map of M is smooth, as is the composition of any two smooth maps. In the special case where M and N are open subsets of Euclidean spaces or half-spaces, it follows from Lemma 1.1 that f is smooth in this new sense if and only if it is smooth in the original sense (continuous partial derivatives of all orders).

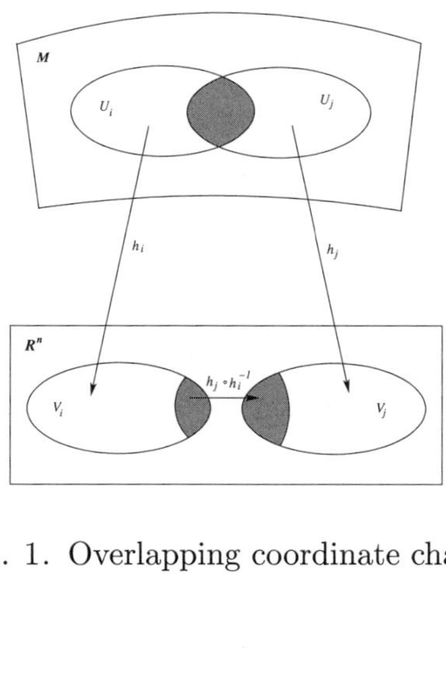

FIG. 1. Overlapping coordinate charts.

LEMMA 1.4. Overlapping coordinate charts. *Let (U_i, V_i, h_i) and (U_j, V_j, h_j) be two coordinate charts for the smooth manifold $(M, \mathcal{S}(M))$, and suppose that $U_i \cap U_j \neq \emptyset$. Then the composition*

$$h_j \circ h_i^{-1} : h_i(U_i \cap U_j) \to h_j(U_i \cap U_j)$$

is a smooth map between open subsets of $\mathbb{R}^n$ or $\mathbb{H}^n$. Conversely, let M be a Hausdorff space and $\{(U_i, V_i, h_i)\}$ a collection of triples where the U_i are open subsets with union equal to M, where each V_i is an open subset of some $\mathbb{R}^{n_i}$ or $\mathbb{H}^{n_i}$, and each h_i is a homeomorphism of U_i onto V_i. If each transition map

$$h_j \circ h_i^{-1} : h_i(U_i \cap U_j) \to h_j(U_i \cap U_j)$$

has continuous partial derivatives of all orders, then there is one and only one smoothness structure $\mathcal{S}(M)$ so that $(M, \mathcal{S}(M))$ is a smooth manifold and so that each h_i is a diffeomorphism.

Thus the definition of smooth manifold as given above, in terms of germs of smooth real valued functions, is completely equivalent to the more customary definition in terms of overlapping coordinate charts. The proof is straightforward. □

1.1. Operations with smooth manifolds. To conclude this section, we describe seven basic constructions involving smooth manifolds.

(1) Submanifolds. Let $(M, \mathcal{S}(M))$ be a smooth manifold and let N be any subset of M. Define the restriction of the smoothness structure $\mathcal{S}(M)$ to N as follows: A germ g of a real valued function at the point $x \in N$ belongs to this restriction $\mathcal{S}(N) = \mathcal{S}(M)|_N$ if and only if there exists a neighborhood U of x in M and a smooth real valued function on U whose restriction to $U \cap N$ represents the given germ g. If this $\mathcal{S}(N)$ is a smoothness structure, or in other words if the pair $(N, \mathcal{S}(N))$ is a smooth manifold, then $(N, \mathcal{S}(N))$ is called a *smooth submanifold* of $(M, \mathcal{S}(M))$. We may also say that $(N, \mathcal{S}(N))$ is *smoothly embedded* in $(M, \mathcal{S}(M))$. Note that this will always be the case if N is an open subset of M.

(2) Boundary. Recall that the boundary of the half-space $\mathbb{H}^n = \{(x_1, \ldots, x_n) ; x_n \geq 0\}$ is defined to be the subset $\partial\mathbb{H}^n = \{(x_1, \ldots, x_n) ; x_n = 0\}$. Similarly, for any open subset $V \subset \mathbb{H}^n$, the boundary is defined to be the set $\partial V = V \cap \partial\mathbb{H}^n$. It is not difficult to check that any diffeomorphism between open subsets of $\mathbb{H}^n$ must carry boundary to boundary.

For any smooth manifold $(M, \mathcal{S}(M))$, the *boundary* $(\partial M, \mathcal{S}(\partial M))$ is a possibly vacuous smooth submanifold. We first define the subset $\partial M \subset M$. If (U, V, h) is a coordinate chart for $(M, \mathcal{S}(M))$, then we set $\partial U = h^{-1}(\partial V)$, and define ∂M to be the union $\bigcup \partial U$ over all such coordinate charts. The closed subset $\partial M \subset M$ defined in this way is always a smooth submanifold. In other words, the restriction $\mathcal{S}(\partial M) = \mathcal{S}(M)|_{\partial M}$ is necessarily a smoothness structure on ∂M. In fact it is not difficult to check that $\partial U = U \cap \partial M$, and that the collection of triples $\{(\partial U_i, \partial V_i, h_i|_{\partial U_i})\}$ constitutes a family of coordinate charts with $\bigcup \partial U_i$ equal to

∂M. The complement $\overset{\circ}{M} = M \smallsetminus \partial M$ is called the *interior* of M. If M is a manifold without boundary (i.e., with vacuous boundary), then every point of M is an interior point.

Since ∂V is an open subset of $\partial \mathbb{H}^n \cong \mathbb{R}^{n-1}$, it follows that ∂V is an $(n-1)$-dimensional manifold without boundary: $\partial(\partial V) = \emptyset$. For any M it follows that ∂M is a manifold without boundary. If M has constant dimension n (for example if M is connected), note that ∂M has constant dimension $n-1$.

(3) Tangent bundle. If p is an interior point of a smooth manifold M, then a *tangent vector* $\mathbf{v}$ at p can be defined for example as an equivalence class $[\phi]$ of smooth curves $\phi : (-\epsilon, \epsilon) \to M$, with $\phi(0) = p$, where $(-\epsilon, \epsilon)$ is some neighborhood of zero in $\mathbb{R}$. By definition, two such curves ϕ and ψ are *equivalent* if, for some coordinate chart (U, V, h) with $p \in U$, the derivatives $dh_i(\phi(t))/dt$ at $t = 0$ are equal to the corresponding derivatives $dh_i(\psi(t))/dt$ at $t = 0$. This condition does not depend on the particular choice of coordinate chart. The set of all such tangent vectors at p form a vector space TM_p. In fact if V has dimension n and if $x(t) = h(\phi(t))$, then the correspondence

$$[\phi] \;\mapsto\; \left.\frac{dx(t)}{dt}\right|_{t=0} = \left.\left(\frac{dx_1}{dt}, \ldots, \frac{dx_n}{dt}\right)\right|_{t=0}$$

maps the tangent vector space TM_p isomorphically onto $\mathbb{R}^n$. Note that any smooth map $f : M \to N$ induces a linear map $\mathcal{D}f_p : TM_p \to TN_{f(p)}$, called the *derivative* of f at p, by the formula $\mathcal{D}f_p[\phi] = [f \circ \phi]$.

If p is a boundary point of M, then we must modify this definition by allowing curves $\phi : [0, \epsilon) \to M$ or $\phi : (-\epsilon, 0] \to M$ which are defined only on a one-sided neighborhood of zero. The corresponding tangent vectors $[\phi]$ are said to point *inward* in the first case and *outward* in the second. *Thus the tangent space TM_p at a boundary point is the union of a half-space of inward vectors together with a half-space of outward vectors.* The intersection of these two half-spaces is the tangent space $T(\partial M)_p$ of the boundary manifold at p.

The *tangent manifold* TM of an n-dimensional smooth manifold M is a $2n$-dimensional smooth manifold which can be expressed as the union of disjoint subsets TM_p, where p ranges over M. This union TM is topologized and given a smoothness structure so that the following two conditions are satisfied:

(a) If V is an open subset of $\mathbb{R}^n$ or $\mathbb{H}^n$, then TV is diffeomorphic to $V \times \mathbb{R}^n \subset \mathbb{R}^{2n}$ under the correspondence $[\phi] \;\leftrightarrow\; \left(\phi(0)\,,\, \frac{d\phi}{dt}(0)\right)$.

(b) If (U, V, h) is a coordinate chart for M, then $(TU, TV, \mathcal{D}h)$ is a coordinate chart for TM. In particular, the tangent manifold TU is an open subset of TM, and the derivative $\mathcal{D}h$ maps TU diffeomorphically onto TV, which can be identified with $V \times \mathbb{R}^n$ as above.

It follows easily that every smooth map $f : M \to N$ gives rise to an induced smooth map $\mathcal{D}f : TM \to TN$ which carries each subset TM_p to the subset $TN_{f(p)}$ by the linear map $\mathcal{D}f_p$. Further details will be left to the reader.

(4) Orientation. By an *orientation* for a finite dimensional vector space W is meant an equivalence class of ordered bases, where two bases $\{e_j\}$ and $\{e'_i = \sum_j a_{ij} e_j\}$ are equivalent (or define the same orientation) if and only if the matrix $[a_{ij}]$ has positive determinant. An orientation for a smooth manifold is a choice of orientation for every tangent vector space TM_p, where these orientations are required to depend continuously on the point p, in the following sense. Each $p_0 \in M$ should have a neighborhood U and a choice of basis $\{p \mapsto e_i(p) \in TM_p\}$ depending continuously on $p \in U$, so that $\{e_i(p)\}$ determines the specified orientation for each $p \in U$. The manifold is said to be *orientable* if it possesses such an orientation. If a connected manifold is orientable, then evidently it can be given exactly two possible orientations.

(5) Disjoint sum. Let $\{(M_i, \mathcal{S}(M_i))\}_{i\in I}$ be an arbitrary family of smooth manifolds, and let $\biguplus M_i$ be the topological sum (that is the disjoint union of the M_i, topologized so that each M_i is embedded as an open subset). We make this into a smooth manifold by setting $\mathcal{S}(\biguplus M_i)$ equal to the union of the $\mathcal{S}(M_i)$. Evidently each $(M_i, \mathcal{S}(M_i))$ is embedded as a smooth open submanifold of this union $\big(\biguplus M_i, \mathcal{S}(\biguplus M_i)\big)$.

(6) Product. If $(M, \mathcal{S}(M))$, $(N, \mathcal{S}(N))$ are smooth manifolds and if at least one of the two has vacuous boundary, then we define $\mathcal{S}(M \times N)$ as the unique smoothness structure on the cartesian product $M \times N$ for which the projections p_1 and p_2 are smooth and which satisfies the following universal property:

Given any smooth manifold $(X, \mathcal{S}(X))$ and smooth maps f_1, f_2, as in the diagram below, the product function $F = (f_1, f_2)$ is also smooth.

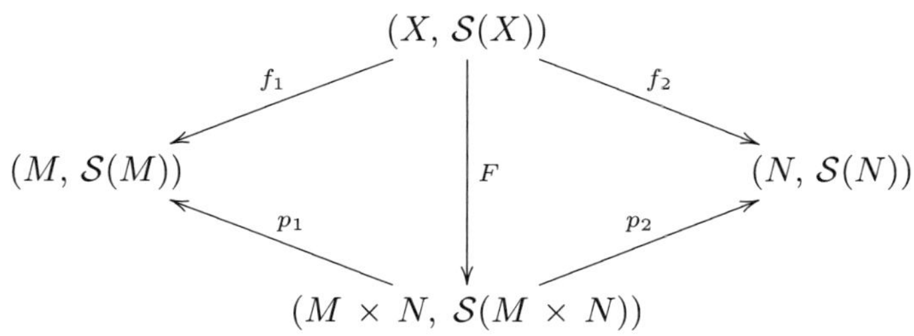

This definition can easily be extended to any finite number of factors, provided that at most one of these factors has non-vacuous boundary.

(7) Quotient manifolds. Let $(M, \mathcal{S}(M))$ be a smooth manifold and let $\stackrel{r}{\sim}$ be an equivalence relation on M. Then we can form the quotient space $M/\stackrel{r}{\sim}$, together with a canonical projection $j : M \to M/\stackrel{r}{\sim}$. If this quotient space is sufficiently well behaved, then it can be given the structure of a smooth manifold, with smoothness structure $\mathcal{S}(M/\stackrel{r}{\sim})$ defined as follows. A germ of a real valued function φ at the point $y \in M/\stackrel{r}{\sim}$ belongs to the ring $\mathcal{S}_y(M/\stackrel{r}{\sim})$ if and only if the germ of the composition $\varphi \circ j$ belongs to $\mathcal{S}_x(M)$ for every x in the equivalence class $j^{-1}(y)$. If $M/\stackrel{r}{\sim}$ is a Hausdorff topological manifold, and if $\mathcal{S}_y(M/\stackrel{r}{\sim})$ is a smoothness structure on this manifold, then $(M/\stackrel{r}{\sim},\ \mathcal{S}(M/\stackrel{r}{\sim}))$ is called the quotient smooth manifold. As an example, if $M = \mathbb{R}^{n+1} \smallsetminus \{0\}$ and if the equivalence class $j^{-1}(j(x))$ is the set $(\mathbb{R} \smallsetminus \{0\})x$ consisting of all non-zero real multiples, then the quotient space $\mathbb{R}^{n+1}/\stackrel{r}{\sim}$ is called the *real projective space* $\mathbb{RP}^n$. If we carry out the same construction using complex numbers in place of real numbers, then we obtain the *complex projective space* $\mathbb{CP}^n$, which is a $2n$-dimensional smooth manifold.

Still another important construction, pasting together the boundaries of two smooth manifolds via a boundary diffeomorphism, will be described at the end of the following section.

2. Manifolds are Homogeneous.

This section will first prove the following.

THEOREM 2.1. Equivalence of disk-embeddings. *Let $f : D^n \to M$ and $g : D^n \to M$ be two smooth embeddings of the closed disk D^n as submanifolds of the interior of a smooth connected n-dimensional manifold M. If M is orientable, we assume also that these two embeddings have compatible orientations. Then there exists a diffeomorphism $h : M \to M$ such that $h \circ f = g$, where h is equal to the identity outside a compact subset of the interior of M.*

The general strategy of the proof can be described as follows. (Compare PALAIS [1960].) We first compose f with a diffeomorphism h_1 so that $h_1 \circ f(0) = g(0)$. Then we compose with a further diffeomorphism h_2 so that the embedding $h_2 \circ h_1 \circ f$ is actually tangent to g at 0. After composing with still another diffeomorphism, we may suppose that the two maps coincide throughout some neighborhood of 0. Finally, by deforming both f and g down into this small neighborhood we can complete the argument.

This proof will depend on a number of lemmas, starting with the following global form of the inverse function theorem.

LEMMA 2.2. Diffeomorphism criterion. *A smooth map $f : M \to N$ is a diffeomorphism if and only the induced map $\mathcal{D}f : TM \to TN$ is bijective (ie., one-to-one and onto).*

In other words, f is a diffeomorphism if and only if (a) f itself is bijective, and (b) for each $p \in M$ the linear map $\mathcal{D}f_p : TM_p \to TN_{f(p)}$ between tangent spaces is also bijective.

Proof. First suppose that $f : V \to V'$ is a map between open subsets of $\mathbb{R}^n$ or $\mathbb{H}^n$. Then the proof proceeds just as for the standard inverse function theorem. It is convenient to identify TV with $V \times \mathbb{R}^n$, and to use the slightly modified notation

$$\mathcal{D}f(x, v) \;=\; \big(f(x)\,,\, \mathcal{D}f_x(v)\big)$$

for the induced map of tangent bundles in this case, where $\mathcal{D}f_x$ can be identified with the $n \times n$ matrix of first derivatives at x. Write the Taylor series at a point as

$$f(x + \Delta x) \;=\; f(x) + \mathcal{D}f_x(\Delta x) + \mathcal{R}\;,$$

where the remainder term satisfies $\|\mathcal{R}\|/\|\Delta x\| \to 0$ as $\Delta x \to 0$. If $y = f(x)$ and $y + \Delta y = f(x + \Delta x)$, then we have

$$f^{-1}(y + \Delta y) \;=\; x + \Delta x \;=\; f^{-1}(y) + (Df_x)^{-1}(\Delta y) + \mathcal{R}'\;,$$

where $\mathcal{R}' = -(\mathcal{D}f_x)^{-1}(\mathcal{R})$. It follows easily that $\|\mathcal{R}'\|/\|\Delta y\| \to 0$ as $\|\Delta y\| \to 0$, so that f^{-1} is differentiable at y, with a derivative $\mathcal{D}(f^{-1})_y = (\mathcal{D}f_x)^{-1}$ which depends continuously on y. Thus f^{-1} is at least C^1-smooth throughout the set $f(V) = V'$.

Now let us apply this same argument to the smooth map $F = \mathcal{D}f : TV \to TV'$. A brief computation shows that the $(2n) \times (2n)$ matrix $\mathcal{D}F_{(x,v)}$ has the form

$$\begin{pmatrix} \mathcal{D}f_x & 0 \\ * & \mathcal{D}f_x \end{pmatrix}$$

where $*$ stands for the matrix of second derivatives of f at x. Thus $\mathcal{D}F$ is also bijective, hence F^{-1} is at least C^1-smooth. Since the matrix of first derivatives of F^{-1} contains the matrix of second derivatives of f^{-1}, this implies that f^{-1} is at least C^2-smooth. Now applying this same argument to F, it follows that F^{-1} is at least C^2-smooth, hence f is at least C^3-smooth. Continuing inductively, it follows that f^{-1} is C^r-smooth for every positive integer r.

This proves Lemma 2.2 for open subsets of $\mathbb{R}^n$ or $\mathbb{H}^n$. The transition to arbitrary smooth manifolds is straightforward, using local coordinate charts. □

Using Lemma 2.2, we will prove that smooth maps of $\mathbb{R}^n$ which are sufficiently close to the identity must necessarily be diffeomorphisms.

LEMMA 2.3. A diffeomorphism criterion in $\mathbb{R}^n$. *Let $f : \mathbb{R}^n \to \mathbb{R}^n$ be a smooth map such that*

$$\left| \frac{\partial f_i}{\partial x_j} - \delta_{ij} \right| \le \frac{1}{2n}$$

for all points in $\mathbb{R}^n$ and all i, j, where $[\delta_{ij}]$ is the identity matrix. Then f is a diffeomorphism.

Proof. Let us write $f(x) = x + a(x)$, with components $f_i(x) = x_i + a_i(x)$. These functions $a_i(x)$ evidently satisfy, $|\partial a_i/\partial x_j| \le 1/2n$, for all i and j. It will be convenient for this proof to use the *max norm* on Euclidean space, defined by the formula $\|x\|_{\max} = \max\{|x_1|, \dots, |x_n|\}$. If we follow a broken path from x to y in $\mathbb{R}^n$, changing just one coordinate at a time, and apply the mean value theorem to each segment of this path, then we see that

$$|a_i(x) - a_i(y)| \le \sum_{j=1}^{n} \tfrac{1}{2n}|x_j - y_j| \le \sum_{j=1}^{n} \tfrac{1}{2n}\|x - y\|_{\max} = \tfrac{1}{2}\|x - y\|_{\max}$$

hence

$$\|a(x) - a(y)\|_{\max} \le \tfrac{1}{2}\|x - y\|_{\max}$$

for all $x, y \in \mathbb{R}^n$. Thus the mapping $x \mapsto a(x)$ contracts distances uniformly.

It follows that f is bijective. For if we choose any base point $z \in \mathbb{R}^n$ and set

$$g_z(x) = z - a(x) ,$$

then clearly the map g_z from $\mathbb{R}^n$ to $\mathbb{R}^n$ also contracts distances uniformly,

$$\|g_z(x) - g_z(y)\| \le \tfrac{1}{2}\|x - y\|.$$

Hence, by a standard argument, g_z has one and only one fixed point $x = g_z(x)$. In fact, if we start with an arbitrary point $y \in \mathbb{R}^n$ and apply the map g_z over and over again, then the resulting orbit $y \mapsto g_z(y) \mapsto g_z(g_z(y)) \mapsto \dots$ will necessarily converge to this unique fixed point x. But clearly x is a solution to the equation

$f(x) = x + a(x) = z$ if and only if $g_z(x) = z - a(x)$ is equal to x. Thus the equation $f(x) = z$ has one and only one solution x.

For any $x \in \mathbb{R}^n$, the linear map $\mathcal{D}f_x : T\mathbb{R}^n_x \to T\mathbb{R}^n_{f(x)}$ also satisfies the hypothesis of Lemma 2.3. Hence the argument above also shows that each $\mathcal{D}f_x$ is bijective. Hence, by Lemma 2.2, f is a diffeomorphism. □

LEMMA 2.4. Moving a point in $\mathbb{R}^n$. *If V is a neighborhood of the point $p \in \mathbb{R}^n$, then for any $q \in \mathbb{R}^n$ which is sufficiently close to p there exists a diffeomorphism $f : \mathbb{R}^n \to \mathbb{R}^n$ with $f(p) = q$ such that f is the identity outside some compact subset of V.*

Without loss of generality, we may assume that $p = 0$ and that V contains the closed unit disk. Let $\sigma : \mathbb{R} \to \mathbb{R}$ be a smooth function such that

$$\sigma(u) = \begin{cases} 1, & \text{if} \quad \mathrm{u} \le 1/2\,, \\ 0, & \text{if} \quad \mathrm{u} \ge 1\,. \end{cases}$$

(Compare Lemma A.1 in the appendix.) Let q be a point in $\mathbb{R}^n$ and $f_i(x) = x_i + \sigma(\|x\|^2)\, q_i$, where $\|x\| = \sqrt{\sum x_i^2}$ is the Euclidean norm. Then

$$\partial f_i/\partial x_j = \delta_{ij} + 2x_j\sigma'\big(\|x\|^2\big)\, q_i\,,$$

from which it follows that

$$\big|\partial f_i/\partial x_j - \delta_{ij}\big| = \big|2x_j\sigma'(\|x\|^2)\, q_i\big|\,.$$

Therefore, it suffices to take $\|q\|$ less than $1/\big(4n \max(|\sigma'(u)|)\big)$ to meet the conditions of Lemma 2.3 and conclude that f is a diffeomorphism which maps the origin to q and equals the identity outside the compact subset $\|x\| \le 1$ of V. This proves Lemma 2.4. □

LEMMA 2.5. Moving a point in a manifold. *Let M be a connected smooth manifold with interior $\mathring{M}$. Given any two points p and q in $\mathring{M}$, there exists a diffeomorphism $f : M \to M$ such that $f(p) = q$. In fact we can always choose f to be a diffeomorphism with compact support contained in $\mathring{M}$.*

That is, we can choose f so that $f(x) = x$ for x outside of some compact subset of the interior of M.

Proof of Lemma 2.5. Let p be a point of M and (U, V, h) a coordinate chart, with $p \in U$ and $h : U \xrightarrow{\cong} V \subset \mathbb{R}^n$. By Lemma 2.4, there is a neighborhood V' of $h(p)$ in V so that we can move $h(p)$ to any point of V' by a diffeomorphism ϕ which is the identity outside of a compact subset of V. Hence we can move p to any point of $h^{-1}(V')$ by a diffeomorphism $h^{-1} \circ \phi \circ h$ which is the identity outside a compact subset of U. Evidently $h^{-1} \circ \phi \circ h$ extends to a diffeomorphism of the whole manifold M.

Now define an equivalence relation on $\mathring{M}$ by setting p equivalent to q whenever there exists a diffeomorphism of M onto itself with compact support in $\mathring{M}$ that maps p to q. By the above arguments, the equivalence classes are open, and therefore closed. Since $\mathring{M}$ is connected, there can be only one equivalence class. □

LEMMA 2.6. Realizing a prescribed first derivative map at a point. *Given an isomorphism (orientation preserving if the manifold is orientable) of the tangent space TM_p of M at p onto the tangent space TM_q, the diffeomorphism of Lemma 2.5 can be chosen such that the derivative $\mathcal{D}f_p : TM_p \to TM_q$ is equal to the given isomorphism.*

Proof. First consider the case $p = q = 0 \in \mathbb{R}^n$. Let

$$f(x) = x + \sigma(\|x\|^2)L(x)$$

where $L(x)$ is a linear transformation of $\mathbb{R}^n$ into itself, and where

$$\sigma(u) = \begin{cases} 1, & \text{if} \quad u \leq 1/2\,, \\ 0, & \text{if} \quad u \geq 1\,, \end{cases}$$

as before. Then, $\partial f_i/\partial x_j \;=\; \delta_{ij} + 2x_j\sigma'(\|x\|^2)L_i(x) + \sigma(\|x\|^2)\partial L_i/\partial x_j$. If the $\partial L_i/\partial x_j$ are sufficiently small, or in other words if the linear transformation L is sufficiently close to the zero map, then, $|\partial f_i/\partial x_j - \delta_{ij}| \;\leq\; 1/2n$. Therefore f is a diffeomorphism, with $f(x) = x$ for $\|x\| \geq 1$, and

$$f(x) \;=\; x + L(x) \qquad \text{for} \qquad \|x\|^2 \leq 1/2\;.$$

Thus $f(0) = 0$, and the derivative $\mathcal{D}f_0$ is equal to $I + L$, where I is the identity map and L can be any linear map sufficiently close to zero.

This implies that the subgroup consisting of all $I + L \in \mathrm{GL}(n, \mathbb{R})$ for which there exists a diffeomorphism f of $\mathbb{R}^n$ with compact support, with $f(0) = 0$, and with $\mathcal{D}f_0 = I + L$, is open. But an open subgroup is necessarily also closed, since its complement is the union of cosets which are open. Hence this subgroup must contain the connected component of the identity.

When the manifold is orientable, Lemma 2.6 follows immediately. If the manifold is not orientable, an argument similar to the proof of Lemma 2.5 shows that there must exist a diffeomorphism with compact support that maps p onto p and reverses the local orientation. The conclusion then follows easily. □

LEMMA 2.7. Extending a given germ to a diffeomorphism. *Given a point $p \in \mathring{M}$, and a diffeomorphism f of some neighborhood of p into $\mathring{M}$ which preserves the orientation, there exists a diffeomorphism h of M onto itself which coincides with f in a smaller neighborhood of p.*

Proof. After composing f with a diffeomorphism of M, we may assume by Lemma 2.6 that $f(p) = p$ and that the derivative $\mathcal{D}f_p$ is the identity map of the tangent space T_pM. Thus, choosing a coordinate chart, it suffices to consider the case where M is a neighborhood of the origin $0 \in \mathbb{R}^n$, with $p = f(p) = 0$ and with $(\partial f_i/\partial x_j)(0) = \delta_{ij}$. Let $f(x) = x + a(x)$, with $a(0) = 0$ and $(\partial a_i/\partial x_j)(0) = 0$. Now choose $r > 0$ sufficiently small so that f is defined throughout the disk $\|x\| < 2r$, and set,

$$h(x) \;=\; x + a(x)\,\sigma(\|x\|^2/r^2)\;,$$

with σ as above. Then h is a smooth map which extends throughout $\mathbb{R}^n$ with

$$h(x) = \begin{cases} f(x), & \text{if} \qquad \|x\| \leq r/\sqrt{2}\,, \\ x, & \text{if} \qquad \|x\| \geq r\,. \end{cases}$$

Furthermore, h is a diffeomorphism if r is sufficiently small. In fact,

$$\partial h_i/\partial x_j \;=\; \delta_{ij} + 2x_j\, a(x)\, \sigma'\big(\|x\|^2/r^2\big)/r^2 + (\partial a_i/\partial x_j)\, \sigma\ ,$$

and as $r \to 0$ we have, $\|a\|/r \to 0$ and $\partial a_i/\partial x_j \to 0$ for all $\|x\| \le r$, with σ, σ' and $|x_j|/r$ bounded. The conclusion follows, using Lemma 2.3. □

LEMMA 2.8. Extending disk embeddings. *Any smooth embedding $f : D^n \to \mathring{M}$ of the closed unit n-disk into the interior of a smooth n-manifold can be extended to an embedding $\hat{f} : D^n(1+\varepsilon) \to M$ of a slightly larger disk.*

Proof. Identifying the tangent space TD^n as usual with the product $D^n \times \mathbb{R}^n$, consider the outward unit vector field $x \mapsto (x, x/\|x\|)$ for $x \in D^n \setminus \{0\}$. Note that the associated differential equation $dx/dt = x/\|x\|$ has a unique integral curve

$$x(t) \;=\; tu \qquad \text{for} \qquad 0 < t \le 1$$

which meets the boundary $\partial D^n = S^{n-1}$ at a specified unit vector u. Using the diffeomorphism $\mathcal{D}f : TD^n \to T(f(D^n))$, we obtain a corresponding vector field $y \mapsto w(y) \in TM_y$ for y belonging to the image disk $f(D^n) \subset M$. Thus the differential equation $dy/dt = w(y)$ has general solution $t \mapsto y = f(tu)$ for $0 < t \le 1$, where u can be any point in the sphere S^{n-1}. Using a partition of unity, it is not difficult to extend the vector field w throughout some neighborhood of $f(D^n)$. (Compare Lemma 2.14 of the Munkres Lecture Notes on page 162.) The solution curves then extend also, so as to be defined say for $0 < t < 1+\varepsilon'$. There is then a unique extension $\hat{f}$ of f to the open $(1+\varepsilon')$-disk so that these extended solution curves are given by $t \mapsto y = \hat{f}(tu)$. Using Lemma 2.2, it follows that $\hat{f}$ is a diffeomorphism on some sufficiently small neighborhood of D^n, say the neighborhood $D^n(1+\varepsilon)$ of radius $1+\varepsilon$. □

Proof of Theorem 2.1. Recall that f and g are embeddings of the closed disk D^n into the interior of M. Choose extensions $\hat{f}$ and $\hat{g}$ as in Lemma 2.8. By Lemma 2.7, after composing one of these two embeddings with a diffeomorphism of M, we may assume that $f(x) = g(x)$ for $\|x\| \le \varepsilon$, provided that ε is small enough. Now let $\sigma_\varepsilon : \mathbb{R} \to \mathbb{R}$ be a monotone smooth function such that

$$\sigma_\varepsilon(r) = \begin{cases} \varepsilon, & \text{if} \quad \mathrm{r} \le 1\,, \\ 1, & \text{if} \quad \mathrm{r} \ge 1+\varepsilon/2\,, \end{cases}$$

and consider the diffeomorphism $S_\varepsilon : \mathbb{R}^n \to \mathbb{R}^n$ defined by $S_\varepsilon(x) = x\sigma_\varepsilon(\|x\|)$. Making use of S_ε, we can define an auxiliary diffeomorphism $F : M \to M$ by the formula

$$F(p) = \begin{cases} p, & \text{if} \quad \mathrm{p} \notin \hat{\mathrm{f}}(\mathrm{D}^n(1+\varepsilon/2))\,, \\ f \circ S_\varepsilon \circ \hat{f}^{-1}(p), & \text{if} \quad \mathrm{p} \in \hat{\mathrm{f}}(\mathrm{D}^n(1+\varepsilon))\,. \end{cases}$$

Then, $F \circ f(x) \;=\; f \circ S_\varepsilon(x) \;=\; f(\varepsilon\, x) \;=\; g(\varepsilon\, x)$, for all $x \in D^n$. Similarly define a diffeomorphism G of M so that $G \circ g(x) = g(\varepsilon x)$ for $x \in D^n$. Then applying G^{-1} to both sides of the equation $F \circ f = G \circ g$, we see that $h = G^{-1} \circ F$ is the required diffeomorphism of M, with $h \circ f = g$. □

To conclude this section, we discuss a way of constructing new manifolds out of old by pasting together diffeomorphic boundary components. Let $(M, \mathcal{S}(M))$ be a smooth manifold with non-vacuous boundary.

Definition. An open neighborhood $U \subset M$ of ∂M is called a *collar neighborhood* of ∂M if $(U, \mathcal{S}(U))$ is diffeomorphic to the product $(\partial M \times [0,1),\ \mathcal{S}(\partial M \times [0,1)))$. Whenever this is so, the diffeomorphism can be chosen in such a way that each point p in $\partial M \subset U$ is mapped to the pair $(p, 0) \in \partial M \times [0,1)$.

THEOREM 2.9. Collar neighborhoods. *If the boundary ∂M is compact, then ∂M has a collar neighborhood within the manifold M.*

A proof can be outlined as follows. (Compare the discussion of tubular neighborhoods in §3 of the Munkres notes, page 168, or in §5 of the Lectures on Differentiable Structures, page 187.) Using a smooth partition of unity, one can construct a smooth Riemannian metric on M. Then any point $p \in M$ which is sufficiently close to ∂M can be joined to ∂M by a unique geodesic of shortest length. (This shortest path is necessarily smooth, and meets the boundary at right angles.) Let $f(p) \in \partial M$ be its boundary endpoint, and let $\ell(p)$ be the length of this shortest path, or in other words the Riemannian distance between p and $f(p)$. Then the correspondence $p \mapsto (f(p), \ell(p)/\varepsilon)$ maps the ε-neighborhood of ∂M diffeomorphically onto $\partial M \times [0,1)$, provided that ε is sufficiently small. □

Now let $(M, \mathcal{S}(M))$, $(N, \mathcal{S}(N))$ be smooth manifolds, and let M_1 and N_1 be connected components of the boundaries ∂M and ∂N, respectively. We suppose that these two boundary components are compact, and that they are diffeomorphic to each other under a diffeomorphism $h : M_1 \to N_1$. Let $P = M \cup_h N$ be the identification space in which M_1 and N_1 are pasted together under h. In other words, let

$$P \;=\; M \cup_h N \;=\; (M \uplus N)/r$$

where the equivalence relation r is such that each $p \in M_1$ is equivalent to $h(p) \in N_1$, but otherwise points are equivalent only to themselves.

COROLLARY 2.10. Pasting together boundaries. *Under these conditions, there exists a smoothness structure $\mathcal{S}(P)$ for P such that $(P, \mathcal{S}(P))$ is a smooth manifold and contains both $(M, \mathcal{S}(M))$ and $(N, \mathcal{S}(N))$ as smooth submanifolds.*

Proof. We need only specify the smoothness structure $\mathcal{S}_x(P)$ at the points of the identified boundary. Choose collar neighborhoods $U \cong M_1 \times [0,1)$ and $U' \cong N_1 \times (-1, 0]$ of M_1 and N_1 respectively. Then the union $U \cup_h U'$ can clearly be given a smoothness structure so that it is diffeomorphic to $M_1 \times (-1,1)$. Further details are straightforward. □

Remark. If we are given explicit collar neighborhoods with explicit diffeomorphisms to $M_1 \times [0,1)$ and $N_1 \times (-1,0]$, then this smoothness structure is unique. In fact, with a little more work, one show that collar neighborhoods with prescribed product structure are unique up to diffeomorphism of the ambient manifold. Hence the resulting smooth manifold $(P, \mathcal{S}(P))$ is also unique up to diffeomorphism.

3. Connected sums and the group $\mathcal{A}_n$ of invertible exotic spheres.

Let M and N be smooth n-dimensional manifolds which are connected and oriented. Then the connected sum $M\#N$ is a new smooth manifold, well defined up to orientation preserving diffeomorphism. Intuitively, it can be constructed by cutting a small disk out of each manifold and then pasting together the resulting boundaries. However, to keep control of orientation and smoothness structure, we proceed more carefully as follows. Let $f : D^n \to \mathring{M}$, $g : D^n \to \mathring{N}$ be smooth embeddings such that f preserves the orientation and g reverses it. Define an orientation preserving diffeomorphism $h : f(\mathring{D}^n \smallsetminus 0) \to g(\mathring{D}^n \smallsetminus 0)$ by

$$h\big(f(ru)\big) \;=\; g\big((1-r)u\big)$$

for every unit vector $u \in S^{n-1} = \partial D^n$ and for every $0 < r < 1$. Then, by definition, the *connected sum*

$$M\#N \;=\; \Big(M \smallsetminus f(0)\Big) \cup_h \Big(N \smallsetminus g(0)\Big)$$

is obtained by removing one point each from M and N and then pasting together the neighborhoods $f(\mathring{D}^n \smallsetminus 0)$ and $g(\mathring{D}^n \smallsetminus 0)$ of these points under h.

Evidently $M\#N$ is also a smooth oriented n-dimensional manifold, compact if both M and N are compact, and connected provided that $n \geq 2$. It follows easily from Theorem 2.1 that this sum is well-defined up to orientation preserving diffeomorphism.

LEMMA 3.1. Properties of connected sums. *If M, N, P are connected, oriented, smooth manifolds of dimension $n \geq 2$, then:*

(1) $M\#S^n \cong M$,
(2) $M\#N \cong N\#M$,
(3) $M\#(N\#P) \cong (M\#N)\#P$,
(4) $M\#\mathbb{R}^n \cong M \smallsetminus$ *(point), and similarly*
(5) $M\#D^n \cong M \smallsetminus f(\mathring{D}^n)$, *where $f(D^n)$ is a smoothly embedded disk. Furthermore:*
(6) *The connected sum of a surface of genus g and a surface of genus h is diffeomorphic to a surface of genus $g + h$.*

Here $\cong$ stands for the relation of orientation preserving diffeomorphism, and S^n stands for the unit sphere in $\mathbb{R}^{n+1}$ with its standard smoothness structure.

Proof. The proof is not difficult, and will be left to the reader. □

It follows that the set of all oriented diffeomorphism classes of smooth oriented n-manifolds forms a *commutative monoid*, that is, a commutative associative semigroup with 2-sided identity element. We will be particularly interested in the submonoid made up out of compact manifolds without boundary:

Definition. Let $\mathfrak{M}_n$ be the set of equivalence classes, under orientation preserving diffeomorphism, of connected, orientable smooth n-manifolds which are compact and without boundary. Evidently the connected sum operation makes

$\mathfrak{M}_n$ into a commutative monoid with the equivalence class of S^n as identity element, for $n \geq 1$. (For compact manifolds without boundary, even the case $n = 1$ makes sense, although it is not very interesting.) As examples:

$\mathfrak{M}_1$ has only the identity element.

$\mathfrak{M}_2$ is a free monoid on one generator, namely the surface $S^1 \times S^1$ of genus one.

$\mathfrak{M}_3$ is a free commutative monoid with infinitely many generators. (See KNESER [1929], MILNOR [1962]. These papers consider only piecewise-linear manifolds. However, in dimensions ≤ 3, it follows from MUNKRES [1960a, 1964] and HIRSCH [1963], together with SMALE [1959b], that there is a one-to-one correspondence between smooth manifolds up to diffeomorphism and piecewise-linear manifolds up to piecewise-linear homeomorphism. Furthermore, MOISE [1952] showed that topological manifolds have an essentially unique piecewise-linear structure, in dimensions ≤ 3.)

However, $\mathfrak{M}_4$ is not free. Indeed, let $\mathbb{CP}^2$ be the oriented complex projective plane and let $\overline{\mathbb{CP}}^2$ be the complex projective plane with opposite orientation. Then

$$(\mathbb{CP}^2 \# \overline{\mathbb{CP}}^2) \# \overline{\mathbb{CP}}^2 \cong (S^2 \times S^2) \# \overline{\mathbb{CP}}^2 , \quad \text{although} \quad \mathbb{CP}^2 \# \overline{\mathbb{CP}}^2 \not\cong S^2 \times S^2 .$$

Thus the cancellation law does not hold in $\mathfrak{M}_4$, so this monoid is certainly not free.

Here is an outline proof of these statements. If M is any complex manifold of complex dimension 2, then the connected sum $M \# \overline{\mathbb{CP}}^2$ can be identified with the manifold obtained by "blowing up" a point in M. (Compare MCDUFF AND SALAMON [1995 p. 216].) On the other hand, identifying S^2 with $\mathbb{CP}^1$, the birational correspondence, $(1:z) \times (1:w) \leftrightarrow (1:z:w)$, between $\mathbb{CP}^1 \times \mathbb{CP}^1$ and $\mathbb{CP}^2$ gives rise to an analytic isomorphism between $\mathbb{CP}^1 \times \mathbb{CP}^1$ with one point $(0:1) \times (0:1)$ blown up and $\mathbb{CP}^2$ with two points $(0:1:0)$ and $(0:0:1)$ blown up. (Compare GRIFFITHS AND HARRIS [1978, pp. 478-480].) However, the two manifolds $\mathbb{CP}^2 \# \overline{\mathbb{CP}}^2$ and $S^2 \times S^2$ are not diffeomorphic; in fact their cohomology rings are different.

We will be particularly interested in elements of $\mathfrak{M}_n$ which admit an inverse under the connected sum operation.

Definition. Let $\mathcal{A}_n$ be the subgroup of $\mathfrak{M}_n$ consisting of all invertible elements. As examples, it follows from the discussion above that $\mathcal{A}_1$, $\mathcal{A}_2$ and $\mathcal{A}_3$ contain only the identity element. On the other hand, we will see that $\mathcal{A}_7$ is non-trivial.

THEOREM 3.2. (Mazur). #-Invertible manifolds are topological spheres. *If M and N are smooth connected n-manifolds, and if $M \# N \cong S^n$, then M as a topological space is homeomorphic to S^n.*

COROLLARY 3.3. *If the group $\mathcal{A}_n$ is non-trivial, then S^n admits two or more essentially distinct differentiable structures.*

In order to prove Theorem 3.2 we introduce the concept of an infinite connected sum.

Definition 3.4. Let M_1, M_2, ... be smooth connected n-manifolds and let

$$\begin{aligned} f_1 : D^n &\longrightarrow M_1 , \\ f_i, g_i : D^n &\longrightarrow M_i , \qquad i = 2, 3, \ldots \end{aligned}$$

be smooth embeddings such that the f_i preserve orientation and the g_i reverse it, with $f_i(D^n) \cap g_i(D^n) = \emptyset$. Consider the disjoint union

$$(1) \qquad \left(M_1 \smallsetminus f_1(0)\right) \uplus \biguplus_{i \geq 2} \left(M_i \smallsetminus \{f_i(0), g_i(0)\}\right)$$

and the equivalence relation,

$$f_i(ru) \sim g_{i+1}((1-r)u)\,, \quad \text{for} \quad \|u\| = 1\,,\ 0 < r < 1\,,\ i = 1\,, 2\,, \ldots$$

The quotient of the disjoint union of Equation (1) with respect to the equivalence relation $\sim$ is by definition the infinite connected sum, $M_1 \# M_2 \# M_3 \# \cdots$.

One then shows, as for finite connected sums, that the infinite sum is well-defined and associative.

Example 3.5. It is easy to see that, $S^n \# S^n \# \cdots \cong \mathbb{R}^n$.

We can now give a proof of the theorem.

Proof of Theorem 3.2. If $M \# N \cong S^n$ then we have,

$$\mathbb{R}^n \cong (M\#N)\#(M\#N)\#\cdots \cong M\#(N\#M)\#\cdots \cong M\#\mathbb{R}^n \cong M \smallsetminus \text{point}\,,$$

which implies that M is homeomorphic to the n-sphere. □

The preceding theorem is contained in the following.

THEOREM 3.6. Characterization of #-invertible manifolds. *If M is an oriented, compact and connected smooth manifold without boundary, then the following conditions are equivalent:*

(1) *$M\#N \cong S^n$ for some N.*

(2) *$M \smallsetminus \{point\} \cong \mathbb{R}^n$.*

(3) *$M = U_1 \cup U_2$ with U_1, U_2 open in M and both diffeomorphic to $\mathbb{R}^n$.*

(4) *If $f(D^n) \subset M$ is an embedded disk, then the complement $M \smallsetminus f(D^n)$ can be smoothly embedded in S^n.*

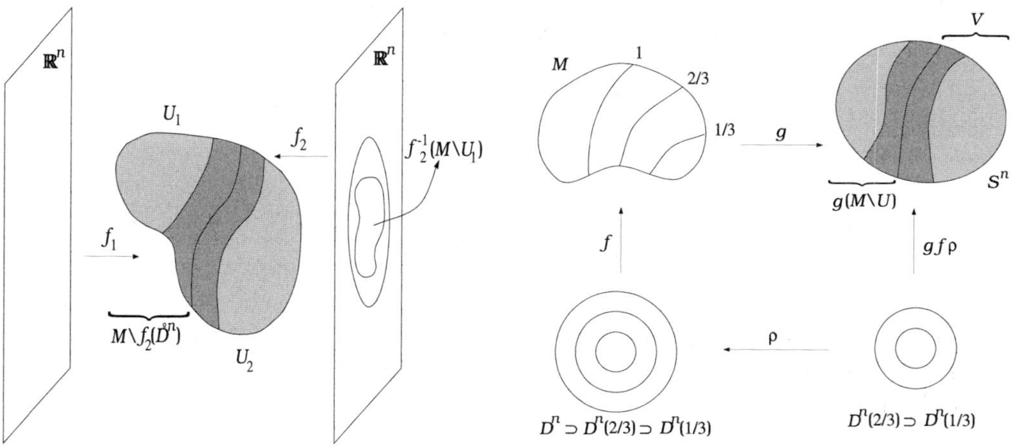

FIG. 2. *On the right: Proof that* (3) ⇒ (4). *On the left: Proof that* (4) ⇒ (1).

Proof. The proof of Mazur's theorem shows that (1) implies (2). To see that (2) implies (3) it is enough to take $U_i = M \setminus p_i$, with distinct points p_1 and p_2. To show that (3) implies (4), first note that $M \setminus U_1$ is compact and is contained in U_2. Let $f_i : \mathbb{R}^n \xrightarrow{\cong} U_i \subset M$. Then $f_2^{-1}(M \setminus U_1)$ is a compact subset of $\mathbb{R}^n$, so it is contained in some disk $D^n(r)$ of large radius. Then $M \setminus f_2(D^n(r)) \subset U_1 \cong \mathbb{R}^n$. Consequently, $M \setminus f_2(D^n(r))$ can be embedded in $\mathbb{R}^n \subset S^n$, as required. (It is enough to prove this for just one smooth embedding of a closed n-disk, since any two such embeddings are equivalent by Theorem 2.1.)

The proof that (4) implies (1) can be outlined as follows. Intuitively, the statement that $M\#N \cong S^n$ means that there exists a smooth embedding $\phi : S^{n-1} \to S^n$, cutting the n-sphere into two regions $\hat{M}$ and $\hat{N}$ with $\phi(S^{n-1})$ as common boundary, so that if we fill the hole in $\hat{M}$ with an n-disk we obtain the manifold M, while if we fill the hole in $\hat{N}$ with an n-disk we obtain N. But condition (4) implies that the manifold $\hat{M}$ (that is, M with an open disk removed) can indeed be embedded in S^n, so that we can construct the required manifold N. To make this argument more precise, let f be a smooth embedding of the disk D^n in M. By (4), there exists a diffeomorphism g from the open set $M \setminus f\big(D^n(1/3)\big)$ onto an open set $U \subset S^n$. Let $A = \mathring{D}^n(2/3) \setminus D^n(1/3)$ be the n-dimensional 'annulus' consisting of vectors $x \in D^n$ with $1/3 < \|x\| < 2/3$. We construct a new manifold N from the disjoint sum of the disk $\mathring{D}^n(2/3)$ and the open set $V = S^n \setminus g\big(M \setminus f(\mathring{D}^n(2/3))\big)$ by pasting $A \subset \mathring{D}^n(2/3)$ onto $g(f(A)) \subset V$ by the diffeomorphism $g \circ f \circ \rho$, where $\rho(ru) = (1-r)u$. Then we claim that the resulting manifold $N = V \cup_{gf\rho} \mathring{D}^n(2/3)$ has the required property that $M\#N \cong S^n$. In fact, let $i : \mathring{D}^n(2/3) \to N$ be the inclusion map. Then $M\#N$ can be obtained from the disjoint sum of $M \setminus f(D^n(1/3)) \cong U$ and $N \setminus i(D^n(1/3)) \cong V$ by pasting together the subsets $f(A)$ and $i(A)$ under $i \circ \rho \circ f^{-1}$ But g on the first summand and the inclusion map $V \to S^n$ on the second induce a diffeomorphism of this identification space with S^n. □

Remarks. Except in dimension 4, we can make the sharper statement that a smooth n-dimensional manifold is homeomorphic to S^n if and only if it is invertible under #. For $n \leq 3$ this is true trivially, since there is only one differentiable structure on the n-sphere, up to diffeomorphism, in these dimensions. (Compare the discussion of $\mathfrak{M}_3$ above.) For the discussion of dimensions $n \geq 5$, see the next section.

4. The group $\Gamma_n \subset \mathcal{A}_n$ of twisted n-spheres, and the homomorphisms $\pi_0\mathrm{Diff}^+(S^{n-1}) \to \Gamma_n$ and $\pi_m(\mathrm{SO}_n) \otimes \pi_n(\mathrm{SO}_m) \to \pi_0\mathrm{Diff}^+(S^{m+n})$.

One particularly easy way of constructing an exotic n-sphere is to take two copies of the closed disk D^n and paste their boundaries together by some diffeomorphism

$$f : S^{n-1} \to S^{n-1} .$$

More precisely, in order to keep track of smoothness structures, let us work with the full Euclidean space $\mathbb{R}^n$ in place of the unit disk.

Definition 4.1. Let $\text{Diff}^+(S^{n-1})$ be the group consisting of all orientation preserving diffeomorphisms of the $(n-1)$-sphere. Given any $f \in \text{Diff}^+(S^{n-1})$ we construct a new smooth manifold

$$\Sigma(f) \;=\; \mathbb{R}^n \cup_F \mathbb{R}^n \,,$$

homeomorphic to the n-sphere, by pasting together the subsets $\mathbb{R}^n \smallsetminus \{0\}$ under the diffeomorphism

$$F(tu) \;=\; f(u)/t \qquad \text{for} \quad u \in S^{n-1} \quad \text{and} \quad 0 < t < \infty \,.$$

This manifold $\Sigma(f)$, with the orientation coming from the first copy of $\mathbb{R}^n$, will be called a *twisted n-sphere*, with *twist f*.

Since $\Sigma(f)$ is homeomorphic to the first copy of $\mathbb{R}^n$, together with a single point at infinity, it is certainly homeomorphic to the standard n-sphere.

Definition 4.2. Two diffeomorphisms $f, g : M \to N$ are said to be *smoothly isotopic* if there is a family of diffeomorphisms $h_t : M \to N$, where t ranges over the real numbers, so that

$$\begin{aligned} h_t &= f \quad \text{for} \quad t \le 0 \\ h_t &= g \quad \text{for} \quad t \ge 1 \,, \end{aligned}$$

and so that the correspondence

$$(t, x) \;\mapsto\; h_t(x)$$

defines a smooth mapping from $\mathbb{R} \times M$ to N.

It is easy to check that this is an equivalence relation between diffeomorphisms. In the special case where $M = N$, the set of all smooth isotopy classes of diffeomorphisms from M to itself will be denoted by $\pi_0\text{Diff}(M)$. The composition of diffeomorphisms defines a group structure in the set $\text{Diff}(M)$ of all diffeomorphisms, and hence in $\pi_0\text{Diff}(M)$. This is a non-abelian group, in general. (As an example, $\pi_0\text{Diff}(S^1 \times S^1)$ can be identified with the group $\text{GL}(2, \mathbb{Z})$ of 2×2 invertible matrices.) However:

LEMMA 4.3. Commutativity. *The group $\pi_0\text{Diff}^+(S^n)$ of isotopy classes of orientation preserving diffeomorphisms of the n-sphere is abelian.*

Proof. We will make use of Theorem 2.1 in a slightly sharper form: Any two smooth orientation preserving embeddings of the n-disk in a connected n-manifold are equivalent under a diffeomorphism of the manifold which is isotopic to the identity. In fact the proof in §2 can easily be modified to provide this sharper statement. Now let f_1, f_2 be any two elements of $\text{Diff}^+(S^n)$. Denote by D^n_+ and D^n_- the upper and lower hemispheres of S^n, respectively. There exists h_1 and h_2, both isotopic to the identity, such that $h_1 \circ f_1 |_{D^n_-}$ is the identity map of D^n_- and $h_2 \circ f_2 |_{D^n_+}$ is the identity map of D^n_+. Then clearly $h_1 \circ f_1$ commutes with $h_2 \circ f_2$. Therefore $f_1 \circ f_2$ is isotopic to $f_2 \circ f_1$. □

LEMMA 4.4. Twisted spheres. *The correspondence* $(f) \mapsto \Sigma(f)$ *gives rise to a homomorphism*

$$\pi_0 \mathrm{Diff}^+(S^{n-1}) \;\to\; \mathcal{A}_n$$

from the commutative group of isotopy classes of orientation preserving diffeomorphisms of S^{n-1} *into the commutative group* $\mathcal{A}_n$ *of invertible exotic* n*-spheres.*

By definition, the image of this homomorphism will be called the group $\Gamma_n \subset \mathcal{A}_n$ of *twisted* n*-spheres*.

Proof of Lemma 4.4. We must first show that if f is isotopic to g, or in other words if $g^{-1} \circ f$ is isotopic to the identity, then $\Sigma(f)$ is diffeomorphic to $\Sigma(g)$. Let $\{h_t\}$ be a smooth family of diffeomorphisms $h_t : S^{n-1} \to S^{n-1}$, where we may assume that h_t is the identity map for $t \le 1$ and that $h_t = g^{-1} \circ f$ for $t \ge 2$. Then the required diffeomorphism from $\Sigma(f) = \mathbb{R}^n \cup_F \mathbb{R}^n$ to $\Sigma(g) = \mathbb{R}^n \cup_G \mathbb{R}^n$ is given by the formula

$$tu \;\mapsto t\, h_t(u)$$

on the first copy of $\mathbb{R}^n$, and by

$$f(u)/t \;\mapsto\; g \circ h_t(u)/t$$

on the second. It is easy to check that this expression is compatible with the identifications, and defines the required diffeomorphism.

Now we must prove that

$$\Sigma(f) \# \Sigma(g) \;\cong\; \Sigma(g \circ f)\ .$$

In fact if $i_1, i_2 : \mathbb{R}^n \to \Sigma(f)$ are the embeddings of the two copies of $\mathbb{R}^n$ into $\Sigma(f)$, and if $j_1, j_2 : \mathbb{R}^n \to \Sigma(g)$ are the corresponding embeddings for $\Sigma(g)$, then we can then form a manifold diffeomorphic to the connected sum from the disjoint union

$$\big(\Sigma(f) \smallsetminus \{i_2(0)\}\big) \;\uplus\; \big(\Sigma(g) \smallsetminus \{j_1(0)\}\big)$$

by identifying the subsets $i_2(\mathbb{R}^n \smallsetminus \{0\})$ and $j_1(\mathbb{R}^n \smallsetminus \{0\})$ under the correspondence $tu \leftrightarrow u/t$. (This is a slight variation on our construction of the connected sum, but is easily seen to be equivalent to it. Compare Theorem 2.9 and its Corollary.) In other words, we form the disjoint sum $i_1(\mathbb{R}^n) \uplus j_2(\mathbb{R}^n)$ and then identify the complements of the origin under the composition

$$tu \;\mapsto\; f(u)/t \;\mapsto t\, f(u) \;\mapsto g(f(u))/t\ .$$

But this is just the required manifold $\Sigma(g \circ f)$. Thus our correspondence

$$\pi_0 \mathrm{Diff}^+(S^{n-1}) \;\to\; \mathcal{A}_n$$

is a well defined group homomorphism. Evidently the image is just the group of oriented diffeomorphism classes of twisted spheres. □

Remarks. There are several other important characterizations of twisted spheres. An argument due to Reeb [1952] shows that a closed n-manifold is a twisted sphere if and only if it admits a Morse function with only two critical points. According to Thom [1959], the manifold M^n is a twisted sphere if and only if a C^1-triangulation $K \to M^n$ yields a simplicial complex K which is piecewise-linearly homeomorphic to the standard piecewise-linear sphere.

In dimensions $n \neq 4$, every smooth manifold homeomorphic to S^n is a twisted sphere, so that $\Gamma_n = \mathcal{A}_n$. In fact, for $n \leq 3$, as noted in the previous section, we have $\Gamma_n = \mathcal{A}_n = 0$. In dimensions ≥ 7, STALLINGS [1960] proved that any compact combinatorial manifold without boundary having the homotopy type of an n-sphere is the union of two open subsets piecewise-linearly homeomorphic to $\mathbb{R}^n$, and hence is invertible and homeomorphic to S^n. (Compare Theorem 3.6.) SMALE [1960, 1961, 1962] proved an even sharper version of this statement, showing that every smooth homotopy n-sphere is a twisted sphere if $n \geq 5$. Thus only dimension four remains open. In fact, CERF [1968] showed that $\Gamma_4 = 0$, and FREEDMAN [1982] showed that every homotopy 4-sphere is a topological 4-sphere. However, the possibility that $\mathcal{A}_4 \neq 0$, and also the possibility that there exist smoothness structures on the 4-sphere which are not #-invertible, remain open, as far as I know.

For these constructions to be useful, we must have some way of constructing non-standard diffeomorphisms of spheres. One easy construction can be described as a bilinear pairing

$$\beta : \pi_m(\mathrm{SO}_n) \otimes \pi_n(\mathrm{SO}_m) \rightarrow \pi_0\mathrm{Diff}^+(S^{m+n}) .$$

Here SO_n is the rotation group, consisting of $n \times n$ orthogonal matrices of determinant $+1$, which acts linearly on the Euclidean space $\mathbb{R}^n$. (As a general reference for this material, see STEENROD [1951, p. 26].) Every element of the homotopy group $\pi_m(\mathrm{SO}_n)$ can be represented by a smooth map

$$\phi : \mathbb{R}^m \rightarrow \mathrm{SO}_n$$

with compact support. That is, we assume that $\phi(x)$ is the identity matrix I_n for $\|x\|$ sufficiently large. The set of all smooth homotopy classes (with uniform compact support) of such maps ϕ forms a group, using the composition operation $\phi_1 \cdot \phi_2$ which comes from the product in SO_n:

$$(\phi_1 \cdot \phi_2)(x) = \phi_1(x) \cdot \phi_2(x) .$$

(This is completely equivalent to the more standard definition in terms of continuous maps from the m-sphere and continuous homotopies.) The resulting group $\pi_m(\mathrm{SO}_n)$ is commutative, since we can always deform one of the two maps ϕ_i by a homotopy, so that the two have disjoint support.

Given $\phi : \mathbb{R}^m \to \mathrm{SO}_n$ and $\psi : \mathbb{R}^n \to \mathrm{SO}_m$, both smooth with compact support, we use the action of the rotation group on the corresponding Euclidean space to construct diffeomorphisms

$$\Phi\,,\ \Psi \in \mathrm{Diff}^+(\mathbb{R}^m \times \mathbb{R}^n)$$

as follows. For each $x \in \mathbb{R}^m$, $y \in \mathbb{R}^n$, set

$$\Phi(x,y) = (x\,,\,\phi(x)y)\,, \qquad \Psi(x,y) = (\psi(y)\,x\,,\,y) .$$

The commutator $\Phi \circ \Psi \circ \Phi^{-1} \circ \Psi^{-1}$ is then a diffeomorphism of $\mathbb{R}^m \times \mathbb{R}^n$. It is easy to check that this commutator has compact support, in the sense that

$$\Phi \circ \Psi \circ \Phi^{-1} \circ \Psi^{-1}(x, y) \;=\; (x, y)$$

whenever either $\|x\|$ or $\|y\|$ is sufficiently large.

Now identify $\mathbb{R}^m \times \mathbb{R}^n \cong \mathbb{R}^{m+n}$ with the complement of a point $p_0 \in S^{m+n}$, under stereographic projection. Then this commutator extends uniquely to a diffeomorphism of S^{m+n} which is the identity in some neighborhood of p_0. Let

$$\beta(\phi, \psi) \;\in\; \pi_0 \mathrm{Diff}^+(S^{m+n})$$

be the isotopy class of this extended commutator.

LEMMA 4.5. The pairing β. *This construction gives rise to a well defined bilinear pairing from $\pi_m(\mathrm{SO}_n) \times \pi_n(\mathrm{SO}_m)$ to the group $\pi_0 \mathrm{Diff}^+(S^{m+n})$.*

Proof. Clearly the isotopy class $\beta(\phi, \psi)$ depends only on the smooth homotopy class of ϕ and ψ. To prove that

$$\beta(\phi_1 \cdot \phi_2\,,\, \psi) \;=\; \beta(\phi_1, \psi) + \beta(\phi_2, \psi)$$

we simply deform ϕ_1 so that its support is contained in the disk $\{x \in \mathbb{R}^m\,;\, \|x\| < 1\}$ and deform ϕ_2 so that its support is contained in $\{x \in \mathbb{R}^m\,;\, 1 < \|x\| < 2\}$. After this modification, the two corresponding diffeomorphisms $\Phi_1 \circ \Psi \circ \Phi_1^{-1} \circ \Psi^{-1}$ and $\Phi_2 \circ \Psi \circ \Phi_2^{-1} \circ \Psi^{-1}$ will have disjoint support. The conclusion then follows easily. □

The final two sections will outline a proof that, for suitable choice of m and n, the composition

$$\pi_m(\mathrm{SO}_n) \otimes \pi_n(\mathrm{SO}_m) \;\to\; \pi_0 \mathrm{Diff}^+(S^{m+n}) \;\to\; \Gamma_{m+n+1}$$

is non-trivial.

5. The invariant $\lambda(M^{4k-1}) \in \mathbb{Q}/\mathbb{Z}$.

This section will first describe the Hirzebruch Signature Theorem. For all details of proof the reader is referred to HIRZEBRUCH [1966], or to MILNOR AND STASHEFF [1974]. We then use this theorem to construct a diffeomorphism invariant λ which can be used to distinguish between different smoothness structures on the sphere S^{4k-1} for suitable k.

Let B be a commutative ring and $x_1, x_2, \ldots$ indeterminates, where by definition the indeterminate x_d is assigned degree d. Consider the ring of polynomials $\mathfrak{P} = B[x_1, x_2, \ldots]$. We define the *degree* of a monomial $x_{i_1} x_{i_2} \cdots x_{i_r}$ as $i_1 + i_2 + \cdots + i_r$ and we denote by $\mathfrak{P}_k$ the finitely generated B-module consisting of polynomials in $\mathfrak{P}$ which are homogeneous of degree k. Thus $\mathfrak{P} = \bigoplus \mathfrak{P}_k$ is a graded B-algebra, with $\mathfrak{P}_0 = B$ and with $\mathfrak{P}_r \mathfrak{P}_s \subset \mathfrak{P}_{r+s}$.

Let $K_0\,,\, K_1\,,\, K_2\,,\, \ldots$ be a sequence of polynomials

$$K_j \;=\; K_j(x_1\,,\, \ldots\,,\, x_j) \;\in\; \mathfrak{P}_j\,,$$

with $K_0 = 1$. Then to every formal power series of the form

$$f(z) = 1 + p_1 z + p_2 z^2 + \cdots$$

we can assign a new formal power series

$$K(f(z)) \;=\; 1 + K_1(p_1)z + K_2(p_1,p_2)z^2 + K_3(p_1,p_2,p_3)z^3 + \cdots ,$$

or in other words

$$K\Big(1+\sum_{j=1}^{\infty} p_j z^j\Big) = \sum_{j=0}^{\infty} K_j(p_1,\ldots,p_j)z^j .$$

We say that $\{K_j\}$ is a *multiplicative sequence* if this correspondence $f(z) \mapsto Kf(z)$ is a multiplicative homomorphism. Equivalently, setting

$$1+\sum_{i=1}^{\infty} x_i^{''} z^i \;=\; \Big(1+\sum_{i=1}^{\infty} x_i z^i\Big)\Big(1+\sum_{i=1}^{\infty} x_i^{'} z^i\Big)$$

where z, x_i, $x_i^{'}$ are indeterminates, the sequence $\{K_j\}$ is multiplicative if

$$\sum_{j=0}^{\infty} K_j(x_1^{''},\ldots,x_j^{''})z^j = \Big(\sum_{j=0}^{\infty} K_j(x_1,\ldots,x_j)z^j\Big)\Big(\sum_{j=0}^{\infty} K_j(x_1^{'},\ldots,x_j^{'})z^j\Big).$$

LEMMA 5.1. Classification of multiplicative sequences. *Given a formal power series $\sum_{j=0}^{\infty} b_j z^j$ with coefficients in B, where $b_0 = 1$, there exists one and only one multiplicative sequence $\{K_j\}$ which satisfies the condition that the coefficient of x_1^j in $K_j(x_1\,,\,\ldots\,,\,x_j)$ is equal to b_j for each j, or in other words the condition that K maps the formal power series $f(z) = 1 + z + 0 + 0 + \cdots$ to*

$$K(1+z) \;=\; 1 + b_1 z + b_2 z^2 + \cdots .$$

By definition, $\sum_{j=0}^{\infty} b_j z^j$ is the *characteristic series* for the multiplicative sequence $\{K_j\}$.

Proof. The proof of this lemma is based on the identity

$$K(1+t_1 z)\cdots K(1+t_n z) \;=\; K(1+\sigma_1 z + \sigma_2 z^2 + \cdots + \sigma_n z^n)\ ,$$

where the t_i are indeterminates of degree one, and where the σ_j are the elementary symmetric functions of the t_i. Details will be omitted. □

Now let us specialize to polynomials with coefficients in the field $\mathbb{Q}$ of rational numbers. To every smooth manifold M^n there are associated the Pontrjagin classes $p_j \in H^{4j}(M^n;\mathbb{Q})$ of its tangent bundle. Thus to any multiplicative sequence $\{K_j\}$ with rational coefficients we can associate cohomology classes

$$K_j(p_1\,,\,\ldots\,,\,p_j) \;\in\; H^{4j}(M^n;\mathbb{Q})\ .$$

In particular, if M^n is compact and oriented, without boundary, of dimension $n = 4k$, then we can integrate the cohomology class $K_k(p_1\,,\,\ldots p_k)$ over the manifold (or in other words evaluate it on the fundamental homology class $[M^{4k}] \in H^{4k}(M^{4k};\,\mathbb{Z})$) to obtain a *characteristic number* $K_k(p_1\,,\,\ldots\,,\,p_j)[M^{4k}]$, or briefly

$$K[M^{4k}] \;\in\; \mathbb{Q}\ .$$

We set $K[M^n] = 0$ if n is not of the form $4k$. The multiplicative property of the sequence $\{K_j\}$ yields the identity

$$K[M^m \times N^n] \;= K[M^m] \cdot K[N^n]\;.$$

Now recall the theory of *cobordism* as defined by THOM [1954]. Two smooth closed oriented manifolds M^n and N^n are *oriented cobordant* if there is a smooth compact oriented $(n+1)$-dimensional manifold whose boundary consists of M^n with its given orientation, together with N^n with the opposite of its given orientation. Thom showed that the collection Ω_* consisting of all cobordism classes of closed oriented manifolds forms a graded ring which, up to torsion, is a polynomial ring

$$\Omega_* \otimes \mathbb{Q} \;\cong\; \mathbb{Q}[\mathbb{CP}^2\,,\, \mathbb{CP}^4\,,\, \mathbb{CP}^6\,,\, \ldots]\;,$$

generated by the complex projective spaces $\mathbb{CP}^{2k}$ of real dimension $4k$. Furthermore, he showed that a manifold M^{4k} represents the zero element of $\Omega_{4k} \otimes \mathbb{Q}$ if and only if all of its Pontrjagin numbers $p_{i_1} \cdots p_{i_r}[M^{4k}]$ are zero.

One important topological invariant of a closed oriented $4k$-manifold is the *signature*. (The term 'index' is also used in the literature.) The definition follows. Given any basis $a_1\,,\,\ldots\,,\,a_\beta$ for the middle dimensional cohomology $H^{2k}(M^{4k};\mathbb{Q})$, the cup products $a_i a_j \in H^{4k}(M^{4k};\mathbb{Q})$ give rise to a symmetric matrix

$$a_i\, a_j\, [M^{4k}]$$

of rational numbers. Choosing the basis so that this matrix is diagonal, the sum of the signs of the diagonal entries $a_i^2[M^{4k}]$ is defined to be the *signature* $\mathrm{sgn}[M^{4k}] \in \mathbb{Z}$. (This choice of notation is supposed to suggest that the signature has properties quite similar to those of a characteristic number.) If the dimension n is not of the form $4k$, then we define $\mathrm{sgn}[M^n]$ to be zero. Thom showed that this signature is a cobordism invariant, with

$$\mathrm{sgn}[M^m \times N^n] \;=\; \mathrm{sgn}[M^m] \cdot \mathrm{sgn}[N^n]\;.$$

He then concluded that the signature can be expressed as a rational linear combination of Pontrjagin numbers. More precisely, following HIRZEBRUCH [1966]:

THEOREM 5.2. The signature formula. *There is one and only one multiplicative sequence $\{L_j\}$ with rational coefficients which satisfies the identity*

$$L[\mathbb{CP}^{2k}] \;=\; 1$$

for every complex projective space $\mathbb{CP}^{2k}$. It follows that

$$L[M^n] \;=\; \mathrm{sgn}[M^n]$$

for every smooth closed oriented manifold. The associated characteristic series is given by the formula

$$L(1+z) = \frac{\sqrt{z}}{\tanh\sqrt{z}} = 1 + \frac{z}{3} - \frac{z^2}{45} + \frac{2z}{945} - + \cdots = 1 + \sum_{k=1}^{\infty}(-1)^{k-1}\frac{2^{2k}}{(2k)!}B_k z^k,$$

where $B_1 = 1/6\,,\ B_2 = 1/30\,,\ B_3 = 1/42\,,\ \ldots$ are Bernoulli numbers.

For example

$$L_1(p_1) = \frac{p_1}{3}\,,\; L_2(p_1,p_2) = \frac{7p_2 - p_1^2}{45}\,,\; L_3(p_1,p_2,p_3) = \frac{62p_3 - 13p_1p_2 + 2p_1^3}{945}\,, \ldots .$$

The proof will be omitted.

Since the polynomial L_k is homogeneous of degree k, we can express it as a sum

$$L_k(p_1,\ldots,p_k) \;=\; L_k(p_1,\ldots,p_{k-1},0) + s_k p_k\,,$$

where the coefficients

$$s_1 = 1/3\,,\; s_2 = 7/45\,,\; s_3 = 62/945\,,\; \ldots\,,\; s_n = 2^{2n}(2^{2n-1}-1)B_n/(2n)!$$

are certain non-zero rational numbers. From this we obtain the following.

COROLLARY 5.3. *The expression*

$$\frac{\mathrm{sgn}[M] - L_k(p_1,\ldots,p_{k-1},0)[M]}{s_k}$$

is an integer, equal to the Pontrjagin number $p_k[M]$.

Now let W be a compact, connected smooth manifold of dimension $4k$ with boundary $\partial W = M$. The cohomology sequence of the pair (W,M) takes the form

$$\cdots \longrightarrow H^{i-1}(M) \xrightarrow{\delta} H^i(W,M) \xrightarrow{j} H^i(W) \xrightarrow{r} H^i(M) \xrightarrow{\delta} \cdots\,,$$

where rational coefficients are to be understood. If M has the rational cohomology of a sphere, that is if $H^i(M) \cong H^i(S^{4k-1})$, then the homomorphism j is an isomorphism for $0 < i < 4k$, and δ is an isomorphism for $i = 4k$. For $0 < i < k$, it follows that the Pontrjagin class $p_i \in H^{4i}(W)$ pulls back to a well defined class $j^{-1}p_i \in H^{4i}(W,M)$.

The orientation of W determines a homomorphism $[W] : H^{4k}(W,M) \to \mathbb{Q}$, still using rational coefficients. We define $\mathrm{sgn}[W]$ to be the signature of the bilinear form determined by the composition

$$H^{2k}(W,M) \otimes H^{2k}(W,M) \xrightarrow{\cup} H^{4k}(W,M) \xrightarrow{[W]} \mathbb{Q}\,.$$

Then we can define a rational number $\Lambda(W,M)$ by the formula

$$\Lambda(W,M) \;=\; \frac{\mathrm{sgn}[W] - L_k(j^{-1}p_1,\ldots,j^{-1}p_{k-1},0)[W]}{s_k} \;\in\; \mathbb{Q}\,.$$

As examples, for $k = 1,2,3$ (simplifying the notation slightly) we obtain the expressions

$$\Lambda \;=\; 3\cdot\mathrm{sgn}[W^4]\,, \qquad \frac{45\cdot\mathrm{sgn} + p_1^2}{7}[W^8]\,, \qquad \frac{945\cdot\mathrm{sgn} - 2\,p_1^3 + 13\,p_1p_2}{62}[W^{12}]$$

respectively.

LEMMA 5.4. The invariant λ. *The residue class of this rational number $\Lambda(W,M)$ modulo $\mathbb{Z}$ depends only of the smooth oriented manifold M, and not on the particular choice of oriented manifold W with boundary M. In other words, if $\partial W_1 = \partial W_2 = M$ then*

$$\Lambda(W_1\,,\,M) \;\equiv\; \Lambda(W_2\,,\,M) \qquad (\mathrm{mod}\ \mathbb{Z})\,.$$

We now define the invariant $\lambda(M) \in \mathbb{Q}/\mathbb{Z}$ to be this common residue class $\Lambda(W, M)$ mod $\mathbb{Z}$. As an example, if M is diffeomorphic to the standard sphere under a diffeomorphism $h : M \to S^{4k-1}$, then we can attach a standard $4k$-disk to W. The resulting union $U = W \cup_h D^{4k}$ can be given a compatible smoothness structure by the Corollary to Theorem 2.9. It then follows easily from the discussion above that $\Lambda(W, M) = p_k[U] \in \mathbb{Z}$. Hence the invariant $\lambda(M) \in \mathbb{Q}/\mathbb{Z}$ is zero.

The proof that $\lambda(M)$ is independent of the choice of W proceeds as follows. Suppose that $M = \partial W_1 = \partial W_2$. Form a closed oriented manifold $U = W_1 \cup_\iota W_2$ from the disjoint sum $W_1 \uplus W_2$ by pasting together boundaries under the identity map ι of M. Here we give U the orientation from W_1, so that the embedding $W_1 \to U$ is orientation preserving but the embedding $W_2 \to U$ is orientation reversing. As above, this union U has a compatible smoothness structure.

We next note that the relative cohomology ring of the pair (U, M) splits naturally as a direct sum

$$H^*(U, M) \;\cong\; H^*(W_1, M) \oplus H^*(W_2, M)\ .$$

In fact we have the following diagram, where the inclusion maps

$$k_1 : (W_2, M) \;\hookrightarrow\; (U, W_1)\,, \qquad k_2 : (W_1, M) \;\hookrightarrow\; (U, W_2)\,,$$

induce isomorphism of cohomology rings. (This is the Eilenberg-Steenrod 'excision' property.)

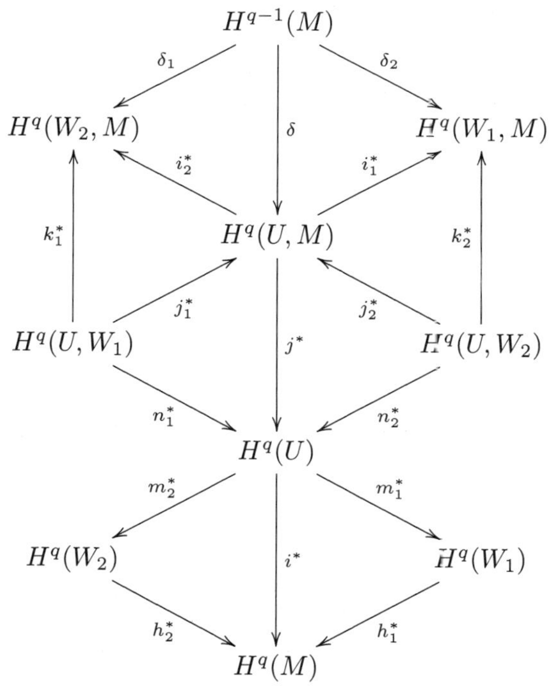

Here each triangle is commutative and each straight line corresponds to an exact sequence, so that

$$\begin{array}{lllllll} \operatorname{Im} \delta & = & \ker j^* , & \operatorname{Im} j^* & = & \ker \ i^* , & \\ \operatorname{Im} j^*_\alpha & = & \ker i^*_\alpha , & \operatorname{Im} n^*_\alpha & = & \ker m^*_\alpha , & \text{for } \alpha = 1, 2 . \end{array}$$

It follows easily from this diagram that the homomorphisms j^*_α and i^*_α induce isomorphisms

$$H^q(U, W_1) \oplus H^q(U, W_2) \xrightarrow{\cong} H^q(U, M) \xrightarrow{\cong} H^q(W_1, M) \oplus H^q(W_2, M) ,$$

as required. Furthermore, these isomorphisms commute with cup products.

Next note that we have a commutative diagram of homomorphisms

$$\begin{array}{ccccc} H^{4k}(U) & \longrightarrow & H^{4k}(U, M) & \xrightarrow{\cong} & H^{4k}(W_1, M) \oplus H^{4k}(W_2, M) \\ \downarrow{\scriptstyle [U]} & & & & \downarrow{\scriptstyle [W_1]-[W_2]} \\ \mathbb{Q} & & \xrightarrow{\quad = \quad} & & \mathbb{Q} \end{array}$$

where the minus sign on the right is used since W_2 and U have opposite orientations. Since

$$H^{2k}(U) \ \cong \ H^{2k}(U, M) \ \cong \ H^{2k}(W_1, M) \oplus H^{2k}(W_2, M) ,$$

it follows easily that

$$\operatorname{sgn}[U] \ = \ \operatorname{sgn}[W_1] - \operatorname{sgn}[W_2] .$$

On the other hand, for every $0 < i < k$ it follows from naturality properties of the Pontrjagin class p_i that the isomorphism

$$H^{4i}(U) \xrightarrow{\cong} H^{4i}(W_1) \oplus H^{4i}(W_2)$$

carries $p_i(U)$ to $p_i(W_1) \oplus p_i(W_2)$. From this, it follows easily that the Pontrjagin numbers of the union U are given by the formula

$$p_{i_1} \cdots p_{i_r}[U] \ = \ p_{i_1} \cdots p_{i_r}[W_1] - p_{i_1} \cdots p_{i_r}[W_2] ,$$

provided that $i_1 , \ldots , i_r < k$. Combining these statements, we see that the difference $\Lambda(W_1, M) - \Lambda(W_2, M)$ is equal to

$$\frac{\operatorname{sgn}[U] - L_k(p_1 , \ldots , p_{k-1} , 0)\,[U]}{s_k} \ = \ p_k[U] .$$

Hence this difference is an integer, which completes the proof of Lemma 5.4. □

This invariant $\lambda(M^{4k-1}) \in \mathbb{Q}/\mathbb{Z}$ is defined for any smooth oriented $(4k-1)$-dimensional manifold which has the rational cohomology $H^*(M^{4k-1}; \mathbb{Q})$ of a sphere, and which represents the zero element of the cobordism group Ω_{4k-1}. In fact we can clarify these conditions by noting the following result.

Assertion 5.5. *If a closed oriented manifold M^n has the mod 2 cohomology of a sphere, then it is a boundary, that is it represents the zero element of Ω_n.*

Proof. In fact, such a manifold M^n must also be a rational cohomology sphere. It then follows from "On the cobordism ring Ω^* and a complex analog, I" (pages 257–273 of this volume), or from AVERBUH [1959] or NOVIKOV [1960, 1962], together with THOM [1954], that M^n can only represent a 2-torsion element in the cobordism group Ω_n; and according to WALL [1960] it must actually represent the zero element. □ In particular, the invariant $\lambda(M^{4k-1})$ is defined for any smooth

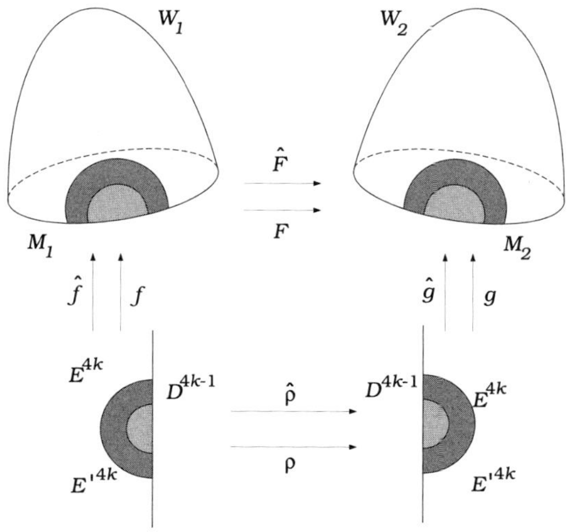

FIG. 3. *The connected sum of boundaries.*

$(4k-1)$-manifold which is a topological sphere.

LEMMA 5.6. The homomorphism λ. *This construction yields a well defined homomorphism from the group $\mathcal{A}_{4k-1}$ of #-invertible manifolds into $\mathbb{Q}/\mathbb{Z}$.*

Proof. We must show that

$$\lambda(M_1 \# M_2) \;=\; \lambda(M_1) + \lambda(M_2)\,.$$

Consider the manifolds $M_1 = \partial W_1$, $M_2 = \partial W_2$. Let f, g be proper embeddings of D^{4k-1} in M_1 and M_2 respectively, f preserving the orientation and g reversing it; let $\rho : (\mathring{D}^{4k-1} \setminus 0) \to (\mathring{D}^{4k-1} \setminus 0)$ be defined by $\rho(ru) = (1-r)u$ and $F = g \circ \rho \circ f^{-1}$. Then with F we obtain $M_1 \# M_2$. We consider D^{4k-1} as contained in the boundary of H^{4k}. Let E^{4k}, E'^{4k} be the half-disks or radius 1 and $1/2$ respectively (see Figure 3). Then ρ, f, g, F can be extended to diffeomorphisms $\hat\rho$, $\hat f$, $\hat g$, $\hat F$ respectively:

$$\begin{aligned}
\hat\rho :\ (\mathring{E}^{4k} \setminus 0) &\;\to\; (\mathring{E}^{4k} \setminus 0)\,,\\
\hat f :\ E^{4k} &\;\to\; W_1\,,\\
\hat g :\ E^{4k} &\;\to\; W_2\,,\\
\hat F = \hat g \circ \hat\rho \circ \hat f^{-1} : \hat f(\mathring{E}^{4k} \setminus 0) &\;\to\; \hat g(\mathring{E}^{4k} \setminus 0)\,.
\end{aligned}$$

Let us set

$$W = (W_1 \setminus f(0)) \cup_{\hat F} (W_2 \setminus g(0))\,.$$

We have then $\partial W = M_1 \# M_2$.

Let $W_1' = W_1 - \hat{f}(\mathring{E}'^{4k})$, $W_2' = W_2 - \hat{g}(\mathring{E}'^{4k})$. Then W_i' is diffeomorphic to W_i, and $W = W_1' \cup_{\hat{F}'} W_2'$, where $\hat{F}'$ is the restriction of $\hat{F}$ to the image under $\hat{f}$ of the boundary of E'^{4k} in H^{4k}. So we conclude that $W_1' \cap W_2'$ is homeomorphic to D^{4k-1}. Therefore it is contractible and when forming the Mayer-Vietoris sequence of the proper triad $(W; W_1', W_2')$ we obtain the isomorphisms

$$H^q(W) \xrightarrow{\varphi} H^q(W_1') \oplus H^q(W_2'), \qquad q > 0\,.$$

In this way we obtain the relation $\sigma(W) = \sigma(W_1) + \sigma(W_2)$, and also the analogous relation for the Pontrjagin numbers. From this it follows that
$\Lambda(W) = \Lambda(W_1) + \Lambda(W_2)$, and the lemma is proved. □

6. Existence of Exotic Spheres (added in 2006).

In §4 we showed how to construct examples of twisted spheres, and in §5 we described an invariant $\lambda(M) \in \mathbb{Q}/\mathbb{Z}$ which can distinguish between different twisted spheres. However, to evaluate $\lambda(M)$ we must present this twisted sphere M as the boundary of a smooth compact manifold. This final section will tie these two chapters together by carrying out this construction. (Compare the discussion in the paper "Differentiable Structures on Spheres", on pages 35-45 of this volume. This will be referred to briefly as [DSS].)

In fact we will describe certain smooth manifolds $M(f_1,\, f_2)$ of the form

$$M(f_1,\, f_2) = \partial W(f_1,\, f_2)\,,$$

where $W(f_1,\, f_2)$ is a smooth compact manifold of dimension $p+q$ having the homotopy type of the union $S^p \vee S^q$ of two spheres intersecting in a single point. Here, for the moment,

$$f_1 : S^{p-1} \to \mathrm{SO}_q \qquad \text{and} \qquad f_2 : S^{q-1} \to \mathrm{SO}_p$$

can be arbitrary smooth maps. To begin the construction, consider the union

$$(D_1^p \times D_1^q) \,\cup\, (D^p \times D^q) \,\cup\, (D_2^p \times D_2^q)$$

of three disjoint copies of the product $D^p \times D^q$. (This is illustrated in Figure 4 for the case $p = q = 1$. Here each dotted line is supposed to be identified with the opposite dotted line.) Now paste the two partial boundaries $S^{p-1} \times D^q$ and $S_1^{p-1} \times D_1^q$ together by the diffeomorphism

$$F_1(x, y) = (x\,,\ f_1(x) \cdot y)\,,$$

where $\|x\| = 1$. Then the union $(D^p \times D^q) \cup_{F_1} (D_1^p \times D_1^q)$ can be described as the D^q-bundle over the p-sphere associated with the homotopy class $(f_1) \in \pi_{p-1}(\mathrm{SO}_q)$. (Compare Steenrod [1951].) I will use the notation $\xi(f_1)$ for this disk bundle.

We will be particularly interested in maps which satisfy any one of the following three equivalent conditions.

(1) The disk bundle $\xi(f_1)$ admits a nowhere zero cross-section.

(2) The associated vector bundle splits as the Whitney sum of an $\mathbb{R}^{q-1}$-bundle and a trivial line bundle.

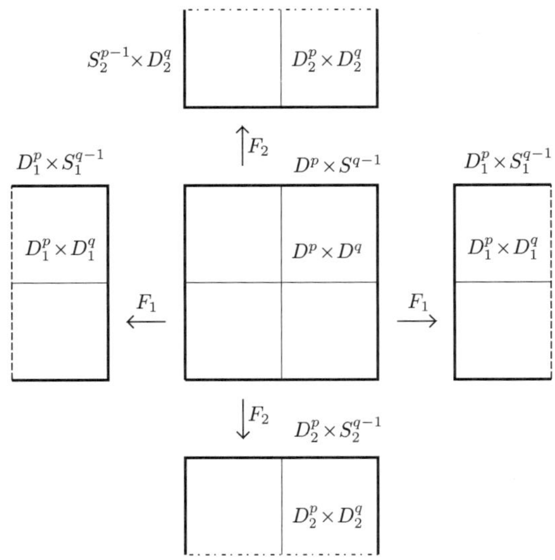

FIG. 4. *Construction of* $W(f_1, f_2)$.

(3) The map f_1 factors as a composition $S^{p-1} \to \mathrm{SO}_{q-1} \subset \mathrm{SO}_q$, up to homotopy.

Similarly, we can paste the two partial boundaries $D^p \times S^{q-1}$ and $D_2^p \times S_2^{q-1}$ together by the diffeomorphism

$$F_2(x, y) = \Big(f_2(y) \cdot x\,,\ y\Big)\,,$$

where $\|y\| = 1$. Then $(D^p \times D^q) \cup_{F_2} (D_2^p \times D_2^q)$ will be a D^p-bundle $\xi(f_2)$ over the q-sphere. The union

$$(D_1^p \times D_1^q) \cup_{F_1^{-1}} (D^p \times D^q) \cup_{F_2} (D_2^p \times D_2^q)$$

will be the required manifold $W(f_1, f_2)$. Evidently the zero-sections of these two disk bundles are spheres of dimension p and q which intersect transversally at the midpoint of $D^p \times D^q$.

There is of course a technical difficulty here, in that $W(f_1, f_2)$, as described, would have a corner along the submanifold $S^{p-1} \times S^{q-1}$ of its boundary. We will assume that the differentiable structure has been modified near this corner, so as to smooth it out. (Compare the discussion in §8 of "Differentiable Manifolds which are Homotopy Spheres", on pages 65–88 of this volume.)

THEOREM 6.1. *If at least one of the two disk-bundles* $\xi(f_1)$ *and* $\xi(f_2)$ *has a nowhere zero cross-section, then the boundary*

$$M(f_1, f_2) = \partial W(f_1, f_2)$$

is a twisted sphere. If both of these bundles have nowhere zero sections, then this twisted sphere can be identified with the image of $\beta(\phi_1, \phi_2)$ *under the homomorphism*

$$\beta : \pi_{p-1}(\mathrm{SO}_{q-1}) \otimes \pi_{q-1}(\mathrm{SO}_{p-1}) \ \to\ \pi_0 \mathit{Diff}^+(S^{p+q-2}) \ \to\ \Gamma_{p+q-1}$$

of Lemma 4.5, where Γ_{p+q-1} is the group of all twisted $(p+q-1)$-spheres. Here ϕ_1 and ϕ_2 are to be homotopy classes which map to (f_1) and (f_2) under the natural homomorphisms

$$\pi_{p-1}(\mathrm{SO}_{q-1}) \to \pi_{p-1}(\mathrm{SO}_q) \qquad \textit{and} \qquad \pi_{q-1}(\mathrm{SO}_{p-1}) \to \pi_{q-1}(\mathrm{SO}_p)\,.$$

Proof. The first statement is an immediate consequence of [DSS, Lemma 1, p. 36]. To prove the second statement, note that the boundary $M(f_1, f_2) = \partial W(f_1, f_2)$ can be obtained from the disjoint union $(D_1^p \times S_1^{q-1}) \cup (S^{p-1} \times D^q)$ by gluing boundaries together under the diffeomorphism

$$F_2 \circ F_1^{-1} \,:\, S_1^{p-1} \times S_1^{q-1} \;\to\; S_2^{p-1} \times S_2^{q-1}\,.$$

However, letting F_1 act as a diffeomorphism of $S_2^{p-1} \times D_2^q$, we see that the gluing map

$$F_1 \circ F_2 \circ F_1^{-1} : S_1^{p-1} \times S_1^{q-1} \to S_2^{p-1} \times S_2^{q-1}$$

will give rise to a diffeomorphic manifold. Similarly, we can compose on the right with the diffeomorphism F_2^{-1} of $D_1^p \times S_1^{q-1}$, so as to obtain the commutator $F_1 \circ F_2 \circ F_1^{-1} \circ F_2^{-1}$ as gluing map.

Now let us use the hypothesis that the image $f_1(S^{p-1})$ lies in the subgroup $\mathrm{SO}_{q-1} \subset \mathrm{SO}_q$, and that $f_2(S^{q-1}) \subset \mathrm{SO}_{p-1} \subset \mathrm{SO}_p$. Intuitively we can think of each $f_1(x)$ as a rotation which fixes the north and south poles of S^{q-1}, and each $f_2(y)$ as a rotation which fixes the poles of S^{p-1}. Furthermore, after deforming f_1 and f_2 by homotopies, we may assume that $f_1(x)$ is the identity rotation, except for x in a small disk N_1 around the north pole, and similarly that $f_2(y)$ is the identity unless $y \in N_2$. It then follows easily that

$$F_1 \circ F_2 \circ F_1^{-1} \circ F_2^{-1}(x,y) \;=\; (x,y) \quad \text{unless} \quad (x,y) \in N_1 \times N_2\,.$$

In other words, a manifold diffeomorphic to $M(f_1, f_2)$ can be obtained from the standard $(p+q-1)$-sphere of radius $\sqrt{2}$ by cutting it open along the set $N_1 \times N_2 \subset S^{p-1} \times S^{q-1}$ and then pasting the two resulting copies of $N_1 \times N_2$ together by $F_1 \circ F_2 \circ F_1^{-1} \circ F_2^{-1}$. The construction of Lemma 4.5 is almost identical, except that we cut along a bi-disk in the equator of S^{p+q-1}. Since it is easy to deform on set onto the other by an isotopy of S^{p+q-1}, this completes the proof. □

Now let us compute that invariant

$$\lambda\Big(M(f_1\,,\, f_2)\Big) \;\in\; \mathbb{Q}/\mathbb{Z}$$

of Lemma 5.4. It is easy to see that this invariant is zero unless both p and q are divisible by 4, so let us assume that $p = 4m$ and $q = 4n$. We will prove the following.

LEMMA 6.2. *Let $\widehat{p}_m(f_1) \in \mathbb{Z}$ be the integer obtained by evaluating the Pontrjagin class $p_m(\xi(f_1))$ on the fundamental homology class of the base space S^{4m}, and define $\widehat{p}_n(f_2)$ similarly. Suppose that at least one of the bundles $\xi(f_1)$ and $\xi(f_2)$ has a non-zero section, so that $M(f_1, f_2)$ is a twisted sphere. Then*

$$\lambda\Big(M(f_1,\, f_2)\Big) \;\equiv\; \pm\widehat{p}_m(f_1)\,\widehat{p}_n(f_2) s_m s_n / s_{m+n} \quad (\mathrm{mod}\ \mathbb{Z})\,,$$

where the sign depends on orientation choices.

Proof. (Compare [DSS, §3].) It is not hard to see that the signature of $W = W(f_1, f_2)$ is zero, and that the only relevant Pontrjagin number is $p_m p_n[W]$. Thus the only problem is to compute the coefficient of $p_m p_n$ in the Hirzebruch polynomial $L_{m+n}(p_1, p_2, \ldots, p_{m+n})$. To do this, consider a closed manifold N^{4n} whose only non-zero Pontrjagin number is $p_n[N^4 n]$, so that the signature formula reduces to

$$\operatorname{sgn}(N^{4n}) = s_n\, p_n[N^{4n}] .$$

If M^{4m} is an analogous manifold of dimension $4m$, then using the identity

$$\operatorname{sgn}(M \times N) = \operatorname{sgn}(M) \cdot \operatorname{sgn}(N) ,$$

a brief computation shows that the coefficient of $p_m p_n$ in the polynomial L_{m+n} is equal to

$$\begin{array}{lll} s_m s_n - s_{m+n} & \text{if} & m \neq n , \\ (s_m s_n - s_{m+n})/2 & \text{if} & m = n , \end{array}$$

where s_{m+n} is the coefficient of p_{m+n}. On the other hand, $p_m p_n[W]$ is equal to either $\widehat{p}_m(f_1)\,\widehat{p}_n(f_2)$ or to twice this number according as $m \neq n$ or $m = n$. The conclusion follows. □

According to HIRZEBRUCH [1966, p. 12], the coefficient s_n is given by the formula

$$s_n = 2^{2n}(2^{2n-1} - 1)B_n/(2n)! ,$$

where the B_n are Bernoulli numbers. (For relevant information about Bernoulli numbers, compare MILNOR AND STASHEFF [1974, Appendix B].)

Here are some numerical values, expressed as quotients of products of primes.

n	1	2	3	4	5	6	7	8
B_n	$\frac{1}{2\cdot3}$	$\frac{1}{2\cdot3\cdot5}$	$\frac{1}{2\cdot3\cdot7}$	$\frac{1}{2\cdot3\cdot5}$	$\frac{5}{2\cdot3\cdot11}$	$\frac{691}{2\cdot3\cdot5\cdot7\cdot13}$	$\frac{7}{2\cdot3}$	$\frac{3617}{2\cdot3\cdot5\cdot17}$
s_n	$\frac{1}{3}$	$\frac{7}{3^2\cdot5}$	$\frac{2\cdot31}{3^3\cdot5\cdot7}$	$\frac{127}{3^3\cdot5^2\cdot7}$	$\frac{2\cdot73}{3^5\cdot5\cdot11}$	$\frac{2\cdot23\cdot89\cdot691}{3^6\cdot5^3\cdot7^2\cdot11\cdot13}$	$\frac{2^2\cdot8191}{3^6\cdot5^2\cdot7\cdot11\cdot13}$	$\frac{31\cdot151\cdot3617}{3^7\cdot5^4\cdot7\cdot11\cdot13\cdot17}$

However, in order to make use of Lemma 6.2 we need to know what values of $\widehat{p}_m(f_1)$ are possible.

LEMMA 6.3. *If $q \leq 2m$ then the homomorphism*

$$\widehat{p}_m : \pi_{4m-1}(\mathrm{SO}_q) \to \mathbb{Z}$$

is zero; but if $q > 2m$ then its image is a non-zero additive group. The generator of this group is a multiple of $(2m-1)!$, and its prime divisors are all $\leq 2m$.

In fact, if $q \geq 4m$ then Bott showed that that this image is generated by either $(2m-1)!$ or $2(2m-1)!$ according as m is even or odd. (Compare BOTT AND MILNOR on page 229–231 of this volume.) The general case follows by using results of Serre on stable homotopy groups of spheres. For details, see [DSS, Lemma 5] on page 43. □

Below is a table showing the denominators of the quotients $s_m s_n/s_{m+n}$, expressed as fractions in lowest terms, for all m and n with $m \leq n < 2m$ and

$m+n \le 8$. Those prime factors which are less than $2n$ are shown in parenthesis, since by Lemma 6.3 they will cancel against factors of $\widehat{p}_n(f_2)$. (In fact, all of the useful factors seem to come from the numerator of s_{m+n}.)

m	n	dimension	denom$(s_m s_n / s_{m+n})$
1	1	7	$\mathbf{7}$
2	2	15	$(3)\cdot\mathbf{127}$
2	3	19	$(5)\cdot\mathbf{73}$
3	3	23	$\mathbf{23}\cdot\mathbf{89}\cdot\mathbf{691}$
3	4	27	$(2\cdot 5\cdot 7)\cdot\mathbf{8191}$
4	4	31	$(7)\cdot\mathbf{31}\cdot\mathbf{151}\cdot\mathbf{3617}$
3	5	31	$(3)\cdot\mathbf{31}\cdot\mathbf{151}\cdot\mathbf{3617}$

As an example, using either of the last two lines together with Lemmas 6.2 and 6.3, we see that there exist at least $31\cdot 151\cdot 3617 = 16931177$ distinct differentiable structures on the 31-dimensional sphere.

Appendix: Construction and Extension of Smooth Real Valued Functions.

The first two lemmas will construct certain smooth functions of one real variable.

LEMMA A.1. *There exists a C^∞-function $\psi : \mathbb{R} \to [0,1]$ such that $\psi(t) = 0$ for $|t| \ge 1$, with $\psi(0) = 1$, and with n-th derivative $\mathcal{D}^n\psi(0) = 0$ for all $n \ge 1$.*

Proof. Start with

$$\phi(t) = \begin{cases} e^{-1/t}, & \text{for} \quad t > 0 \\ 0, & \text{for} \quad t \le 0, \end{cases}$$

the standard example of a function which is C^∞ but not real analytic. Evidently $\phi(t) > 0$ if and only if $t > 0$. It follows that $\phi(t) + \phi(1-t) > 0$ everywhere. The ratio

$$\eta(t) = \phi(1-t)/\big(\phi(t) + \phi(1-t)\big)$$

is then a C^∞-function satisfying $0 \le \eta(t) \le 1$, with $\eta(t) = 0$ for $t \ge 1$ and $\eta(t) = 1$ for $t \le 0$. Setting $\psi(t) = \eta(t^2)$, we obtain a function with the required properties. □

LEMMA A.2. *Given a completely arbitrary sequence of real numbers $a_0, a_1, a_2, \ldots$ there exists a C^∞ map $f : \mathbb{R} \to \mathbb{R}$ whose n-th derivative at the origin is equal to a_n for each $n \ge 0$.*

Proof. Choose numbers $b_k \ge |a_k| + 1$, and set

$$f(t) = \sum_{k\ge 0} a_k\, \psi(b_k t)\, t^k/k!\,, \qquad (2)$$

with ψ as above. If the k-th term of this series is non-zero, then we must have $|b_k t| < 1$ hence

$$\big|\, a_k\, \psi(b_k t)\, t^k/k! \,\big| \;<\; b_k\cdot 1\cdot (1/b_k)^k/k! \;\le\; 1/k!$$

for $k \geq 1$. Thus the series of Equation (2) converges uniformly, hence f is a well defined continuous function. Note also that $f(t) = 0$ for $|t| \geq 1$, since $b_k \geq 1$ for all k.

It follows by induction on n that the n-th derivative of the k-th term in the series can be expressed as a sum

$$(3) \qquad \mathcal{D}^n\big(\,a_k\,\psi(b_k t)\,t^k/k!\,\big) \;=\; \sum_{j=0}^{\min(k,n)} \binom{n}{j}\, a_k b_k^{\,n-j}\psi^{(n-j)} t^{k-j}/(k-j)!\;,$$

where $\psi^{(n-j)}$ stands for the derivative $\mathcal{D}^{n-j}\psi$ evaluated at the point $b_k t$, where again we may assume that $|b_k t| \leq 1$. From this, we easily obtain an upper bound of the form

$$\big|\mathcal{D}^n\big(\,a_k\,\psi(b_k t)\,t^k/k!\,\big)\big| \;\leq\; C_n\, b_k^{1+n-k}/(k-n)!$$

for $k \geq n$, where C_n is a constant which depends only on n. Now sum over $k \geq n$ for fixed n. Since the resulting series converges uniformly for each n, it follows that f has continuous derivatives of all orders.

Finally, for the special case $t = 0$, note that the j-th term of the summation of Equation (3) is equal to a_n if $j = k = n$, and is zero otherwise. It follows that $\mathcal{D}^n f(0) = a_n$, as required. □

We now apply this lemma to study a local smooth function on the closed half-space $\mathbb{H}^n \subset \mathbb{R}^n$.

LEMMA A.3. *Let V be a neighborhood of the origin in $\mathbb{R}^n$, and suppose that $f : V \cap \mathbb{H}^n \to \mathbb{R}$ has continuous partial derivatives of all orders. Then there exists a smaller neighborhood V' of the origin in $\mathbb{R}^n$ and a smooth function $g : V' \to \mathbb{R}$ which coincides with f throughout the intersection $V' \cap \mathbb{H}^n$.*

Proof. It is convenient to identify points of $\mathbb{H}^n$ with pairs $(x, t) \in \mathbb{R}^{n-1} \times \mathbb{R}$ where $t \geq 0$. Let $a_k(x)$ be the k-th partial derivative of $f(x, t)$ with respect to t at $t = 0$. Let D_ϵ be the closed ϵ-disk centered at the origin of $\mathbb{R}^{n-1} \times \mathbb{R}$, where ϵ is small enough so that $D_\epsilon \subset V$, and choose real numbers b_k so that $b_k \geq 1 + |a_k(x)|$ whenever $\|x\| \leq \epsilon$. For every (x, t) in D_ϵ set

$$g(x,t) \;=\; \begin{cases} f(x,t), & \text{if} \quad t \geq 0 \\[2ex] \sum_{k\geq 0} a_k(x)\,\psi(b_k t)\,t^k/k!, & \text{if} \quad t \leq 0 \end{cases}$$

Then it is not difficult to check that g is smooth throughout the interior of D_ϵ. Since g clearly coincides with f whenever both functions are defined, the conclusion follows. □

For a much more general statement about smooth extensions of real valued functions defined on subsets of Euclidean space, the reader is referred to WHITNEY [1936].

PART 3. RELATIONS WITH ALGEBRAIC TOPOLOGY

This section will consist of the following four papers:

On the parallelizability of the spheres (with R. Bott), Bulletin American Mathematical Society **64** (1958) 87–89.

Some consequences of a theorem of Bott, Annals of Mathematics **68** (1958) 444–449.

On the Whitehead homomorphism J, Bulletin American Mathematical Society **64** (1958) 79–82.

Bernoulli numbers, homotopy groups, and a theorem of Rohlin (with M. Kervaire), in "Proceedings International Congress of Mathematics 1958", Cambridge Univ. Press (1960) 454–458.

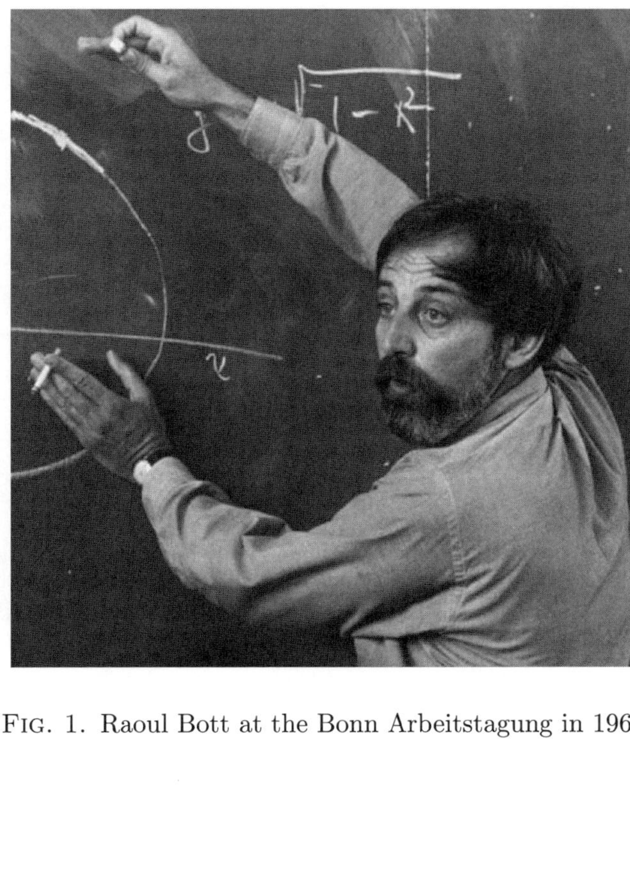

FIG. 1. Raoul Bott at the Bonn Arbeitstagung in 1969

Introduction

Although these papers present results which seem to have nothing to do with exotic sphere, the proofs are in fact very closely related to the proofs in Part 1. One fundamental result which ties all of these papers together is the following remarkable theorem of Raoul BOTT [1958, 1959], which he proved by arguments in Morse Theory.

> **Theorem.** *For any vector bundle ξ over the sphere S^{4m}, the Pontrjagin class $p_m(\xi) \in H^{4m}(S^{4m}; \mathbb{Z}) \cong \mathbb{Z}$ is divisible by* $(2m-1)!$ *(and in fact by* $2(2m-1)!$ *when m is odd).*

Wu Wen-Tsün, in his study of the cohomology of Grassmann manifolds, had described relations between the Pontrjagin classes $p_m(\xi) \in H^{4m}(B; \mathbb{Z})$ of a vector bundle ξ over any base space B and the Stiefel Whitney classes $w_k(\xi) \in H^k(B; \mathbb{Z}/2)$. For the special case of a vector bundle over S^{4m}, these Wu relations imply that w_{4m} is zero if and only if p_m is divisible by four. Combining these theorems of Bott and Wu with earlier known results, we easily obtain the following statement.[1]

> *There exists a vector bundle over S^n with $w_n \neq 0$ only in the classical cases studied by Hopf, with $n = 1, 2, 4,$ or 8.*

This result is described in the two papers which follow: **On the Parallelizability of the Spheres** (written with Bott), and **Some Consequences of a Theorem of Bott**. As easy consequences, it follows that there exists a division algebra over the real numbers only in these dimensions, and that the sphere S^{n-1} is parallelizable only for these values of n.

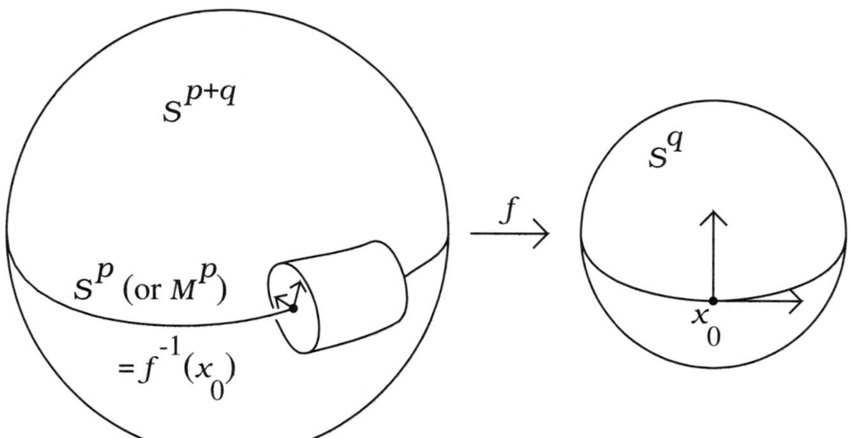

The next paper, **On the Whitehead homomorphism** J describes a relationship between differential topology, algebraic topology, and number theory which I found quite surprising. In particular, it provides a lower bound for the size of the

[1]This was proved independently, with a very different argument, by ADAMS [1958]. See also ATIYAH AND HIRZEBRUCH [1961].

stable homotopy group $\Pi_{4k-1} = \pi_{4k-1+q}(S^q)$, with q large, in terms of the denominator of the Bernoulli number B_k. This can be contrasted with results in Part 1 which estimate the size of the group Θ_{4k-1} of homotopy spheres in terms of the numerator of B_k. In fact the two estimates are closely related. Their relationship is based on the J-homomorphism

$$J \ : \ \pi_p(\mathrm{SO_q}) \ \to \pi_{\mathrm{p+q}}(\mathrm{S^q})$$

of Heinz Hopf and George Whitehead, which provides a fundamental connection between homotopy groups of rotation groups and homotopy groups of spheres, and is also intimately related to the study of differentiable manifolds. In particular, the J-homomorphism generalizes to the ***Pontrjagin-Thom construction*** which assigns to each embedded $M^p \subset S^{p+q}$ with framed normal bundle a map $f : S^{p+q} \to S^q$ which carries each normal q-disk in a tubular neighborhood onto S^q, with M^p mapping to a base point x_0, with normal frame mapping to a fixed frame at x_0, and with the complement of the tubular neighborhood mapping to the antipodal point $-x_0$. Taking M^p to be the standard $S^p \subset S^{p+q}$ with an arbitrary normal framing, we recover the J-homomorphism. In particular, taking q large, we obtain the ***stable*** J-homomorphism $J : \pi_p(\mathrm{SO}) \to \Pi_{\mathrm{p}}$.

Consider a homotopy class in the kernel of this J-homomorphism. Application of the Pontrjagin-Thom construction shows that S^p, with its normal framing, is the boundary of a normally framed submanifold Q^{p+1} in the disk D^{p+q+1}. Gluing a standard $(p+1)$-disk onto the sphere $S^p = \partial Q^{p+1}$, we obtain a closed differentiable manifold M^{p+1}.

Now suppose that $p+1 = 4k$. Since Q^{4k} is parallelizable, it follows that all of the Pontrjagin classes p_i of M^{4k} are zero for $i < k$. Hence the signature formula reduces to the simple form

$$\text{signature} = s_k\, p_k[M^{4k}] \ .$$

Here the coefficient s_k is a rational number which Hirzebruch computed explicitly as a multiple of the k-th Bernoulli number,

$$s_k = 2^{2k}(2^{2k-1} - 1)B_k/(2k)! \ ,$$

for example

$$s_1 = 1/3 \,, \quad s_2 = 7/45 \,, \quad s_3 = 2/945 \,.$$

Since the signature is an integer,[2] it follows that the Pontrjagin number $p_k[M^{4k}]$ is always divisible by the denominator of the rational number s_k. For example, taking $k = 2$, the number $p_2[M^8]$ must be divisible by $\mathrm{denom}(s_2) = 45$.

The structure of the stable homotopy groups $\pi_{n-1}(\mathrm{SO}) \cong \pi_{\mathrm{n}}(\mathrm{B_{SO}})$ has been worked out by Bott. In particular, for $n = 4k$, the group $\pi_{4k-1}(\mathrm{SO})$ is free cyclic. Furthermore, there is a canonical embedding which I will denote by

$$p_{k*} \ : \ \pi_{4k-1}(\mathrm{SO}) \ \to \ \mathbb{Z} \ ,$$

[2] In fact the signature is an integer divisible by 8, since Q^{4k} is parallelizable. However, this extra information is not very useful (except when $k \leq 2$) because of the large factor 2^{2k} in the formula for s_k.

defined as follows. To each $\eta \in \pi_{4k-1}(\mathrm{SO})$ there is associated an SO-bundle ξ_η over the sphere S^{4k}, with Pontrjagin class $p_k(\xi_\eta) \in H^{4k}(S^{4k};\mathbb{Z}) \cong \mathbb{Z}$. Now set $p_{k*}(\eta)$ equal to the corresponding integer $p_k(\xi_\eta)[S^{4k}]$. Bott showed that the image $p_{k*}(\pi_{4k-1}(\mathrm{SO})) \subset \mathbb{Z}$ is generated by $a_k(2k-1)!$, where a_k is either 1 or 2 according as k is even or odd. (See his description on p. 230 below.)

If η belongs to the kernel of the J-homomorphism, then it is not hard to show that $p_{k*}(\eta)$ can be identified with the Pontrjagin number $p_k[M^{4k}]$ of the manifold M^{4k} constructed above, up to sign. Hence, in this case, the product $s_k\, p_{k*}(\eta)$ must be an integer.

Since the stable homotopy groups of spheres are finite, by a theorem of Serre, the image $J\pi_{4k-1}(\mathrm{SO}) \subset \Pi_{4k-1}$ must be a finite cyclic group. Let j_k be its order. Then a generator η_0 of the kernel of the J-homomorphism will map to the integer

$$p_{k*}(\eta_0) = j_k a_k (2k-1)!\ .$$

Hence it follows from the argument above that the product $s_k j_k a_k(2k-1)!$ must be an integer. *In other words, the order j_k of the group $J\pi_{4k-1}(\mathrm{SO})$ must be a multiple of the denominator of the rational number*

$$s_k a_k (2k-1)! = 2^{2k} a_k (2^{2k-1}-1) B_k/2k\ .$$

Using classical results about Bernoulli numbers,[3] we see that the factor of $2^{2k-1}-1$ never causes any cancellation, so this is just the odd primary part of the denominator of B_k/k. Furthermore, an odd prime power p^a divides this denominator if and only if k is a multiple of $p^{a-1}(p-1)/2$. As examples, for $k = 1, 2, \ldots, 6$ this denominator takes the values

$$3\,, \qquad 3\cdot 5\,, \qquad 3^2\cdot 7\,, \qquad 3\cdot 5\,, \qquad 3\cdot 11\,, \qquad 3^2\cdot 5\cdot 7\cdot 13$$

respectively.

This estimate is sharpened to include the prime 2 in the paper **Bernoulli Numbers, Homotopy Groups, and a Theorem of Rohlin**, written with Kervaire, which shows that the order of the cyclic group $J\pi_{4k-1}(\mathrm{SO})$ is at least equal to the denominator of the rational number $B_k/4k$. As examples, for $k = 1, 2, \ldots, 6$ the ratio $B_k/4k$ takes the values

$$\frac{1}{2^3\cdot 3}\,, \quad \frac{1}{2^4\cdot 3\cdot 5}\,, \quad \frac{1}{2^3\cdot 3^2\cdot 7}\,, \quad \frac{1}{2^5\cdot 3\cdot 5}\,, \quad \frac{1}{2^3\cdot 3\cdot 11}\,, \quad \frac{691}{2^4\cdot 3^2\cdot 5\cdot 7\cdot 13}$$

respectively. The proof is based on the $\widehat{A}$-*genus* of a closed oriented smooth $4k$-manifold. This is a rational combination of Pontrjagin numbers of the form

$$\widehat{A}(M^{4k}) = \frac{-B_k}{2\,(2k)!}\, p_k[M^{4k}] \;+\; (\text{terms in } p_1\,, \ldots\,, p_{k-1})$$

which was introduced by Hirzebruch.[4] The basic result, proved by Borel and Hirzebruch [1960], is that $\widehat{A}(M^{4k})$ is an integer whenever $w_2(M^{4k}) = 0$. (In fact

[3]For the number theory of Bernoulli numbers, compare Milnor and Stasheff [1974, Appendix B].

[4]See Hirzebruch [1959, 1966], and compare Problem 7 of Hirzebruch [1954]. In terms of the A-genus, which he had introduced earlier, the $\widehat{A}$-genus is defined to be $A(M^{4k})/2^{4k}$.

it is an even integer if k is odd.) The rest of the proof is completely analogous to the one outlined above, simply making use of $\widehat{A}(M^{4k})$ in place of the signature. For discussion of the theorem of Rohlin (= Rokhlin or Rochlin), see GUILLOU AND MARIN [1986].

Further Developments

In fact, it turns out that the lower bound described above is optimal. The much more difficult problem of finding an upper bound for the order of $J\pi_{4k-1}(\mathrm{SO})$ was solved a few years later by Frank ADAMS [1966], who showed that:

> *The order of $J\pi_{4k-1}(\mathrm{SO})$ is precisely equal to the denominator of $B_k/4k$.*

For k equal to 1 through 6, this order is respectively

$$2^3\cdot 3\,, \qquad 2^4\cdot 3\cdot 5\,, \qquad 2^3\cdot 3^2\cdot 7\,, \qquad 2^5\cdot 3\cdot 5\,, \qquad 2^3\cdot 3\cdot 11\,, \qquad 2^4\cdot 3^2\cdot 5\cdot 7\cdot 13\,.$$

In general, the order of $J\pi_{4k-1}(\mathrm{SO})$ can be expressed as a product $\prod p^{a+1}$ where p ranges over all primes such that $2k$ is a multiple of $p-1$, and where p^a is the highest power of p dividing $4k$ (taking $a=0$ if p does *not* divide $4k$). As an example, if $p=2n+1$ is an odd prime, then p divides this denominator if and only if k is a multiple of n, and some higher power of p divides the denominator only if k is also a multiple of p.

ON THE PARALLELIZABILITY OF THE SPHERES

BY R. BOTT AND J. MILNOR

Communicated by H. Samelson, February 13, 1958

(The following note consists of excerpts from two letters.)
(*Milnor to Bott*; December 23, 1957.)

. . . Hirzebruch tells me that you have a proof of his conjecture that the Pontrjagin class p_k of a GL_m-bundle over the sphere S^{4k} is always divisible by $(2k-1)!$. I wonder if you have noted the connection of this result with classical problems, such as the existence of division algebras, and the parallelizability of spheres.

According to Wu the Pontrjagin classes of any GL_m-bundle, reduced modulo 4, are determined by the Stiefel-Whitney classes of the bundle. (See *On the Pontrjagin classes* III, Acta Math. Sinica vol. 4 (1954) in Chinese.) The proof makes use of the Pontrjagin squaring operation, together with the coefficient homomorphism $i: Z_2 \to Z_4$. Although I do not know the exact formula which Wu obtains, the following special case is not hard to prove:

LEMMA. *If the Stiefel-Whitney classes $w_1, w_2, \cdots, w_{4k-1}$ of a GL_m-bundle are zero then the Pontrjagin class p_k, reduced modulo* 4, *is equal to $i_* w_{4k}$.*

For a bundle over S^{4k} this means that w_{4k} is zero if and only if p_k is divisible by 4. Now if you can prove that p_k is divisible by $(2k-1)!$ it will follow that w_{4k} must be zero, whenever $k \geqq 3$.

THEOREM. *There exists a GL_m-bundle over S^n with $w_n \neq 0$ only if n equals* 1, 2, 4 *or* 8.

PROOF. Wu has shown that such a bundle can only exist if n is a power of 2. But the above remarks show that the cases $n = 16, 32, \cdots$ cannot occur.

COROLLARY 1. *The vector space R^n possesses a bilinear product operation without zero divisors only for n equal to* 1, 2, 4 *or* 8.

PROOF. Given such a product operation the map $S^{n-1} \to GL_n$ defined by $x \to$(left multiplication by x) gives rise to a GL_n-bundle over S^n for which it can be shown that $w_n \neq 0$.

COROLLARY 2. *The sphere S^{n-1} is parallelizable only for $n-1$ equal to* 1, 3 *or* 7.

PROOF. Given linearly independent vector fields $v_1(x), \cdots, v_{n-1}(x)$, on S^{n-1}, the correspondence

$$x \to (x, v_1(x), \cdots, v_{n-1}(x))$$

carries S^{n-1} into the Stiefel manifold of n-frames in R^n. Identifying this space with GL_n, we again obtain a GL_n-bundle over S^n with $w_n \neq 0 \cdots$

(*Bott to Milnor*, January 6, 1958.)

... Here is what I can show:

THEOREM. *Let $B = B_U$ be the universal base-space of the infinite unitary group. Then the image of $\pi_{2n}(B)$ in $H_{2n}(B)$ is divisible by precisely $(n-1)!$.*

This then refines the result of Borel-Hirzebruch that these classes are divisible by $(n-1)!$ except for the prime 2, [3], and confirms their conjecture. Because the Pontryagin classes are in the last analysis pre-images of classes in B_U, it follows that for any $GL_n(R)$ bundle over S^{4k}, p_k is divisible by $(2k-1)!$. This is all you needed.

The precise divisibility of p_k, for a real bundle over S^{4k}, is actually given by:

$$p_k = 0, \quad \begin{matrix} \operatorname{mod} (2k-1)!, & k \text{ even}, \\ \operatorname{mod} (2k-1)!2, & k \text{ odd}. \end{matrix}$$

This is seen by considering the fibering $U/O \to B_O \to B_U$.

The theorem follows from the fact, that if $\Omega = \Omega SU$ is the loopspace on SU, then there exists a homotopy equivalence $f: B \to \Omega$, as was announced in [1] and is proved in [2]. By standard theory the double suspension, S, from Ω into B, defines a homomorphism $\pi_k(\Omega) \to \pi_{k+2}(B)$ which is bijective for dimensions ≥ 1.

Let $\lambda = f_* \circ S$. It is then clear that:

$$\pi_{2k}(\Omega) = \lambda^{k-1}(\pi_2(\Omega)). \tag{1}$$

Now in [2] the Hopf algebra $H_*(\Omega)$ is described. It turns out that $H_*(\Omega) = Z[\sigma_1, \sigma_2, \cdots]$, $\dim \sigma_i = 2i$, the diagonal map being: $\Delta_* \sigma_i = \sum \sigma_s \otimes \sigma_t$; $s + t = i$; $\sigma_0 = 1$. Hence the primitive subspace, P_*, is generated by elements $\{p_n\}$, $n = 1, 2, \cdots$, which are inductively determined by the relation:

$$p_n - p_{n-1} \cdot \sigma_1 + p_{n-2} \cdot \sigma_2 - \cdots \pm n\sigma_n = 0, \qquad n = 1, 2, \cdots. \tag{2}$$

Let λ_* be the homomorphism corresponding to λ in homology. It will preserve spherical classes, and annihilate decomposable elements. It therefore follows from (2) that $\lambda_* p_i = \pm i \lambda_* \sigma_i$. As the spherical classes generate P_* (over the rationals, SU is a product of odd spheres!) this relation implies that $\lambda_*^{k-1}(H_2(\Omega))$ is divisible by at least $(k-1)!$. By (1) it follows that the spherical classes in dimension $2k$ are divisible by at least $(k-1)!$. This is the best bound because it is not hard to see that λ_* is *not* divisible on all of $H_*(\Omega)$.

An easy corollary of the theorem is that $\pi_{2n}(U_n)=Z/n!Z$. Kervaire also has decided the parallelizability question.* He uses this formula as his starting point . . . .

REFERENCES +

1. R. Bott, *The stable homotopy of the classical groups*, Proc. Nat. Acad. Sci. U.S.A. vol. 43 (1951) pp. 933–935.

2. ———, *The Pontryagin ring of* ΩG (to be published in Michigan Math. J.).

3. A. Borel and F. Hirzebruch: *Characteristic classes and homogeneous spaces* (to be published in Amer. J. Math.).

UNIVERSITY OF MICHIGAN AND
OXFORD UNIVERSITY

---

* Unpublished.

\+ See BOTT [1958] as well as BOREL and HIRZEBRUCH [1960] in the Bibliography.

ANNALS OF MATHEMATICS
Vol. 68, No. 2, September, 1958
Printed in Japan

SOME CONSEQUENCES OF A THEOREM OF BOTT

BY JOHN MILNOR[1]

(Received February 11, 1958)

It will be shown that the following theorem, due to R. Bott [3], can be used to solve several well known problems ; including the problem of the existence of division algebras, and the parallelizability of spheres[2]. (Independent solutions of these problems, also based on Bott's work, have been given by Kervaire and Hirzebruch.)

THEOREM OF BOTT. *For any O_m-bundle ξ over the sphere S^{4k}, the Pontrjagin class $p_k(\xi) \in H^{4k}(S^{4k}; Z)$ is divisible by* $(2k-1)!$. (This result was conjectured, and proved up to powers of 2, by Borel and Hirzebruch [1]).

The following result, which follows from Wu Wen-Tsün [19], will also be needed. Since Wu's paper is in Chinese a proof is included in the appendix. The epimorphism $Z \to Z_4$ induces a homomorphism $H^*(K; Z) \to H^*(K; Z_4)$ which will be denoted by $\alpha \to (\alpha)_4$. Let $i: Z_2 \to Z_4$ denote the inclusion homomorphism.

THEOREM OF WU. *For any O_m-bundle ξ over a complex K, the class $(p_k(\xi))_4 \in H^{4k}(K; Z_4)$ is determined by the Stiefel-Whitney classes $w_i(\xi) \in H^i(K; Z_2)$. In particular if the Stiefel-Whitney classes $w_1(\xi), \cdots, w_{4k-1}(\xi)$ are zero then $(p_k(\xi))_4 = i_* w_{4k}(\xi)$.*

Combining these two results, the following is obtained.

THEOREM 1. *There exists an O_m-bundle ξ over the sphere S^n with $w_n(\xi) \neq 0$ only for $n = 1, 2, 4$ or 8.*

(Examples of such bundles can be given as follows : for $n = 1$ the 2-fold covering of the circle, and for $n = 2$, 4 or 8 the O_n-bundle over S^n associated with the Hopf fibering $S^{2n-1} \to S^n$.)

PROOF. According to Wu [16] such a bundle can exists only if n is a power of 2. Hence it is certainly sufficient to consider the case $n = 4k$, $k > 2$. The identity

$$(p_k(\xi))_4 = i_* w_{4k}(\xi) \in H^{4k}(S^{4k}; Z_4) = Z_4$$

is valid, since the lower Stiefel-Whitney classes must be zero. In other words the class $w_{4k}(\xi)$ is zero if and only if $p_k(\xi)$ is divisible by 4. But

[1] The author holds a Sloan fellowship.

[2] A preliminary account of this work has been given in [20].

$p_k(\xi)$ is known to be divisible by $(2k-1)!$. For $k > 2$ this proves that $w_{4k}(\xi) = 0$.

THEOREM 2. *The sphere S^r is parallelizable only for $r = 1, 3, 7$.* (Compare Steenrod and Whitehead [10].)

PROOF. The fibering $SO_r \longrightarrow SO_{r+1} \xrightarrow{f} S^r$ associated with the tangent bundle of S^r has the following homotopy sequence :

$$\longrightarrow \pi_r(SO_{r+1}) \xrightarrow{f_*} \pi_r(S^r) \xrightarrow{\partial} \pi_{r-1}(SO_r) \longrightarrow \pi_{r-1}(SO_{r+1}) \longrightarrow 0 .$$

The group $\pi_r(S^r)$ will be identified with the integers. Then $\partial(1) \in \pi_{r-1}(SO_r)$ is the element which corresponds to the tangent bundle of S^r. (See Steenrod [9, § 18]).

For each $\lambda \in \pi_r(SO_{r+1})$ let ξ denote the corresponding SO_{r+1}-bundle over S^{r+1}, and let $X(\xi)$ denote its Euler class (=top Stiefel-Whitney class with integer coefficients). Let μ be the standard generator of $H_{r+1}(S^{r+1}; Z)$. Then $f_*(\lambda)$ is equal to the negative of the " Euler number " $\langle X(\xi), \mu\rangle$. [Proof. Let $\mathfrak{o}(\xi) \in H^{r+1}(S^{r+1}; \pi_r(SO_{r+1}))$ denote the obstruction to the existence of a cross-section of ξ. Then $X(\xi)$, the obstruction to the existence of a cross-section in the associated sphere bundle, is equal to $f_*(\mathfrak{o}(\xi))$. According to Steenrod [9, p. 180] the identity $\langle \mathfrak{o}(\xi), \mu\rangle = -\lambda$ is satisfied. Therefore $\langle X(\xi), \mu\rangle = -f_*(\lambda)$].

Now if S^r is parallelizable then $\partial(1) = 0$, hence there exists $\lambda \in \pi_r(SO_{r+1})$ with $f_*(\lambda) = 1$. For the corresponding bundle ξ, the class $X(\xi)$ generates $H^{r+1}(S^{r+1}; Z)$; hence the class $w_{r+1}(\xi) = (X(\xi))_2$ is non-zero. Together with Theorem 1 this complete the proof.

It follows immediately that the real projective space P^r is parallelizable only for $r = 1, 3, 7$. (For consequences concerning the immersion of P^r in euclidean space see Milnor, Comm. Math. Helv. 30 (1956), p. 284).

COROLLARY 1. *There exists a division algebra of rank n over the real numbers only for $n = 1, 2, 4, 8$.*

PROOF. The existence of a bilinear product operation without zero-divisors in the vector space R^n implies that the projective space P^{n-1} is parallelizable. (See Stiefel [11, p. 216]). [ALTERNATIVE PROOF. Suppose that such a product operation in R^n is given. Then the correspondence $S^{n-1} \to GL_n$ defined by $x \to$ (left multiplication by x) gives rise to a GL_n-bundle ξ over S^n. It is not hard to verify that $w_n(\xi) \neq 0$].

COROLLARY 2. *For $r \geqq 8$ the groups $\pi_{r-1}(SO_r)$ are as follows :*

r modulo 8:	0	1	2	3	4	5	6	7
$\pi_{r-1}(SO_r)$:	$Z+Z$	*group of order* 4	$Z+Z_2$	Z_2	$Z+Z$	Z_2	Z	Z_2

PROOF. This follows from Bott's computation [2] of the stable groups $\pi_{r-1}(SO_{r+1})$, together with the exact homotopy sequence used to prove Theorem 2.

THEOREM 3. *Let M^{2n} be a simply-connected differentiable manifold such that the cohomology group $H^i(M^{2n}; Z)$ is infinite cyclic for $i = 0, n, 2n$, and zero otherwise. Then n must be* 2, 4, *or* 8.

(Examples are provided by the complex, quaternion, and Cayley projective planes. It will be shown in a later paper [7] that the condition of simple-connectivity can be eliminated. This will give an answer to Problem 5 of [5]).

PROOF. If α generates $H^n(M^{2n}; Z)$, then the Poincaré duality theorem implies that $\alpha \smile \alpha$ generates $H^{2n}(M^{2n}; Z)$. Hence

$$Sq^n : H^n(M^{2n}; Z_2) \longrightarrow H^{2n}(M^{2n}; Z_2)$$

in non-zero. The formulas of Wu [15] now imply that the Stiefel-Whitney class w_n of the tangent bundle θ is non-zero. Choose a map $g : S^n \to M^{2n}$ which, under the Hurewicz homomorphism, corresponds to a generator of $H_n(M^{2n}; Z)$. Then the bundle θ' over S^n induced from θ by g will satisfy $w_n(\theta') \neq 0$. Therefore n must be 1, 2, 4 or 8. Since the case $n = 1$ is easily excluded, this completes the proof.

Bott's theorem is related to the question of the existence of maps with Hopf invariant 1 as follows. Let $J : \pi_{n-1}(SO_m) \to \pi_{m+n-1}(S^m)$ be the homomorphism of G. W. Whitehead [13], and let $\gamma_n : \pi_{m+n-1}(S^m) \to Z_2$ be the generalized Hopf invariant of Steenrod [8], which is defined using the functional Sq^n operation. For each odd prime q let

$$\gamma_{q,i} : \pi_{m+2i(q-1)-1}(S^m) \longrightarrow Z_q$$

denote the corresponding homomorphism based on the reduced q^{th} power $\mathscr{P}^i$.

THEOREM 4a. *The image $J\pi_{n-1}(SO_m)$, $m \geqq n$, contains an element $J\lambda$ with generalized Hopf invariant $\gamma_n(J\lambda)$ different from zero only if n equals* 2, 4, *or* 8.

THEOREM 4b. *The image $J\pi_{2i(q-1)-1}(SO_m)$, $m \geqq 2i(q-1)$, contains an element $J\lambda$ with $\gamma_{q,i}(J\lambda)$ different from zero only if $i = 1$.*

PROOF OF 4b. Let ξ be the SO_m-bundle over S^n associated with λ, where $n = 2i(q-1)$. Let E be the total space of the associated bundle having the unit ball B^m as fibre, so that the boundary $\dot{E}$ is the total space of the associated sphere bundle. According to [7, Theorem 3, Corollary 1], the collapsed space $E/\dot{E}$ can be obtained from the sphere S^m by attaching an $(m+n)$-cell, using on attaching map in the homotopy class $J\lambda$. Thus the generalized Hopf invariant $\gamma_{q,i}(J\lambda)$ is non-zero if and only if the homomorphism

$$\mathscr{P}^i : H^m(E, \dot{E}\,; Z_q) \longrightarrow H^{m+n}(E, \dot{E}\,; Z_q)$$

is non-zero.

Let $\phi : H^j(S^n\,; Z_q) \to H^{j+m}(E, \dot{E}\,; Z_q)$ denote the isomorphism of Thom [12]. According to Wu [18, § IV] the class $\phi^{-1}\mathscr{P}^i\phi(1) \in H^n(S^n\,; Z_q)$ can be expressed as a polynomial in the Pontrjagin classes of ξ, reduced modulo q. But these Pontrjagin classes are zero, except for $p_{i(q-1)/2}(\xi)$ which is divisible by $(i(q-1)-1)!$. For $i > 1$, since the number $(i(q-1)-1)!$ is divisible by q, it follows that the operation $\mathscr{P}^i$ must be zero.

Theorem 4a is proved in a similar way, using Theorem 1 together with Thom's definition of the Stiefel-Whitney classes. (See [12]).

Appendix

PROOF OF THE THEOREM OF WU. Following Hirzebruch [6] define the Pontrjagin class p_k of an O_m-bundle as $(-1)^k$ times the Chern class c_{2k} of the U_m-bundle induced by the inclusion $O_m \to U_m$. This is slightly different from the Pontrjagin class as defined by Pontrjagin and Wu. (Compare [17, Theorem 4]).

Consider the exact sequence of cohomology group corresponding to the coefficient sequence

$$0 \longrightarrow Z_2 \xrightarrow{\ i\ } Z_4 \xrightarrow{\ j\ } Z_2 \longrightarrow 0\,;$$

as well as the Pontrjagin squaring operation

$$\mathfrak{P} : H^{2k}(K\,; Z_2) \longrightarrow H^{4k}(K\,; Z_4)\ .$$

(See for example Whitehead [14]).

LEMMA 1. *The Pontrjagin class p_k of any O_m-bundle is related to the Stiefel-Whitney classes $w_1, \cdots, w_{4k}$ by an identity*

$$(p_k)_4 = \mathfrak{P}(w_{2k}) + i_* f_k(w_1; \cdots, w_{4k})$$

where f_k is a polynomial with coefficients in Z_2.

PROOF. It is clearly sufficient to consider the case of the universal bundle over the Grassmann space $G_m(R)$, with m large. The identity $j_*\mathfrak{P}(w) = w \smile w$ holds for any cohomology class w. Comparing this with the relation[3]

[3] See Wu [17, Theorem 3].

$$j_*((p_k)_4) = (p_k)_2 = w_{2k} \smile w_{2k}$$

it follows that

$$(p_k)_4 - \mathfrak{P} w_{2k} \in (\text{kernel } j_*) = i_* H^{4k}(G_m(R)\,;Z_2)\ .$$

Since the cohomology ring $H^*(G_m(R)\,;Z_2)$ is generated by the Stiefel-Whitney classes, this proves Lemma 1.

To prove the theorem it is only necessary to show that the coefficient of w_{4k} in f_k is non-zero. Let Υ denote the universal U_m-bundle over the complex Grassmann space $G_m(C)$. Recall that the cohomology ring $H^*(G_m(C)\,;Z)$ is a polynomial ring[4] generated by the Chern classes of Υ. The inclusion $U_m \to O_{2m}$ induces an O_{2m}-bundle over $G_m(C)$ which will be denoted by Υ_R. Applying Lemma 1 to this bundle Υ_R, the relations[5]

$$p_k(\Upsilon_R) = c_k(\Upsilon)^2 - 2c_{k-1}(\Upsilon)c_{k+1}(\Upsilon) + - \cdots \pm 2c_0(\Upsilon)c_{2k}(\Upsilon)$$

and

$$w_{2r+1}(\Upsilon_R) = 0\ , \qquad w_{2r}(\Upsilon_R) = (c_r(\Upsilon))_2$$

show that the polynomial f_k must satisfy

$$f_k(0, w_2, 0, w_4, \cdots, w_{4k}) = w_{2k-2}w_{2k+2} + w_{2k-4}w_{2k+4} + \cdots + w_0 w_{4k}\ .$$

Therefore $f_k(0, 0, \cdots, 0, w_{4k}) = w_{4k}$; which completes the proof.

PRINCETON UNIVERSITY

REFERENCES

1. A. BOREL and F. HIRZEBRUCH, *Characteristic classes and homogeneous spaces*, to appear (Amer. J. Math.). *
2. R. BOTT, *On the stable homotopy of the classical groups*, Proc. Nat. Acad. Sci. U.S.A. 43 (1957), 933–935.
3. ———, *The space of loops on a Lie-group*, to appear (Mich. J. Math.). +
4. S. S. CHERN, *Characteristic classes of Hermitian manifolds*, Ann. of Math. 47 (1946), 85–121.
5. F. HIRZEBRUCH, *Some problems on differentiable and complex manifolds*, Ann. of Math. 60 (1954), 213–236.
6. ———, Neue topologische Methoden in der algebraischen Geometrie, Springer, 1956.
7. J. MILNOR, *On spaces with a gap in cohomology*, to appear. ++
8. N. E. STEENROD, *CohＯlomogy invariants of mappings*, Ann. of Math. 50 (1949), 954–988.
9. ———, The topology of fibre bundles, Princeton, 1951.
10. ——— and J.H.C. WHITEHEAD, *Vector fields on the n-sphere*, Proc. Nat. Acad. Sci. U.S.A. 37 (1951), 58–63.
11. E. STIEFEL, *Über Richtungsfelder in den projektiven Räumen und einen Satz aus der reelen Algebra*, Comm. Math. Helv. 13 (1940), 201–218.
12. R. THOM, *Espaces fibrés en sphères et carrés de Steenrod*, Ann. sci. Ecole norm. sup 69 (1952), 109–182.

[4] Chern [4].

[5] See Hirzebruch [6, p. 68] and Steenrod [9, p. 212].

* See BOREL and HIRZEBRUCH [1960] in the Bibliography.

+ See BOTT [1958] in the Bibliography.

++ Never published. (Compare pages 7-8 in this volume.)

13. G. W. WHITEHEAD, *On the homotopy groups of spheres and rotation groups*, Ann. of Math. 43 (1942), 634–640.
14. J.H.C. WHITEHEAD, *On simply connected 4-dimensional polyhedra*, Comm. Math. Helv. 22 (1949), 48-92.
15. WU WEN-TSÜN, *Classes caractéristiques et i-carrés d'une variété*, C. R. Acad. Sci. Paris 230 (1950), 508-511.
16. ———, *Les i-carrés dans une variété grassmannienne*, C. R. Acad. Sci. Paris 230 (1950), 918-920.
17. ———, *On Pontrjagin classes I*, Scientia Sinica 3 (1954), 353–367.
18. ———, *On Pontrjagin classes II*, Scientia Sinica 4 (1955), 455–490.
19. ———, *On Pontrjagin classes III*, Acta Math. Sinica 4 (1954), 323–346.
20. R. BOTT and J. MILNOR, *On the parallelizability of the spheres*, Bull. Amer. Math. Soc. 64 (1958), 87–89.

Reprinted from the
BULLETIN OF THE AMERICAN MATHEMATICAL SOCIETY
Vol. 64, No. 3, Part I, pp. 79-82
May, 1958

ON THE WHITEHEAD HOMOMORPHISM J

BY JOHN MILNOR[1]

Communicated by R. H. Fox, February 3, 1958

Consider the homomorphism $J: \pi_{r-1}(SO_n) \to \pi_{n+r-1}(S^n)$ of G. W. Whitehead[2] in the stable range $n > r$. The object of this note is to prove:

THEOREM 2. *Let q be an odd prime and let r be any multiple of $2(q-1)q^i$, $i \geqq 0$. Then for $n > r$ the image $J\pi_{r-1}(SO_n) \subset \pi_{n+r-1}(S^n)$ contains a cyclic subgroup of order q^{i+1}.*

According to recent work of Adams (as yet unpublished) the stable group $\pi_{n+r-1}(S^n)$ has the following q-primary components:

Z_q for $r = 2i(q-1)$, $i < q$ (this result is due to Cartan);

Z_q for $r = 2q(q-1) - 1$;

Z_{q^2} for $r = 2q(q-1)$; and zero for other values of r less than $2q(q-1)$. Comparing this with Theorem 2 we have:

COROLLARY. *For $r < 2q(q-1) - 1$, and for $r = 2q(q-1)$, the image $J\pi_{r-1}(SO_n)$ contains the q-primary component of the stable group $\pi_{n+r-1}(S^n)$.*

The corresponding assertion for $r = 2q(q-1) - 1$ is false, since the group $\pi_{r-1}(SO_n)$ is zero[3] in this case.

The proof will be based on work of Thom, Hirzebruch, Borel and von Staudt.

THEOREM 1. *Let ξ be the SO_n-bundle over S^r corresponding[4] to an element λ of $\pi_{r-1}(SO_n)$. If $J\lambda = 0$ then there exists an oriented manifold M^r, differentiably imbedded in the sphere S^{n+r}, and having the following property: Some map $g: M^r \to S^r$ of degree $+1$ is covered by a bundle map of the normal bundle of M^r into the given bundle ξ.*

(It is not asserted that M^r is connected.) The manifolds M^r constructed in this way will be further studied in a later paper.[5]

PROOF OF THEOREM 1. Let E be the total space of the n-cell bundle over S^r associated with ξ; so that the boundary $\dot{E}$ is the total space

[1] The author holds a Sloan fellowship.

[2] G. W. Whitehead, Ann. of Math. vol. 43 (1942), pp. 634–640.

[3] See R. Bott, Proc. Nat. Acad. Sci. U.S.A. vol. 43 (1957) pp. 933–935.

[4] See Steenrod, *The topology of fibre bundles*, 1951, p. 99.

[5] J. Milnor, *A generalization of a theorem of Rohlin*, to appear.

of the associated $(n-1)$-sphere bundle. Consider the identification space $E/\dot{E}$ obtained by collapsing $\dot{E}$ to a point e^0. This space has a cell subdivision with cells e^0, e^n, and e^{n+r}; where e^n corresponds to the inverse image of a base point in S^r. The Thom isomorphism[6] $\Phi: H^i(S^r) \to H^{i+n}(E, \dot{E})$ can be used to give specific orientations to the cells e^n, e^{n+r}. In any such cell complex the cell e^{n+r} is attached to the sphere $e^0 \cup e^n$ by means of an attaching map $S^{n+r-1} \to e^0 \cup e^n$ which is well defined up to homotopy.

LEMMA 1. *The homotopy class of the attaching map, considered as an element of* $\pi_{n+r-1}(S^n)$, *is equal to* $J\lambda$.

The proof, which is not difficult, will be given in a subsequent paper.[7]

Now suppose that $J\lambda = 0$. Then the complex $E/\dot{E}$ has the homotopy type of the union $S^n \vee S^{n+r}$ with a single point in common; hence there exists a map $f: S^{n+r} \to E/\dot{E}$ of degree $+1$.

The complement $E/\dot{E} - e^0$ can be considered as a differentiable manifold, with submanifold S^r corresponding to the trivial cross-section of the cell bundle. Following Thom[8] the map f can be approximated by a map f_1 which is t-regular on S^r. The inverse image $M^r = f_1^{-1}(S^r)$ is then a differentiable manifold with a canonical orientation. Furthermore, if $g: M^r \to S^r$ denotes the restriction of f_1, then g is covered by a bundle map of the normal bundle of M^r into the given bundle ξ.

Let T_1 and T_2 be suitably chosen open tubular neighborhoods of M^r and S^r. From the commutativity of the diagram

$$\begin{array}{ccccc} H^r(S^r) & \xrightarrow{\Phi} & H^{n+r}(E/\dot{E}, E/\dot{E} - T_2) & \to & H^{n+r}(E/\dot{E}) \\ \downarrow g^* & & \downarrow & & \downarrow f_1^* \\ H^r(M^r) & \xrightarrow{\Phi} & H^{n+r}(S^{n+r}, S^{n+r} - T_1) & \to & H^{n+r}(S^{n+r}) \end{array}$$

it follows that g has degree $+1$. This completes the proof of Theorem 1.

Now suppose that r is equal to $4k$. The Pontrjagin classes p_i of the manifold M^r are clearly zero for $i < k$. Furthermore the Pontrjagin number $p_k[M^r]$ is equal to $-\langle p_k(\xi), \mu \rangle$, where μ denotes the standard generator of $H_r(S^r; Z)$. For such a manifold the Hirzebruch index

---

[6] R. Thom, Ann. Sci. Ecole Norm. Sup. vol. 69 (1952) pp. 109–181.

[7] J. Milnor, *On spaces with a gap in cohomology*, Theorem 3, Corollary 1, to appear.

[8] R. Thom, Comment. Math. Helv. vol. 28 (1954) pp. 17–86.

formula[9] reduces to $\tau(M^r)=s_k p_k[M^r]$; where $s_1=1/3$, $s_2=7/45, \cdots$, and in general s_k equals $2^{2k}(2^{2k-1}-1)/(2k)!$ times the Bernoulli number B_k. Since $\tau(M^r)$ is an integer by definition this implies:

COROLLARY 1. *Let ξ and λ be as above, with $r=4k$. The condition $J\lambda=0$ implies that $s_k\langle p_k(\xi), \mu\rangle$ is an integer.*

Borel and Hirzebruch[10] have constructed an example of an SO_n-bundle ξ_0 over S^{4k}, $n>4k$, such that the number $\langle p_k(\xi_0), \mu\rangle$ is equal to $2(2k-1)!$. Let λ_0 be the corresponding element of $\pi_{4k-1}(SO_n)$ and let h be the order of its image $J\lambda_0$. (h is a positive integer since the stable homotopy groups of spheres are known to be finite groups.) Then for the bundle ξ_1 corresponding to $h\lambda_0$ the number

$$s_k\langle p_k(\xi_1), \mu\rangle = s_k h 2(2k-1)!$$

must be an integer. An immediate consequence is the following.

COROLLARY 2. *The order h of the element $J\lambda_0$ of $\pi_{n+4k-1}(S^n)$ is a multiple of the denominator of the rational number $2(2k-1)!\, s_k$, expressed as a fraction in lowest terms.*

The author is indebted to Hirzebruch for calling his attention to the following two theorems, which will be used to compute the above denominator. An odd prime q is said to be "of rank k" if $2k\equiv 0$ (modulo $q-1$). Let b_k denote the product of all odd primes of rank k.

First theorem of von Staudt.[11] *The denominator of the Bernoulli number B_k, expressed as a fraction in lowest terms, is equal to $2b_k$.*

Thus, setting $B_k=a_k/2b_k$, the integer a_k is odd and has no prime factor of rank k. Any positive integer k can be expressed as a product $k=2^t k_1 k_2$ with $k_1 k_2$ odd, where all of the prime factors of k_1 are of rank k, while none of the prime factors of k_2 is of rank k.

Second theorem of von Staudt.[11] *The numerator a_k of B_k is congruent to zero modulo k_2.*

Now the number $2(2k-1)!\, s_k=2^{2k}(2^{2k}-2)B_k/2k$ can be written as

$$2^{2k-t-2}(2^{2k}-2)(a_k/k_2)/b_k k_1.$$

The numerator and denominator of this expression are clearly integers. Furthermore the prime factors of the denominator are all odd primes of rank k. But no such prime divides the numerator. (The

[9] F. Hirzebruch, *Neue topologische Methoden in der algebraischen Geometrie*, 1956, p. 85.

[10] A. Borel and F. Hirzebruch, *Characteristic classes and homogeneous spaces*, to appear.

[11] See for example N. Nielsen, *Traité élémentaire des nombres de Bernoulli*, 1923, pp. 240–250.

condition $2k \equiv 0 \pmod{q-1}$ implies that $2^{2k}-2 \equiv -1 \pmod{q}$.) Thus we have proved:

LEMMA 2. *The number* $2(2k-1)!\, s_k$, *when expressed as a fraction in lowest terms, has denominator* $b_k k_1$.

PROOF OF THEOREM 2. If $r=4k$ is a multiple of $2(q-1)q^i$, it follows that q is of rank k. Hence q divides b_k and q^i divides k_1; so that q^{i+1} divides the denominator of $b_k k_1$. Together with Theorem 1, Corollary 2 this completes the proof.

BERNOULLI NUMBERS, HOMOTOPY GROUPS, AND A THEOREM OF ROHLIN

By JOHN W. MILNOR AND MICHEL A. KERVAIRE

A homomorphism $J: \pi_{k-1}(\mathbf{SO}_m) \to \pi_{m+k-1}(S^m)$ from the homotopy groups of rotation groups to the homotopy groups of spheres has been defined by H. Hopf and G. W. Whitehead[16]. This homomorphism plays an important role in the study of differentiable manifolds. We will study its relation to one particular problem: the question of possible Pontrjagin numbers of an 'almost parallelizable' manifold.

Definition. A connected differentiable manifold M^k with base point x_0 is *almost parallelizable* if $M^k - x_0$ is parallelizable. If M^k is imbedded in a high-dimensional Euclidean space R^{m+k} $(m \geqslant k+1)$ then this is equivalent to the condition that the normal bundle ν, restricted to $M^k - x_0$, be trivial (compare the argument given by Whitehead[17], or Kervaire ([5], § 8)).

The following theorem was proved by Rohlin in 1952 (see Rohlin[11, 12], Kervaire[6]).

Theorem (Rohlin). Let M^4 be a compact oriented differentiable 4-manifold with Stiefel-Whitney class w_2 equal to zero. Then the Pontrjagin number $p_1[M^4]$ is divisible by 48.

Rohlin's proof may be sketched as follows. It may be assumed that M^4 is a connected manifold imbedded in R^{m+4}, $m \geqslant 5$.

Step 1. It is shown that M^4 is almost parallelizable.

Let f be a cross-section of the normal $\mathbf{SO}_m$-bundle ν restricted to $M^4 - x_0$. The obstruction to extending f is an element

$$\mathfrak{o}(\nu, f) \in H^4(M^4; \pi_3(\mathbf{SO}_m)) \approx \pi_3(\mathbf{SO}_m).$$

Step 2. It is shown that $J\mathfrak{o}(\nu, f) = 0$.

Since J carries the infinite cyclic group $\pi_3(\mathbf{SO}_m)$ onto the cyclic group $\pi_{m+3}(S^m)$ of order 24, this implies that $\mathfrak{o}(\nu, f)$ is divisible by 24. Now identify the group $\pi_3(\mathbf{SO}_m)$ with the integers.

Step 3. It is shown that the Pontrjagin class $p_1(\nu)$ is equal to $\pm 2\mathfrak{o}(\nu, f)$.

Since by Whitney duality $p_1(\nu) = -p_1$ (tangent bundle), it follows that $p_1[M^4]$ is divisible by 48.

The first step in this argument does not generalize to higher dimensions. However Step 2, the assertion that $J\mathfrak{o}(\nu, f) = 0$, generalizes immediately. In fact we have:

Lemma 1. *Let* $\alpha \in \pi_{k-1}(\mathbf{SO}_m)$; *then* $J\alpha = 0$ *if and only if there exists an almost parallelizable manifold* $M^k \subset R^{m+k}$ *and a cross-section* f *of the induced normal* $\mathbf{SO}_m$*-bundle* ν *over* $M^k - x_0$ *such that* $\alpha = \mathfrak{o}(\nu, f)$.

Step 3 can be replaced by the following. Identify the group $\pi_{4n-1}(\mathbf{SO}_m)$, $m > 4n$, with the integers (compare Bott[2]). *Define* a_n to be equal to 2 for n odd and 1 for n even.

Lemma 2. *Let* ξ *be a stable* $\mathbf{SO}_m$*-bundle over a complex* K $(\dim K < m)$, *and let* f *be a cross-section of* ξ *restricted to the skeleton* $K^{(4n-1)}$. *Then the obstruction class* $\mathfrak{o}(\xi, f) \in H^{4n}(K; \pi_{4n-1}(\mathbf{SO}_m))$ *is related to the Pontrjagin class* $p_n(\xi)$ *by the identity* $p_n(\xi) = \pm a_n . (2n-1)!\, \mathfrak{o}(\xi, f)$.

Combining Lemmas 1 and 2, we obtain the following theorems.

Define j_n as the order of the finite cyclic group $J\pi_{4n-1}(\mathbf{SO}_m)$ in the stable range $m > 4n$.

Theorem 1. *The Pontrjagin number* $p_n[M^{4n}]$ *of an almost parallelizable* $4n$*-manifold is divisible by* $j_n a_n (2n-1)!$.

(For $n = 1$, this gives Rohlin's assertion, since $j_1 = 24$, $a_1 = 2$.)

Proof. This follows since $\mathfrak{o}(\nu, f)$ must be divisible by j_n.

Conversely:

Theorem 2. *There exists an almost parallelizable manifold* M_0^{4n} *with*

$$p_n[M_0^{4n}] = j_n a_n . (2n-1)!.$$

The proof is clear.

Proof of Lemma 1. Given an imbedding $i: V^{k-1} \to R^{m+k-1}$ of a compact differentiable manifold V^{k-1} into Euclidean space, and given a cross-section f of the normal $\mathbf{SO}_m$-bundle over V^{k-1}, a well-known procedure due to Thom associates with i and f a sphere mapping $\phi: S^{m+k-1} \to S^m$ (compare Kervaire[5], p. 223).

The map ϕ is homotopic to zero if and only if there exists a bounded manifold Q^k with boundary V^{k-1} imbedded in R^{m+k} on one side of R^{m+k-1} such that:

(i) the restriction to V^{k-1} of the imbedding of Q^k is the given imbedding of V^{k-1} in R^{m+k-1};

(ii) Q^k meets R^{m+k-1} orthogonally so that the restriction to V^{k-1} of the normal bundle of Q^k is just the normal bundle of V^{k-1} in R^{m+k-1}; and

(iii) the cross-section f can be extended throughout Q^k as a cross-section f' of the normal $\mathbf{SO}_m$-bundle.

These facts follow from Thom [15], ch. I, § 2 and Lemmas IV, 5, IV.5′.

To obtain Lemma 1 above, take $V^{k-1} = S^{k-1}$ and take $i(S^{k-1})$ to be the unit sphere in $R^k \subset R^{m+k-1}$. Since the normal m-plane at each point of $i(S^{k-1})$ in R^{m+k-1} admits a natural basis (consisting of the radius vector followed by the vectors of a basis for R^{m+k-1}/R^k), the cross-section f

provides a mapping $a\colon S^{k-1} \to \mathbf{SO}_m$. Let $\alpha \in \pi_{k-1}(\mathbf{SO}_m)$ be its homotopy class. It is easily seen (compare Kervaire[7], § 1.8) that the map $\phi\colon S^{m+k-1} \to S^m$ associated with i and f represents $J\alpha$ up to sign.

If $J\alpha = 0$ then there exists a bounded manifold $Q^k \subset R^{m+k}$ satisfying conditions (i), (ii) and (iii). Let $M^k \subset R^{m+k}$ denote the unbounded manifold obtained from Q^k by adjoining a k-dimensional hemisphere, which lies on the other side of R^{m+k-1} and has the same boundary $i(S^{k-1})$. Since the normal bundle ν restricted to Q^k has a cross-section f', it follows that M^k is almost parallelizable. Clearly the obstruction class $\mathfrak{o}(\nu, f')$ is equal to α.

Conversely, let M^k be a manifold imbedded in R^{m+k} and let f be a cross-section of the normal bundle ν restricted to $M^k - x_0$. After modifying this imbedding by a diffeomorphism of R^{m+k} we may assume that some neighborhood of x_0 in M^k is a hemisphere lying on one side of the hyperplane R^{m+k-1}, and that the rest of M^k lies on the other side. Removing this neighborhood we obtain a bounded manifold $Q^k \subset R^{m+k}$ just as above, having the unit sphere $S^{k-1} \subset R^k \subset R^{m+k-1}$ as boundary. The cross-section f restricted to S^{k-1} gives rise to a map $a\colon S^{k-1} \to \mathbf{SO}_m$ which represents the homotopy class $\mathfrak{o}(\nu, f)$. The argument above shows that $J\mathfrak{o}(\nu, f) = 0$; which completes the proof of Lemma 1.

Remark. Lemma 1 could also be proved using the interpretation of J given in Milnor[9].

Proof of Lemma 2. (Compare Kervaire[8].) The $\mathbf{SO}_m$-bundle ξ induces a $\mathbf{U}_m$-bundle ξ' and hence a $\mathbf{U}_m/\mathbf{U}_{2n-1}$-bundle ξ''. Similarly, the partial cross-section f induces partial cross-sections f' and f''. By definition the obstruction class $\mathfrak{o}(\xi'', f'')$ is equal to the Chern class $c_{2n}(\xi')$ and hence to the Pontrjagin class $\pm p_n(\xi)$. Therefore $p_n(\xi)$ equals $\pm q_* h_* \mathfrak{o}(\xi, f)$, where

$$h\colon \pi_{4n-1}(\mathbf{SO}_m) \to \pi_{4n-1}(\mathbf{U}_m) \quad \text{and} \quad q\colon \pi_{4n-1}(\mathbf{U}_m) \to \pi_{4n-1}(\mathbf{U}_m/\mathbf{U}_{2n-1})$$

are the natural homomorphisms and h_*, q_* are the homomorphisms in the cohomology of K induced by the coefficient homomorphisms h, q.

Using the following computations of Bott[2]:

$$\pi_{4n-1}(\mathbf{U}_m) \approx \mathbf{Z}, \quad \pi_{4n-1}(\mathbf{U}_m/\mathbf{SO}_m) \approx \mathbf{Z}_{a_n}, \quad \pi_{4n-2}(\mathbf{SO}_m) = 0,$$

it follows that h carries a generator into a_n times a generator. Similarly, using the fact that

$$\pi_{4n-2}(\mathbf{U}_{2n-1}) \approx \mathbf{Z}_{(2n-1)!} \text{ (see [3])} \quad \text{and} \quad \pi_{4n-2}(\mathbf{U}_m) = 0,$$

it follows that q carries a generator into $(2n-1)!$ times a generator. Therefore $p_n(\xi) = \pm a_n(2n-1)!\, \mathfrak{o}(\xi, f)$. This completes the proof of Lemma 2.

Hirzebruch's index theorem[4] states that the index $I(M^{4n})$ of any $4n$-manifold is equal to

$$2^{2n}(2^{2n-1}-1)\,B_n p_n[M^{4n}]/(2n)! + (\text{terms involving lower Pontrjagin classes}).$$

Here B_n denotes the nth Bernoulli number. For an almost parallelizable manifold the lower Pontrjagin classes are zero. Therefore

Corollary. The index $I(M_0^{4n})$ *is equal to* $2^{2n-1}(2^{2n-1}-1)\,B_n j_n a_n/n$; *and the index of any almost parallelizable* $4n$*-manifold is a multiple of this number.*

The fact that $I(M_0^{4n})$ is an integer can be used to estimate the number j_n (compare Milnor[9]). However, a sharper estimate, which includes the prime 2, can be obtained as follows, using a new generalization of Rohlin's theorem.

Borel and Hirzebruch ([1], §§ 23.1 and 25.4) define a rational number

$$\hat{A}[M^{4n}] = -B_n p_n[M^{4n}]/2(2n)! + (\text{terms involving } p_1, \ldots, p_{n-1});$$

and prove that the denominator of $\hat{A}[M^{4n}]$ is a power of 2.

Theorem 3. If the Stiefel–Whitney class w_2 *of* M^{4n} *is zero then* $\hat{A}[M^{4n}]$ *is actually an integer.*†

The proof will be given in a subsequent paper by Borel and Hirzebruch. It is based on the methods of [1], together with the assertion that the Todd genus of a generalized almost complex manifold is an integer (Milnor[10]).

Applying this theorem to the manifold M_0^{4n} of Theorem 2 it follows that $B_n j_n a_n/4n$ is an integer. Therefore:

Theorem 4. The order j_n *of the stable group* $J\pi_{4n-1}(\mathbf{SO}_m)$ *is a multiple of the denominator of the rational number* $B_n a_n/4n$.

As examples, for $n = 1, 2, 3$, the number $B_n a_n/4n$ is equal to $1/12$, $1/240$, and $1/252$ respectively. Since $\pi_{m+7}(S^m)$ is cyclic of order 240, it follows that $j_2 = 240$. Since $\pi_{m+11}(S^m)$ is cyclic of order 504, it follows that j_3 is either 252 or 504. It may be conjectured that j_n is always equal to the denominator of $B_n/4n$.

The theorems of von Staudt[13, 14] can be used to compute such denominators (compare Milnor[9]).

Lemma 3. The denominator of $B_n/2n$ *can be described as follows. A prime power* p^{i+1} *divides this denominator if and only if*

$$2n \equiv 0 \mod p^i(p-1).$$

† See note at the end of the paper.

Combining Lemma 3 with Theorem 4, we see that the stable homotopy groups of spheres contain elements of arbitrary finite order. In fact:

Corollary. If $2n$ is a multiple of the Euler Φ function $\Phi(r)$, then the stable group $\pi_{m+4n-1}(S^m)$ contains an element of order r.

REFERENCES

[1] Borel, A. and Hirzebruch, F. Characteristic classes and homogeneous spaces. *Amer. J. Math.* 80, 458–538 (1958).

[2] Bott, R. The stable homotopy of the classical groups. *Proc. Nat. Acad. Sci., Wash.*, 43, 933–935 (1957).

[3] Bott, R. and Milnor, J. W. On the parallelizability of the spheres. *Bull. Amer. Math. Soc.* 64, 87–89 (1958).

[4] Hirzebruch, F. *Neue topologische Methoden in der algebraischen Geometrie.* Springer, 1956.

[5] Kervaire, M. A. Courbure intégrale généralisée et homotopie. *Math. Ann.* 131, 219–252 (1956).

[6] Kervaire, M. A. On the Pontryagin classes of certain SO(n)-bundles over manifolds. *Amer. J. Math.* 80, 632–638 (1958).

[7] Kervaire, M. A. An interpretation of G. Whitehead's generalization of Hopf's invariant. *Ann. Math.* (2), 69, 345–365 (1959).

[8] Kervaire, M. A. A note on obstructions and characteristic classes. (In preparation.) *

[9] Milnor, J. W. On the Whitehead homomorphism *J. Bull. Amer. Math. Soc.* 64, 79–82 (1958).

[10] Milnor, J. W. On the cobordism ring Ω^*, and a complex analogue. (In preparation.) +

[11] Rohlin, V. A. Classification of mappings of S^{n+3} onto S^n. (Russian.) *Dokl. Akad. Nauk, SSSR*, 81, 19–22 (1951).

[12] Rohlin, V. A. New results in the theory of 4-dimensional manifolds. (Russian.) *Dokl. Akad. Nauk, SSSR*, 84, 221–224 (1952).

[13] von Staudt. *J. Math.* 21, 372–374 (1840).

[14] von Staudt. *De numeris Bernoullianis commentatio.* Erlangen, 1845.

[15] Thom, R. Quelques propriétés globales des variétés différentiables. *Comment. math. Helvet.* 28, 17–86 (1954).

[16] Whitehead, G. W. On the homotopy groups of spheres and rotation groups. *Ann. Math.* (2), 43, 634–640 (1942).

[17] Whitehead, J. H. C. On the homotopy type of manifolds. *Ann. Math.* (2), 41, 825–832 (1940).

[*Added in proof.*] For the case n odd, Hirzebruch has since sharpened Theorem 3, showing that $\hat{A}[M^{4n}]$ is an even integer. Thus the factor a_n in Theorem 4 can be cancelled.

* See Amer. J. Math. **81**, 1959, pp. 773-784.

+ This volume pp. 257-273.

PART 4: COBORDISM.

This section will consist of the following:

- On the cobordism ring Ω^* (research announcement), Notices American Mathematical Society **5** (1958) 457.
- On the cobordism ring Ω^* and a complex analogue, Part I, American Journal of Mathemtics **82** (1960) 505–521.
- Travaux de Milnor sur le cobordisme, by R. Thom, Séminaire Bourbaki 1958/59, no. **180** Fév. 1959.
- A survey of cobordism theory, Enseignement Mathématique **8** (1962) 16–23; together with an added Erratum.
- Spin structures on manifolds, Enseignement Mathématique **9** (1963) 198–203.
- Remarks concerning spin manifolds, in "Differential and Combinatorial Topology, A Symposium in honor of Marston Morse", Princeton Univ. Press (1965) 55–62.
- On the Stiefel-Whitney numbers of complex manifolds and of spin manifolds, Topology **3** (1965) 223–230.

Introduction

The research announcement **On the Cobordism Ring Ω^*** describes some of the structure of the cobordism ring for oriented manifolds. Proofs were to appear in a pair of papers. The first of these, **On the Cobordism Ring Ω^* and a Complex Analogue, Part I**, combines the techniques introduced by René Thom with work of Frank Adams to show that the oriented cobordism ring (nowadays denoted by Ω_*) has no odd torsion. It also shows that the analogous ring Ω_*^{U} for manifolds with a complex structure on the stable normal bundle has no torsion at all. These results were also obtained in the Soviet Union by B. G. AVERBUH [1959] in the real case, and by Sergei NOVIKOV [1960] in the complex case. (See also NOVIKOV [1962].)

The projected Part II of this paper was never written. In fact I am chagrined to discover that I have never published any details about some of the announced results which were intended to appear in it. I was very grateful when Thom gave me permission to reprint his 1959 Bourbaki lecture **Travaux de Milnor sur le Cobordisme**, which gives a better account of this work than anything which I have published. In particular, it sketches a proof that the complex cobordism ring Ω_*^{U} is a polynomial ring of the form $\mathbb{Z}[Y_1', Y_2', Y_3', \ldots]$, which is non-zero only in even dimensions. Here the Y_k' are certain smooth complex algebraic varieties (not necessarily connected) of complex dimension k. Furthermore, the oriented cobordism ring Ω_* modulo 2-torsion is a polynomial ring of the form $\mathbb{Z}[Y_1, Y_2, Y_3, \ldots]$, where each Y_k can be identified with the complex manifold Y_{2k}', of real dimension $4k$. The full structure of the oriented cobordism ring Ω_*, including 2-torsion, was worked out shortly afterwards by Terry WALL [1960].

The key to recognizing which manifolds Y_k' can serve as generators is provided by an additive characteristic class $\mathbf{s}_k$, introduced by Thom.[1] The total Chern class

$$c = 1 + c_1 + c_2 + \cdots$$

of a Whitney sum of complex vector bundles satisfies the identity

$$c(\xi \oplus \eta) = c(\xi) \cdot c(\eta)\,.$$

Therefore its logarithm, defined by the power series expansion

$$\log(1+c_1+c_2+\cdots) \;=\; (c_1+c_2+\cdots) - \frac{1}{2}(c_1+c_2+\cdots)^2 + \frac{1}{3}(c_1+c_2+\cdots)^3 - + \cdots,$$

satisfies

$$\log\, c(\xi \oplus \eta) = \log\, c(\xi) + \log\, c(\eta)\,.$$

It will be convenient to set $\log c = \mathbf{s}_1 - \frac{1}{2}\mathbf{s}_2 + \frac{1}{3}\mathbf{s}_3 - + \cdots$ where

$$\mathbf{s}_1 = c_1\,, \qquad \mathbf{s}_2 = c_1^{\,2} - 2c_2\,, \qquad \mathbf{s}_3 = c_1^{\,3} - 3c_1c_2 + 3c_3\,, \qquad \cdots.$$

Each $\mathbf{s}_k = \mathbf{s}_k(c_1, c_2, \ldots, c_k)$ is also an additive characteristic class, expressed as a polynomial with integer coefficients in the Chern classes, with leading term $c_1^{\,k}$. It has the marvelous property that the associated characteristic number satisfies

$$\mathbf{s}_{m+n}[M \times N] = 0$$

for any product $M \times N$ of compact complex manifolds of complex dimensions $m, n > 0$.

Now let $\mathcal{H}_{m,n}$ be a hypersurface of degree (1, 1) in the product $\mathbb{P}\mathbb{C}^m \times \mathbb{P}\mathbb{C}^n$.

Lemma. *If $m, n > 1$, then the characteristic number $\mathbf{s}_{m+n-1}[\mathcal{H}_{m,n}]$ is equal to*

$$-\binom{m+n}{m} = -\frac{(m+n)!}{m!\,n!}\,.$$

The proof can be outlined as follows. The total Chern class for the tangent bundle of $\mathbb{P}\mathbb{C}^m$ is equal to $(1+x)^{m+1}$, where x is the standard 2-dimensional generator. Hence $\log c = (m+1)\log(1+x)$ and $\mathbf{s}_k(c) = (m+1)x^k$. It follows that

[1] As background for the following material, see MILNOR AND STASHEFF [1974]. Caution: These additive characteristic classes $\mathbf{s}_k$ have nothing to do with the rational coefficients s_k introduced by Hirzebruch and mentioned on page 225 above.

FIG. 1. *Relaxing at the 1959 Arbeitstagung in Bonn. From the left: Thom, Grothendieck, von Viereck, Milnor, Dombrowski.*

the corresponding characteristic class for the tangent bundle of the product of two projective spaces has the form

$$\mathbf{s}_k = (m+1)x^k \ + \ (n+1)y^k \,,$$

where x and y are the respective two-dimensional generators. Taking $k = m+n-1$ equal to the complex dimension of the hypersurface, this will be zero whenever $m,\, n > 1$ so that $k > m,\, n$.

On the other hand, the total Chern class for the normal line bundle ν of the hypersurface can be computed as $1+(x+y)$, restricted to the hypersurface. Hence $\mathbf{s}_k(\nu)$ is the restriction of $(x+y)^k$. If τ is the tangent bundle of $\mathcal{H}_{m,n}$, then using the identity $\mathbf{s}_k(\tau)+\mathbf{s}_k(\nu) = \mathbf{s}_k(\tau\oplus\nu) = 0$ we see that $-\mathbf{s}_{m+n-1}(\tau)$ is equal to

$$\mathbf{s}_{m+n-1}(\nu) = (x+y)^{m+n-1} = \begin{pmatrix} m+n-1 \\ m-1 \end{pmatrix} x^{m-1}y^n + \begin{pmatrix} m+n-1 \\ m \end{pmatrix} x^m y^{n-1}\,,$$

restricted to $\mathcal{H}_{m,n}$. Finally, both $x^{m-1}y^n$ and $x^m y^{n-1}$, evaluated on the fundamental homology class of the hypersurface, yield $+1$. Thus we obtain

$$-\mathbf{s}_k[\mathcal{H}_{m,n}] \ = \ \begin{pmatrix} m+n-1 \\ m-1 \end{pmatrix} + \begin{pmatrix} m+n-1 \\ m \end{pmatrix} \ = \ \begin{pmatrix} m+n \\ m \end{pmatrix},$$

as asserted. □

On the other hand, for the projective space $\mathbb{PC}^k$ we have

$$\mathbf{s}_k[\mathbb{PC}^k] = k+1 = \binom{k+1}{1}.$$

For fixed $k+1 = m+n$, it is not difficult to see that the greatest common divisor of the numbers $\binom{k+1}{m}$ for $1 \le m \le k$ is given by

$$\mathrm{GCD} = \begin{cases} p, & \text{if} \quad k+1 \quad \text{is a power of some prime} \quad p, \\ 1, & \text{if} \quad k+1 \quad \text{is not a prime power.} \end{cases}$$

For each k, we can choose Y'_k to be a disjoint union of complex varieties $\mathcal{H}_{m,n}$ and $\mathbb{PC}^k$ of complex dimension k, so that $\mathbf{s}_k[Y'_k]$ will be equal to this greatest common divisor. These Y'_k are the required generators for the complex cobordism ring Ω^{U}_*.

The situation for real oriented cobordism, modulo 2-torsion, is completely analogous. In fact if ξ is a complex vector bundle, then the additive characteristic class $\mathbf{s}_{2k}(c_1, c_2, \ldots, c_{2k})$ for ξ depends only on the underlying real vector bundle $\xi_{\mathbb{R}}$, with complex structure forgotten. In fact it coincides with the class $\mathbf{s}_k(p_1, p_2, \ldots, p_k)$ of $\xi_{\mathbb{R}}$. Those generating manifolds Y'_{2k} for Ω^{U}_* which have even complex dimension serve also as generators for the oriented cobordism ring $\Omega_*/(\text{torsion})$.

The expository paper **A Survey of Cobordism Theory** begins with an outline of the results described above, and ends with a very tentative exploration of more general cobordism theories, for example for topological or piecewise linear manifolds. However, its main emphasis is on cobordism for smooth Riemannian manifolds with an "X-structure" on the stable tangent bundle, where X is some space on which the infinite orthogonal group $\mathrm{O} = \varinjlim \mathrm{O}_n$ acts. By definition, an X-*structure* on M is a section of the X-bundle associated with the tangent bundle of M. For example, if $X = \mathrm{O}/G$ where G is a subgroup of O, then the X-structures on M correspond uniquely to reductions of the structural group of the stable tangent bundle to this subgroup G. Thus an O/SO-structure is just an orientation, while an O/U-structure is a stable almost complex structure, and an O/{1}-structure is a stable framing of the manifold.[2] For each choice of X, there is an associated cobordism theory, which could be denoted by $\mathcal{N}^G_*$, or better by Ω^G_* when $G \subset \mathrm{SO}$. (The notations which are actually used in the paper, $H_*(\mathcal{M})$ for the cobordism groups associated with some class $\mathcal{M}$ of compact manifolds, or $\mathcal{N}_*(X)$ for the cobordism groups for smooth manifolds with X-structure, seem rather confusing, and are not recommended. Similarly, the notation J_n for the integers mod n was soon forgotten.)

This construction is also used to describe spin structures on a manifold; however, the exposition is not quite right. See the **Erratum** for a corrected version.

[2]The examples which are cited all have the important property that an X-structure on the stable tangent bundle determines an X-structure on the stable normal bundle up to homotopy, and conversely. Compare R. K. LASHOF [1963].

Alternatively, a spin structure on a principal SO(n)-bundle $P \twoheadrightarrow B$ can be specified much more simply by choosing a cohomology class in $H^1(P;\,\mathbb{Z}/2)$ whose restriction to each fiber is non-zero. This idea, due to Moe Hirsch, is used in the next paper **Spin Structures on Manifolds**, which contains some preliminary computations of the spin cobordism ring $\Omega_*^{\rm Spin}$. This paper also contains the curious conjecture that two different spin structures on an SO(n)-bundle necessarily give rise to isomorphic Spin(n)-bundles, provided that the dimension of the base space B is strictly less than n. I am not sure about the status of this conjecture in general; but in the special case of a trivial SO(n)-bundle it is easily verified as follows. In this special case, it is equivalent to the statement that any element of $H^1(B;\,\mathbb{Z}/2)$ is the image of the non-zero class in $H^1(\mathrm{SO};\,\mathbb{Z}/2)$ under the homomorphism induced by some map $B \to \mathrm{SO(n)}$. Recall that there is a canonical embedding of the projective space $\mathbb{RP}^{n-1}$ into SO(n). In fact each pair of unit vectors $\pm\mathbf{u}$ in the sphere S^{n-1} gives rise to a rotation $\rho_{\mathbf{u}} \circ \rho_{\mathbf{u}_0} \in \mathrm{SO(n)}$, where $\rho_{\mathbf{u}}$ is the reflection in the hyperplane orthogonal to $\mathbf{u}$, and where $\mathbf{u}_0 \in S^{n-1}$ is a base point. Since every element of $H^1(B;\,\mathbb{Z}/2)$ is induced by a corresponding map from B to $\mathbb{RP}^{n-1}$, the conclusion follows.

The next paper, **Remarks Concerning Spin Manifolds**, contains two unrelated results. After a review of spin structures, section 2 uses a diffeomorphism invariant of Eells and Kuiper [1962] to construct a non-standard smooth involution of the 7-sphere. (For other free actions of finite groups on spheres, see these Collected Papers, Volume 2, "The Fundamental Group".)

The last section uses a construction of Atiyah in topological K-theory (or more explicitly KO-theory) to study the relationship between spin cobordism and exotic spheres. If

$$D^{8k} \hookrightarrow E \twoheadrightarrow B$$

is a $8k$-disk bundle with spin structure, then there is a canonical KO-*orientation class*, belonging to the group $KO(E,\dot E) \cong \widetilde{KO}(E/\dot E)$, where $\dot E$ is the boundary sphere-bundle. Applying this construction to the universal Spin($8k$)-bundle, where $8k > n$ so that

$$\pi_{8k+n}(E/\dot E) \cong \Omega_n^{\rm Spin}$$

is the n-th spin cobordism group, then we see that this orientation class gives rise to a homomorphism

$$\Omega_n^{\rm Spin} \to \widetilde{KO}(S^{8k+n}) \cong \widetilde{KO}(S^n)\ .$$

Now specialize to the case $n \equiv 1$ or $n \equiv 2 \pmod 8$, so that $\widetilde{KO}(S^n)$ is cyclic of order 2, and consider the corresponding group of homotopy spheres Θ_n. For n equal to 9, 10, 17, or 18, it is shown that the composition

$$\Theta_n \to \Omega_n^{\rm Spin} \to \widetilde{KO}(S^n) \cong \mathbb{Z}/2$$

is non-zero. *Thus, in these dimensions, there exists an exotic sphere which does not bound any spin manifold.* (Using Adams [1966], which appeared one year later, the proof extends immediately to all higher values of n which are congruent to 1 or 2 mod 8. Compare Lawson and Michelsohn [1989, p. 93].)

The last paper in this section, **On the Stiefel-Whitney Numbers of Complex Manifolds and Spin Manifolds**, shows that the image of the natural homomorphism

$$\Omega_*^{\mathrm{U}} \to \Omega_* \to \mathcal{N}_*$$

from complex cobordism to non-oriented cobordism is precisely equal to the set of all squares in $\mathcal{N}_*$, and gives partial results towards identifying the image of the corresponding homomorphism $\Omega_*^{\mathrm{Spin}} \to \Omega_* \to \mathcal{N}_*$.

Further Developments

The full structure of the spin cobordism ring was worked out shortly afterwards by ANDERSON, BROWN AND PETERSON [1967], making use of KO characteristic numbers. (Compare the exposition in R. STONG [1968].) In particular, their Corollaries 2.4 and 2.5 answer two open questions from my last mentioned paper as follows:

> **Assertion:** *An unoriented cobordism class contains a spin manifold if and only if all Stiefel-Whitney numbers involving w_1 or w_2 are zero.*
>
> **Assertion:** *There exist unoriented cobordism classes which contain spin manifolds but do not contain the square of any orientable manifold. The first such example occurs in dimension 24.*

Another quite different form of cobordism is the study of cobordism between knots in 3-space. ROBERTELLO [1965] showed that the knot cobordism group has at least two elements, while FOX AND MILNOR [1966] proved that it is not finitely generated by showing that the Alexander polynomial of a null-cobordant (=slice) knot must be equivalent to a polynomial of the form $p(t)p(1/t)$. (This result had been announced in 1958, and a partial proof was published by TERASAKA [1959].) The FOX-MILNOR paper also notes that for any piecewise linear embedding of the 2-sphere in 4-space, the sum of the knots which characterize the various singular points must be null-cobordant.

The study of knot cobordism, also in higher dimensions, has been developed much further by CASSON AND GORDON [1986].

On the cobordism ring Ω^*

(Research announcement 547-26, Notices AMS 5 (1958), 457)

It is proved that the cobordism groups Ω^n (defined by R. Thom, Comm. Math. Helv. **28** (1954)) have no odd torsion. The quotient ring, Ω^* modulo torsion elements, is a polynomial ring on generators of dimension 4, 8, 12, ... A given manifold M^{4k} qualifies as a generator if and only if it satisfies the following condition. Let $s_k(\sigma_1, \ldots, \sigma_k)$ denote the polynomial which expresses the symmetric function $t_1^k + \ldots + t_m^k$ in terms of the elementary symmetric functions σ_i; and let $p_1, \ldots, p_k$ denote the Pontrjagin classes of M^{4k}. Then the characteristic number $s_k(p_1, \ldots, p_k)[M^{4k}]$ must be equal to $\pm q$ if $2k + 1$ is a power of the prime q; and must be equal to ± 1 if $2k + 1$ is not a prime power. The proof makes use of a spectral sequence due to F. Adams. (Received March 26, 1958.)

ON THE COBORDISM RING Ω^* AND A COMPLEX ANALOGUE, PART I.*

By J. MILNOR.

This paper will prove that the cobordism groups Ω^i, defined by Thom [15], have no odd torsion.[1] Furthermore, it is shown that certain related groups $\pi_{i+2n}M(U_n)$ have no torsion at all; providing that n is large. The proofs are based on a spectral sequence due to J. F. Adams [1, 2].

The following is a brief summary of Thom's constructions. Let G be a subgroup of the orthogonal group O_n. (More generally one could start with any Lie group G, together with a specified representation into O_n.) Beginning with a universal bundle for G we can form:

1) The weakly associated bundle having the disk D^n as fibre. Let $\pi\colon E \to B(G)$ denote the projection map of this bundle.

2) The weakly associated bundle having the sphere S^{n-1} as fibre. Let $\partial E \subset E$ denote the total space.

The *Thom space* $M(G)$ is now defined as the identification space obtained from E by collapsing ∂E to a point.

Taking G to be the rotation group $SO_n \subset O_n$, Thom showed that the homotopy group $\pi_{i+n}M(SO_n)$ is independent of n, providing that n is large. He showed that this group is isomorphic to the "cobordism group" Ω^i; and determined its structure up to torsion. The 2-torsion subgroup of Ω^i has recently been determined by C. T. C. Wall. Hence the assertion that Ω^i has no odd torsion completes the description of this group.

Let $M(U_n)$ denote the Thom space for the unitary group $U_n \subset O_{2n}$. In Part II of this paper it will be shown that the stable homotopy group $\pi_{i+2n}M(U_n)$ can be interpreted as a "complex cobordism group." Part I will determine the structure of this group without attempting to interpret it.

* Received July 27, 1959.

[1] Added in proof. This result has been obtained independently by B. G. Averbuch, *Doklady Akademii Nauk SSSR*, vol. 125 (1959), pp. 11-14. The results on complex cobordism have been obtained independently by Novikov.

The first section proves several lemmas concerning the Steenrod algebra, which are needed later. The second section describes the Adams spectral sequence, which relates the cohomology module of any space to the stable homotopy groups of the space. Sections 3 and 4 complete the argument by computing the cohomology modules of $M(U_n)$ and $M(SO_n)$ respectively.

1. Lemmas concerning the Steenrod algebra. Let A denote the Steenrod algebra corresponding to a fixed prime p. (See Cartan [6], Adem [3].) The Bochstein coboundary operation will be denoted by $Q_0 \in A^1$. The two-sided ideal generated by Q_0 in A will be denoted by (Q_0).

LEMMA 1. *The Steenrod algebra contains a subalgebra A_0 with the following properties.*

(i) *A_0 is a Grassmann algebra over Z_p with generators $Q_0, Q_1, \cdots$ of odd dimension.*

(ii) *A is free as a right A_0-module.*

(iii) *The identity map of A induces an isomorphism between the left A-modules $A \otimes_{A_0} Z_p$ and $A/(Q_0)$.*

[*Explanation of* (iii). The field Z_p is considered as a left A_0-module with $Q_i Z_p = 0$. Hence $A \otimes_{A_0} Z_p$ is the quotient of A by the left ideal $AQ_0 + AQ_1 + AQ_2 + \cdots$.]

Proof for the case p odd. We will first prove the corresponding statements with left and right interchanged. According to Milnor [10, Theorem 4a]:

(1) There is a basis for A over Z_p consisting of elements $Q_0^{e_0} Q_1^{e_1} \cdots \mathcal{P}^R$. Here the integers $e_0, e_1, \cdots$ should be 0 or 1, and almost all zero. The letter R stands for a sequence $(r_1, r_2, \cdots)$ of non-negative integers, almost all zero.

[*Explanation.* The element $\mathcal{P}^R$ is a complicated polynomial in the Steenrod operations, with dimension $\sum r_j(2p^j - 2)$. For the special case $R = (r, 0, 0, \cdots)$ the element $\mathcal{P}^R$ is equal to the Steenrod operation $\mathcal{P}^r$. The element Q_i of dimension $2p^i - 1$ can be defined inductively by the rule $Q_{i+1} = \mathcal{P}^{p^i} Q_i - Q_i \mathcal{P}^{p^i}$.]

Furthermore:

(2) The elements Q_i are odd dimensional, and satisfy $Q_i Q_j + Q_j Q_i = 0$, $Q_i Q_i = 0$.

Thus the Q_i generate a Grassmann algebra which may be denoted by $A_0 \subset A$. Clearly A is free as a left A_0-module, with basis $\{\mathcal{P}^R\}$.

Consider the right ideal $Q_0A + Q_1A + Q_2A + \cdots$. The following identity (see [10, Theorem 4a]) proves that this is also a left ideal. Define $p^i\Delta_j$ as the sequence $(0, \cdots, 0, p^i, 0, \cdots)$ with p^i in the j-th place.

(3) $\mathcal{P}^R Q_i$ is equal to $Q_i\mathcal{P}^R + \sum Q_{i+j}\mathcal{P}^{R-p^i\Delta_j}$, to be summed over all $j > 0$ for which $R - p^i\Delta_j$ is a sequence of non-negative integers. (That is, all j for which $r_j \geqq p^i$.)

Thus $Q_0A + Q_1A + \cdots$ is a two-sided ideal which contains Q_0, and therefore contains (Q_0).

As a special case of (3), the identity $\mathcal{P}^{\Delta_j}Q_0 = Q_0\mathcal{P}^{\Delta_j} + Q_j$ is valid. Thus the elements Q_j belong to the ideal (Q_0). This proves that the ideal $Q_0A + Q_1A + \cdots$ is equal to (Q_0). Dividing A by these ideals, it follows that $Z_p \otimes_{A_0} A$ is isomorphic to $A/(Q_0)$.

This proves Lemma 1 for p odd, except that right and left have been interchanged. To complete the proof it is only necessary to recall:

(4) There exists an anti-automorphism of A; that is, a Z_p-isomorphism $c\colon A \to A$ satisfying

$$c(xy) = (-1)^{\dim x \dim y} c(y)c(x).$$

Furthermore, c carries Q_i into $-Q_i$.

This is proved in [10, § 7]. Clearly Lemma 1 follows (for p odd).

LEMMA 2. *The elements $\mathcal{P}^R \in A$ yield a basis over Z_p for the quotient algebra $A/(Q_0)$.*

Proof for p odd. Recall that $\{\mathcal{P}^R\}$ forms a basis for A, considered as a left A_0-module. Hence it forms a basis for $Z_p \otimes_{A_0} A = A/(Q_0)$ over Z_p, which completes the proof.

Conventions. The sum $R + R'$ of two sequences is defined as the term by term sum, and nR denotes the sequence $(nr_1, nr_2, \cdots)$. The binomial coefficient (R, R') is defined as the product over i of $(r_i + r'_i)!/r_i!r'_i!$. The symbol Δ_j stands for a sequence with 1 in the j-th place and zero elsewhere.

Proof of Lemmas 1 *and* 2 *for the case* $p = 2$. The Steenrod algebra over Z_2 has a basis consisting of elements Sq^R of dimension $r_1 + 3r_2 + 7r_3 + \cdots$. (See [10, Appendix 1].) Define $\mathcal{P}^R$ to be Sq^{2R} and define Q_{i-1} to be Sq^{Δ_i}. (For example $Q_0 = \mathrm{Sq}^{\Delta_1} = \mathrm{Sq}^1$ which checks with the definition of Q_0 as the

Bochstein coboundary operator.) Then we will prove Assertions (1), (2), (3) and (4) above. Using these, the proof of Lemmas 1 and 2 can be carried out just as for p odd.

The formula for products $\mathrm{Sq}^{R'}\mathrm{Sq}^{R}$ is rather complicated; however the following special case will suffice.

(5) If E is a sequence satisfying $e_i \leqq 1$, then $\mathrm{Sq}^E\mathrm{Sq}^R$ is equal to $(E,R)\,\mathrm{Sq}^{E+R}$.

For a proof see [10, Corollary 4 and Appendix 1]. As examples, taking $E = \Delta_{i+1}$, $R = \Delta_{j+1}$ we find that $Q_iQ_j = Q_jQ_i$, and that $Q_iQ_i = 0$. This proves Assertion (2) for the case $p = 2$.

By induction the product $Q_0^{e_1}Q_1^{e_2}\cdots$ is equal to Sq^E. Furthermore, a binomial coefficient of the form $(E, 2R)$ is always odd, hence $\mathrm{Sq}^E\mathcal{P}^R = \mathrm{Sq}^E\mathrm{Sq}^{2R}$ is equal to Sq^{E+2R}. Since every sequence can be written uniquely in the form $E + 2R$, it follows that these elements form a basis for A over Z_2. This proves Assertion (1).

Proof of Assertion (3) *for* $p = 2$. Direct application of the general product rule [10, Theorem 4b] shows that

$$\mathrm{Sq}^{2R}\mathrm{Sq}^{\Delta_{i+1}} = \mathrm{Sq}^{\Delta_{i+1}}\mathrm{Sq}^{2R} + \sum \mathrm{Sq}^{2R-2^{i+1}\Delta_j+\Delta_{i+1+j}},$$

to be summed over all $j \geqq 1$ for which $r_j \geqq 2^i$. On the other hand, using Assertion (5), the j-th term on the right can be written as

$$\mathrm{Sq}^{\Delta_{i+1+j}}\mathrm{Sq}^{2R-2^{i+1}\Delta_j} = Q_{i+j}\mathcal{P}^{R-2^i\Delta_j}.$$

Thus $\mathcal{P}^RQ_i = Q_i\mathcal{P}^R + \sum Q_{i+j}\mathcal{P}^{R-2^i\Delta_j}$, as required.

Since Assertion (4) is also true for $p = 2$, this completes the proof of Lemmas 1 and 2.

[*Remark.* There is one essential difference between the case p odd and the case $p = 2$. For p odd the elements $\mathcal{P}^R$ span a subalgebra of A isomorphic to $A/(Q_0)$; but for $p = 2$ there is no such subalgebra. This can be seen using the identity $\mathrm{Sq}^2\mathrm{Sq}^2 = \mathrm{Sq}^1\mathrm{Sq}^2\mathrm{Sq}^1 \neq 0$.]

The symbol Δ_0 will denote the sequence $(0, 0, \cdots)$.

LEMMA 3. *If p is odd, then the cohomology operations $\mathcal{P}^R$ have the following properties.*

(1) *For $x, y \in H^*(X; Z_p)$ the element $\mathcal{P}^R(xy)$ is equal to*

$$\sum_{R_1+R_2=R} (\mathcal{P}^{R_1}x)(\mathcal{P}^{R_2}y).$$

(2) *For a 2-dimensional cohomology class $t \in H^2(X; Z_p)$, the element $\mathcal{P}^R t$ is equal to t^{p^i} if $R = \Delta_i$; and is zero if R is not equal to one of the sequences $\Delta_0, \Delta_1, \Delta_2, \cdots$.*

Proof. The first assertion follows from [10, Lemma 9]. For the special case $R = r\Delta_1$, the second assertion is well known. That is:

$$\mathcal{P}^0 t = t, \quad \mathcal{P}^1 t = t^p, \quad \mathcal{P}^r t = 0 \text{ for } r > 1.$$

But every $\mathcal{P}^R$ is a "polynomial" in the Steenrod operations $\mathcal{P}^r$. Proceeding by induction on the complexity of this polynomial, we see that $\mathcal{P}^R t$ must have the form kt^i, where $k \in Z_p$ is some constant, and $2i$ is the dimension.

To evaluate k it is sufficient to consider one example. As example, let X be the $2i$-skeleton of the Eilenberg-MacLane complex $K(Z_p, 1)$. According to [10, Lemmas 4, 6] we have:

$$\lambda(t) = t \otimes \xi_0 + t^p \otimes \xi_1 + \cdots;$$

hence

$$\mathcal{P}^R t = \sum_i \langle \mathcal{P}^R, \xi_i \rangle t^{p^i}.$$

Using the definition of $\mathcal{P}^R$, this is equal to t^{p^i} if $R = \Delta_i$ and is zero otherwise. This completes the proof.

For the prime $p = 2$, both assertions of Lemma 3 would be false. However the following modified assertions are proved by the same method:

(1′) $\mathrm{Sq}^R(xy) = \sum_{R_1 + R_2 = R} (\mathrm{Sq}^{R_1} x)(\mathrm{Sq}^{R_2} y)$.

(2′) If $a \in H^1(X; Z_2)$, then $\mathrm{Sq}^{\Delta_i} a = a^{2^i}$; and $\mathrm{Sq}^R a = 0$ for R not of the form Δ_i.

Using these statement the following result will be proved.

LEMMA 3′. *Let $p = 2$ and let $H^*(X; Z_2)$ be a cohomology ring which is annihilated by the operation $Q_0 = \mathrm{Sq}^1$. The assertions* (1) *and* (2) *of Lemma* 3 *are valid as originally stated.*

Proof of (1). If R_1 is a sequence containing some odd integer, then Sq^{R_1} belongs to the ideal (Q_0) (compare the proof of Lemma 1), and therefore annihilates the cohomology of X. Thus in formula (1′) above, it is sufficient to consider sequences R_1 and R_2 which are "even." This proves assertion (1).

Proof of (2). It will be convenient to weaken the hypothesis on X, and assume only that $\mathrm{Sq}^1 t = 0$. Then just as in the proof of Lemma 3, it follows that $\mathfrak{P}^R t$ has the form kt^i. In order to determine the constant $k \in Z_2$, it is sufficient to consider the example of a real projective space X, with $t = a^2$, Using (1′) and (2′) it is seen that $\mathfrak{P}^R t$ equals t^{2^i} for $R = \Delta_i$ and equals zero otherwise. This completes the proof of Lemma 3′.

2. The spectral sequence of Adams. Let X, Y be finite CW-complexes with base point denoted by o; and let A be the Steenrod algebra for some fixed prime p. Thus the cohomology group $H^*(X \bmod o; Z_p)$ is a graded left A-module.

The m-fold *suspension* $S^m X$ is obtained from the product $X \times I^m$ by collapsing $(X \times \partial I^m) \cup (o \times I^m)$ to a point. Here I^m denotes the unit m-cube. The *stable track group* $\{X, Y\}_n$ is the direct limit under suspension of the group of homotopy classes of maps $S^{m+n} X \to S^m Y$. (The integer n may be positive or negative.)

THEOREM OF ADAMS. *There exists a spectral sequence $\{E_r^{st}, d_r\}$ determined by X, Y and p such that*

$$E_2^{st} = \mathrm{Ext}_A^{st}(H^*(Y \bmod o; Z_p), H^*(X \bmod o; Z_p))$$

and such that

$$E_\infty^{st} = B^{st}/B^{s+1\, t+1},$$

where $\{X, Y\}_n = B^{0n} \supset B^{1\, n+1} \supset B^{2\, n+2} \supset \cdots$ is a certain filtration. The intersection $\bigcap_s B^{s\, n+s}$ of these groups is equal to the subgroup of $\{X, Y\}_n$ consisting of elements whose order is finite and prime to p. Each succeeding term E_{r+1} of the spectral sequence is equal to the homology of E_r with respect to the differential operator

$$d_r \colon E_r^{st} \to E_r^{s+r\, t+r-1};$$

and E_∞ is the limit as $r \to \infty$ of E_r.

The functor Ext_A^{st} is defined as follows. If M and N are graded left A-modules let $\mathrm{Hom}_A{}^t(M, N) = \mathrm{Ext}_A^{0t}(M, N)$ denote the group of A-homomorphisms $M \to N$ of degree $-t$. Choose a projective resolution

$$\cdots \to P_2 \xrightarrow{d} P_1 \xrightarrow{d} P_0 \to M \to 0,$$

where the A-homomorphisms d have degree zero. Then $\mathrm{Ext}_A^{st}(M, N)$ is defined as the homology group (kernel modulo image) of the sequence

$$\mathrm{Hom}_A{}^t(P_{s-1}, N) \xrightarrow{d^*} \mathrm{Hom}_A{}^t(P_s, N) \xrightarrow{d^*} \mathrm{Hom}_A{}^t(P_{s+1}, N).$$

It will be convenient to add an E_1 term to the spectral sequence by defining $E_1^{st} = \mathrm{Hom}_A{}^t(P_s, N)$, $d_1 = d^*$.

For the special case $X = S^0$ this theorem is proved in Adams [1]. The more general case is proved by the same argument. It is only necessary to replace the homotopy group $\{S^0, \ \}_n$ by the track group $\{X, \ \}_n$ throughout. See Adams [2].

More generally the finite complex Y may be replaced by a "spectrum" in the sense of Lima [9] and Spanier [13]; or by an "object in the stable category" in the sense of Adams [2]. For our purpose the following definition will be convenient. A *stable object* $\mathbf{Y}$ is a sequence of CW-complexes $(Y_0, Y_1, \cdots)$ such that each suspension SY_i is a subcomplex of Y_{i+1}. The imbedding $SY_i \subset Y_{i+1}$ must be explicitly given.

Given such an object, define the chain group $C_n(\mathbf{Y})$ as the direct limit under suspension of the chain groups $C_{n+i}(Y_i \bmod o)$. Homology and cohomology groups are then defined as usual. Similarly, for any finite complex X define $\{X, \mathbf{Y}\}_n = \text{dir.lim.}\{S^iX, Y_i\}_n$. The abbreviation $\pi_n\mathbf{Y}$ will sometimes be used for $\{S^0, Y\}_n$.

Remark. The suspension homomorphism of chain groups should be defined by the correspondence

$$\alpha \to \alpha \times \iota, \text{ for } \alpha \in C_*(Y_i \bmod o), \ \iota \in C_1(I \bmod \partial I),$$

so as to commute with boundary homomorphisms.

Examples. Any finite complex Y may be defined with the stable object

$$\mathbf{Y} = (Y, SY, S^2Y, \cdots).$$

We will see later that the suspension of the Thom space $M(SO_n)$ is imbedded naturally as a subcomplex of $M(SO_{n+1})$. Hence the *stable Thom object*

$$\mathbf{M}(SO) = (o, M(SO_1), M(SO_2), \cdots)$$

is defined. Note that the track group

$$\{S^0, \mathbf{M}(SO)\}_n = \text{dir. lim. } \pi_{n+i}(M(SO_i))$$

is isomorphic to the cobordism group Ω^n.

Assertion. The theorem of Adams remains valid if the finite complex Y is replaced by any stable object $\mathbf{Y}$; providing that the following *finiteness condition* is satisfied. The groups $C_n(\mathbf{Y}; Z)$ should be finitely generated, and should vanish for n less than some constant.

This can be proved in two ways. One can simply take the direct limit of the spectral sequences for the "finite sub-objects" of $\mathbf{Y}$; or the theorem can be proved from the beginning in the stable category. See Adams [2]. The second approach is preferable, since the proof is much easier in the stable category. Details will not be given.

Using the Adams spectral sequence we will prove the following key result. Let $\mathbf{Y}$ be an object such that $H^n(\mathbf{Y};Z_p)$ is zero for n odd. Then $H^*(\mathbf{Y};Z_p)$ is annihilated by the element Q_0, and hence can be considered as a graded module over the quotient algebra $A/(Q_0)$.

THEOREM 1. *If $H^*(\mathbf{Y};Z_p)$ is a free $A/(Q_0)$-module with even dimensional generators, and if $C_*(\mathbf{Y};Z)$ satisfies the finiteness condition, then the stable homotopy group $\{S^0,\mathbf{Y}\}_n$ contains no p-torsion.*

The idea of the proof is to compute the spectral sequence for the track group $\{X,\mathbf{Y}\}_n$, where X is a "co-Moore space" having cohomology groups $H^i(X \bmod o;Z)$ equal to Z_p for $i=k$ and equal to zero for $i\neq k$.

The following universal coefficient theorem has been proved by Peterson [11]. *There exists an exact sequence*

$$0\to\{S^k,\mathbf{Y}\}_n\otimes Z_p\to\{X,\mathbf{Y}\}_n\to \operatorname{Tor}(\{S^k,\mathbf{Y}\}_{n-1},Z_p)\to 0.$$

An immediate consequence is the following.

LEMMA 4. *If $\{S^0,\mathbf{Y}\}_n$ contains p-torsion, then $\{X,\mathbf{Y}\}_m$ must be non-trivial for two consecutive values of m.*

On the other hand, assuming that $H^*(\mathbf{Y};Z_p)$ is a free $A/(Q_0)$-module on even dimensional generators, we will see that $\{X,\mathbf{Y}\}_m$ is zero for m odd. This will prove Theorem 1.

Construction of an A-free resolution for $H^*(\mathbf{Y};Z_p)$.

First consider the Grassmann algebra A_0 and the A_0-module Z_p. According to Cartan's theory of constructions, to each Grassmann algebra A_0 there corresponds a twisted polynomial algebra P and a differential operator d on $A_0\otimes P$ so that this tensor product becomes acyclic. If A_0 has generators $Q_0,Q_1,\cdots$, then P has a basis over Z_p consisting of elements $b(r_0,r_1,\cdots)$ of dimension $\sum r_i(\dim Q_i+1)$. The integers $r_0,r_1,\cdots$ should be non-negative and almost all zero. The differential operator d is defined as follows. (In order to make the signs come out correctly, we let d act on the right.) For any $a\in A_0$:

$$a \otimes b(r_0, r_1, \cdots) d = \sum a Q_i \otimes b(r_0, \cdots, r_i - 1, r_{i+1}, \cdots),$$

summed over all i for which $r_i > 0$.

Proof that $A_0 \otimes P$ is acyclic. For a Grassmann algebra on one generator, see Cartan [5, p. 704, I]. But a Grassmann algebra with finitely many generators in each dimension can be considered as a tensor product of Grassmann algebras with one generator. Hence the conclusion follows by applying the Künneth theorem.

This conclusion can be formulated as follows. Let F_s be the free A_0-module generated by those symbols $b(r_0, r_1, \cdots)$ for which $r_0 + r_1 + \cdots = s$. Then $A_0 \otimes P$ can be considered as the direct sum $F_0 + F_1 + \cdots$. The augmentation $\epsilon \colon F_0 \to Z_p$ is the A_0-homomorphism defined by $b(0, 0, \cdots) \to 1$. It follows that *the sequence*

$$\cdots \xrightarrow{d} F_1 \xrightarrow{d} F_0 \xrightarrow{\epsilon} Z_p \to 0$$

is an A_0-free resolution of Z_p.

Now apply the functor $A \otimes_{A_0}$ to this exact sequence. Since A is free as a right A_0-module, we obtain an exact sequence

$$\cdots \to A \otimes_{A_0} F_1 \to A \otimes_{A_0} F_0 \to A \otimes_{A_0} Z_p \to 0$$

of left A-modules. Furthermore, each $A \otimes_{A_0} F_s$ is a free A-module. Thus we have constructed an A-free resolution of $A \otimes_{A_0} Z_p$.

According to Lemma 1, the A-module $A/(Q_0)$ is isomorphic to $A \otimes_{A_0} Z_p$. Hence in order to form an A-free resolution of any $A/(Q_0)$-free module, it is sufficient to take the direct sum of a number of copies of the above resolution. This proves the following.

LEMMA 5. *Let $H^*(\mathbf{Y}; Z_p)$ be a free module over $A/(Q_0)$ with basis $\{y_\alpha\}$. Then there exists an A-free resolution*

$$\cdots \to F_1' \to F_0' \to H^*(\mathbf{Y}; Z_p) \to 0,$$

where each F_s' has a basis consisting of elements $b_\alpha(r_0, r_1, \cdots)$, with $r_0 + r_1 + \cdots = s$. The dimension of such a basis element is equal to $\dim y_\alpha + \sum 2r_i(p^i - 1) + s$.

[*Explanation.* The integer s has been added to the dimension of $b_\alpha(r_0, r_1, \cdots)$ so that the homomorphisms $d' \colon F_s' \to F_{s-1}'$ will have degree zero.]

Now consider the complex X consisting of a circle with a 2-cell attached by a map of degree p. Let

$$x \in H^1(X \bmod o; Z_p), \qquad Q_0 x \in H^2(X \bmod o; Z_p)$$

be generators. Then the term

$$E_1^{st} = \mathrm{Hom}_A{}^t(F_s', H^*(X \bmod o; Z_p))$$

of the spectral sequence for $\{X, \boldsymbol{Y}\}$ has a basis consisting of the following elements.

(1) For each $b_\alpha(r_0, r_1, \cdots)$ of dimension $t+1$, the homomorphism $h_\alpha(r_0, r_1, \cdots)$ which carries this basis element into x and carries the other basis elements into zero.

(2) For each $b_\alpha(r_0, r_1, \cdots)$ of dimension $t+2$, the homomorphism $h_\alpha'(r_0, r_1, \cdots)$ which carries this basis element into $Q_0 x$ and carries the other basis elements into zero.

The boundary operator $d_1 \colon E_1^{st} \to E_1^{s+1\,t}$ is given by

$$d_1 h_\alpha(r_0, r_1, \cdots) = h_\alpha'(r_0 + 1, r_1, \cdots),$$

and

$$d_1 h_\alpha'(r_0, r_1, \cdots) = 0.$$

Thus E_2^{st} has as basis the set of elements $h_\alpha'(0, r_1, r_2, \cdots)$, with total dimensions $t - s$ equal to $\dim y_\alpha + \sum 2r_i(p^i - 1) - 2$.

If the integers $\dim y_\alpha$ are all even, then everything in the spectral sequence is even dimensional. It follows that $\{X, \boldsymbol{Y}\}_m$ is zero for m odd. Together with Lemma 4, this completes the proof of Theorem 1.

3. Computation of $\boldsymbol{H^*(B(U_n); Z_p)}$ and $\boldsymbol{H^*(M(U); Z_p)}$. This section will complete the study of $M(U_n)$ by constructing a stable object

$$\boldsymbol{M}(U) = (o, o, M(U_1), SM(U_1), M(U_2), SM(U_2), \cdots);$$

and showing that $H^*(\boldsymbol{M}(U); Z_p)$ is a free module over $A/(Q_0)$, with even dimensional generators, for any prime p.

The proof of this assertion is an immediate generalization of the argument which Thom used to compute the non-orientable cobordism group. In our terminology, Thom showed that $H^*(\boldsymbol{M}(O); Z_2)$ is a free A-module. (See [15, pp. 39-42].)

First a description of $H^*B(U_n)$. *The coefficient group Z_p is to be* *understood, where p is some fixed prime.* (However, integer coefficients could equally well be used.) Let $T_n \subset U_n$ be the n-torus consisting of diagonal unitary matrices. There is a natural map $B(T_n) \to B(U_n)$ of classifying spaces. The cohomology algebra $H^*B(T_n)$ is a polynomial algebra on generators $t_1, \cdots, t_n$ of dimension 2. According to Borel and Serre [4] we may identify $H^*B(U_n)$ with the subalgebra consisting of all symmetric polynomials.

A basis for $H^{2r}B(U_n)$ over Z_p is given as follows. Let $\omega = i_1 \cdots i_k$ range over all partitions of r such that the "length" k is less than or equal to n. (A partition of r is an unordered sequence of positive integers with sum r.) *Define* $s(\omega)$ as the "smallest" symmetric polynomial which contains the term $t_1^{i_1} \cdots t_k^{i_k}$.

[The notation $\sum t_1^{i_1} \cdots t_k^{i_k}$ is commonly used. A more precise definition would be the following. Consider all distinct monomials which can be obtained from $t_1^{i_1} \cdots t_k^{i_k}$ by permuting the n variables; and let $s(\omega)$ denote their sum. It is clear that these elements $s(\omega)$ from a basis for the vector space of symmetric polynomials.]

Next we must study the Thom complex $M(U_n)$. For a group $G \subset SO_m$ recall that $M(G)$ is the quotient space $E/\partial E$, where E is an oriented m-disk bundle over $B(G)$. Any CW-cell subdivision of $B(G)$ induces a cell subdivision of $M(G)$ as follows. For each open i-cell e of $B(G)$, the inverse image e' in $E - \partial E$ is an $(i+m)$-cell. Clearly, $M(G)$ is the disjoint union of these cells e', together with the base point. It is not difficult to verify that $M(G)$ thus becomes a CW-complex.

Let $G \times 1$ denote the group G, considered as a subgroup of SO_{m+1}. The CW-complex $M(G \times 1)$ can be identified with the suspension $SM(G)$ as follows. Let D^m denote the m-disk and I the unit interval. Map $D^m \times I$ onto D^{m+1} by the correspondence

$$(x_1, \cdots, x_m), y \to x_1, \cdots, x_m, (2y-1)(1-x_1^2 - \cdots - x_m^2)^{\frac{1}{2}}.$$

This correspondence gives rise to a map f of $E \times I$ onto the total space E_1 of the associated $(m+1)$-disk bundle. Since f carries $(\partial E \times I) \cup (E \times \partial I)$ onto the boundary ∂E_1, it follows that f gives rise to a map $f'\colon SM(G) \to M(G \times 1)$. But f is a relative homeomorphism, hence f' is a homeomorphism.

The Thom isomorphism

$$\phi\colon H^iB(G) \to H^{i+m}(M(G) \bmod o)$$

is defined as follows. (see [14, Théorème I.4]). The cohomology of $M(G) \bmod o$ will be identified with the cohomology of $E \bmod \partial E$. It can be verified that $H^m(E \bmod \partial E; Z)$ is an infinite cyclic group, with standard generator u. The isomorphism ϕ is now defined by the formula $\phi(a) = \pi^*(a)u$, where $\pi\colon E \to B(G)$ denotes the projection map. It follows from this definition that the following diagram is commutative:

$$\begin{array}{ccc} H^{i+m}(M(G) \bmod o) & \xrightarrow{S} & H^{i+m+1}(M(G\times 1) \bmod o) \\ \uparrow \phi & & \uparrow \phi \\ H^iB(G) & = & H^iB(G\times 1). \end{array}$$

Here S denotes the cohomology suspension, defined using the cohomology cross product.

Now let us specialize to the case $G = U_n \subset SO_{2n}$. The classifying space $B(U_n)$ has a standard cell subdivision due to Ehresmann [7] and $B(U_n)$ is a subcomplex of $B(U_{n+1})$. Hence $M(U_n)$ is a CW-complex and the two-fold suspension

$$S^2M(U_n) = M(U_n \times 1 \times 1)$$

is a subcomplex of $M(U_{n+1})$. Thus

$$\boldsymbol{M}(U) = (0, 0, M(U_1), SM(U_1), M(U_2), \cdots)$$

is a stable object. The track group $\{S^0, \boldsymbol{M}(U)\}_k$ is clearly isomorphic to the stable homotopy group $\pi_{k+2n}(M(U_n))$, with n large.

On the other hand the complexes $B(U_1) \subset B(U_2) \subset \cdots$ have a union $B(U)$ which is again a CW-complex. The isomorphisms

$$\phi\colon H^iB(U_n) \to H^{i+2n}(M(U_n) \bmod o)$$

give rise, in the limit, to an isomorphism

$$\phi\colon H^iB(U) \to H^i\boldsymbol{M}(U).$$

It follows that $H^*\boldsymbol{M}(U)$ has a basis over Z_p consisting of the elements $\phi s(\omega)$, where ω ranges over all partitions.

THEOREM 2. *The cohomology $H^*\boldsymbol{M}(U)$ with coefficient group Z_p is a free module over $A/(Q_0)$, having as basis the elements $\phi s(\lambda)$, where λ ranges over all partitions which contain no integer of the form $p^j - 1$.*

Together with Theorem 1, and the fact that $\boldsymbol{M}(U)$ has no odd dimensional cohomology, this clearly implies the following.

THEOREM 3. *The groups $\{S^0, \boldsymbol{M}(U)\}_m$ have no torsion.*

The full structure of these stable homotopy groups can now be determined, using the fact that the stable Hurewicz homomorphism

$$\{S^0, \boldsymbol{Y}\}_m \to H_m(\boldsymbol{Y}; Z)$$

is a $\mathcal{C}$-isomorphism, where $\mathcal{C}$ denotes the class of finite groups. (See Serre [12] for definitions. This particular assertion is not in Serre's paper, but is well known.)

COROLLARY. *The group $\{S^0, \boldsymbol{M}(U)\}_m = \pi_m \boldsymbol{M}(U)$ is zero for m odd, and is free abelian for $m = 2n$, the number of generators being equal to the number of partitions of n.*

The proof of Theorem 2 will be based on a peculiar partial ordering of partitions, due to Thom. Given a sequence $R = (r_1, r_2, \cdots)$, define ω_R as the partition of $\sum r_j(p^j - 1)$ consisting of r_j copies of $p^j - 1$ for each $j \geqq 1$. Thus every partition ω can be written uniquely in the form $\lambda\omega_R$, where $\lambda = h_1 \cdots h_l$ contains no integer of the form $p^j - 1$. Let l denote the length of λ and let $\Sigma = h_1 + \cdots + h_l$ denote the sum of the integers in λ. Similarly, given a second partition ω', define l' and Σ'.

Definition. ω' is *less than* ω if $l' < l$, or if $l' = l$ and $\Sigma' > \Sigma$. (Note that integers of the form $p^j - 1$ are completely ignored in this definition.)

LEMMA 6. *The cohomology operation $\mathfrak{P}^R$ carries $\phi s(\lambda) \in H^*\boldsymbol{M}(U)$ into $\phi s(\lambda\omega_R)$ plus a linear combination of elements $\phi s(\omega')$ with ω' less than $\lambda\omega_R$.*

Proof. It is clearly sufficient to prove the corresponding assertion for $H^*M(U_n)$, where n is large (say $n \geqq l + r_1 + r_2 + \cdots$), but finite. Consider the cross-section

$$f \colon B(U_n) \to E, \partial E$$

of the $2n$-disk bundle, determined by the center points of the disks. The induced cohomology homomorphism f^* carries the fundamental cohomology class $u \in H^{2n}(E \bmod \partial E)$ into the characteristic class

$$c_n = t_1 \cdots t_n = s(1 \cdots 1) \in H^{2n}B(U_n).$$

(See Thom [14], Borel and Serre [4].) Hence f^* carries the general element $\phi(a) = \pi^*(a)u \in H^{i+2n}(E \bmod \partial E)$ into the cup product $ac_n \in H^{i+2n}B(U_n)$. But the correspondence $a \to ac_n$ is a monomorphism; hence f^* is a monomorphism. Thus in order to prove Lemma 6 it is sufficient to prove the following.

9

Assertion. *$\mathcal{P}^R(s(\lambda)c_n)$ is equal to $s(\lambda\omega_R)c_n$ plus a linear combination of elements $s(\omega')c_n$ with ω' less than $\lambda\omega_R$.*

Consider a typical monomial $t_1^{a_1}\cdots t_n^{a_n}$ of the sum $s(\lambda)c_n$. Here l of the integers $a_1,\cdots,a_n$ are equal to the integers $1+h_1,\cdots,1+h_l$ in some order; while the remaining $n-l$ integers a_i are equal to 1. According to Lemma 3 we have

$$\mathcal{P}^R(t_1^{a_1}\cdots t_n^{a_n}) = \sum_{R_1+\cdots+R_n=R} (\mathcal{P}^{R_1}t_1^{a_1})\cdots(\mathcal{P}^{R_n}t_n^{a_n}).$$

This formula is valid even for the case $p=2$, since $B(U_n)$ has no odd dimensional cohomology. (See Lemma 3′.) The expression $\mathcal{P}^{R_i}t_i^{a_i}$ is equal to some constant k_i times $t_i^{b_i}$, where $b_i \geqq a_i$. The case $b_i = a_i$ can occur only if $R_i = 0$.

Each such monomial $(k_1\cdots k_n)t_1^{b_1}\cdots t_n^{b_n}$ contributes to a symmetric polynomials $s(\omega')c_n$, where ω' denotes the partition obtained from the sequence $b_1-1,\cdots,b_n-1$ by deleting zero. We wish to choose $R_1,\cdots,R_n$ so that this partition ω' is as "large" as possible, in the sense of the partial ordering. The first requirement is that as few as possible of the integers b_i-1 should be of the form p^j-1. But if $a_i=1$, and if the constant k_i is non-zero, then $\mathcal{P}^{R_i}t_i^{a_i}$ is necessarily of the form $t_i^{p^j}$. (See Lemma 3.) Thus the best we can do is to *choose $R_1,\cdots,R_n$ so that b_i is a power of p only if $a_i=1$.*

The second requirement in order to make ω' "large" is that the sum of all b_i-1 for which b_i is not a power of p should be as small as possible. Evidently, the best we can do in this direction is to *choose $R_i=0$ whenever $a_i>1$*; so that b_i will be equal to a_i whenver $a_i>1$.

Now consider the sum of all terms $(\mathcal{P}^{R_1}t_1^{a_1})\cdots(\mathcal{P}^{R_n}t_n^{a_n})$ for which this last condition (that R_i must be equal to zero whenever $a_i>1$) is satisfied. Each such term has the form $t_1^{b_1}\cdots t_n^{b_n}$, where l of the integers $b_1,\cdots,b_n$ are equal to $1+h_1,\cdots,1+h_l$ in some permutation; and the remaining $n-l$ integers b_i are powers of p. Recall that $\mathcal{P}^{R_i}t_i$ is equal to $t_i^{p^j}$ if $R_i=\Delta_j$ and is zero otherwise. Hence the relation $R_1+\cdots+R_n=R=(r_1,r_2,\cdots)$ implies that a given power p^j, $j\geqq 1$, must occur exactly r_j times in the sequence $b_1,\cdots,b_n$. The integer 1 must therefore occur $n-l-r_1-r_2-\cdots$ times in the sequence $b_1,\cdots,b_n$. Taking the sum of all monomials $t_1^{b_1}\cdots t_n^{b_n}$ which satisfy these conditions, we obtain exactly the polynomial $s(\lambda\omega_R)c_n$. This completes the proof of Lemma 6.

Proof of Theorem 2. The equations

$$\mathcal{P}^R\phi s(\lambda) = \phi s(\lambda\omega_R) + \sum(\text{constant})\phi s(\lambda'\omega_{R'}),$$

with all λ' less than λ, can be solved inductively, giving rise to equations:

$$\phi s(\lambda\omega_R) = \mathcal{P}^R\phi s(\lambda) + \sum(\text{constant})\mathcal{P}^{R'}\phi s(\lambda'),$$

with all λ' less than λ. (Only a finite number of terms are involved, since $H^*\boldsymbol{M}(U)$ is finitely generated in each dimension.) But the elements $\phi s(\lambda\omega_R)$ are known to form a Z_p-basis for $H^*\boldsymbol{M}(U)$. Therefore the elements $\mathcal{P}^R\phi s(\lambda)$ also form a Z_p-basis for $H^*\boldsymbol{M}(U)$. Since $\{\mathcal{P}^R\}$ is a basis for the vector space $A/(Q_0)$ over Z_p, this implies that the elements $\phi s(\lambda)$ form an $A/(Q_0)$-basis for $H^*\boldsymbol{M}(U)$. This completes the proof of Theorem 2, and hence Theorem 3.

4. Cohomology computations for $B(SO_{2n})$ and $M(SO)$. Consider the torus $T_n \subset U_n \subset SO_{2n}$, and the corresponding homomorphism

$$H^*(B(SO_{2n});Z_p) \to H^*(B(T_n);Z_p).$$

According to Borel and Serre [4], if p is odd, then the first algebra may be identified with the subalgebra of the second consisting of all polynomials $a + t_1 \cdots t_n b$, where a and b are symmetric polynomials in the elements $t_1^2, \cdots, t_n^2$. Thus a basis for $H^*(B(SO_{2n});Z_p)$ over Z_p is given by the elements $s(\omega)$ and $s(\omega)t_1 \cdots t_n$, where $\omega = i_1 \cdots i_k$, $k \leqq n$, is a partition into even integers. Letting n tend to infinity, *a Z_p-basis for $H^*(B(SO);Z_p)$ is given by the elements $s(\omega)$, where ω ranges over all partitions into even integers.*

Carrying out an argument completely analogous to that in Section 3, we construct a stable object

$$\boldsymbol{M}(SO) = (o, M(SO_1), M(SO_2), \cdots),$$

and prove the following.

THEOREM 4. *Let p be an odd prime, and let $\lambda = h_1 \cdots h_l$ range over all partitions into integers h_i which are even and not of the form $p^j - 1$. Then $H^*(\boldsymbol{M}(SO);Z_p)$ is the free $A/(Q_0)$-module having as basis the elements $\phi s(\lambda)$.*

Together with Theorem 1 this proves the following

THEOREM 5. *The cobordism groups $\Omega^i = \pi_i(\boldsymbol{M}(SO))$ contain no odd torsion.*

C. T. C. Wall has recently proved that an element in the 2-torsion subgroup of Ω^i is completely determined by its Stiefel-Whitney numbers. Together with Theorem 5, this proves the following conjecture of Thom.

COROLLARY 1. *If the Stiefel-Whitney numbers and the Pontrjagin numbers of a compact, oriented, differentiable manifold V^i are all zero, then V^i is a boundary.*

As special cases:

COROLLARY 2. *Suppose that V^i can be imbedded in euclidean space so as to have trivial normal bundle. Then V^i is a boundary.*

The proof is clear.

COROLLARY 3. *Suppose that $H_*(V^i;Z_2)$ is isomorphic to $H_*(S^i;Z_2)$. Then V^i is a boundary.*

Proof. The Stiefel-Whitney number $w_i[V^i]$ is equal to the Euler characteristic reduced modulo 2; hence is zero. If $i=4n$, then the Pontrjagin number $p_n[V^i]$ is zero by the index theorem (Hirzebruch [8]). Since the other characteristic numbers of V^i are trivially zero, it follows that V^i is a boundary.

Concluding Remarks. There are other homotopy groups which may be accessible, using the Adams spectral sequence. For example, the symplectic groups $Sp(n)\subset SO_{4n}$ give rise to a stable object

$$\boldsymbol{M}(Sp)=(o,o,o,o,M(Sp(1)),SM(Sp(1)),$$
$$S^2M(Sp(1)),S^3M(Sp(1)),M(Sp(2)),\cdots).$$

Assertion. The groups $\pi_i\boldsymbol{M}(Sp)$ have no odd torsion.

This can be proved directly from the spectral sequence; or can be derived from Theorem 5, using the natural map $\boldsymbol{M}(Sp)\rightarrow\boldsymbol{M}(SO)$.

Problem. Can one compute the spectral sequence for $\pi_*\boldsymbol{M}(Sp)$ corresponding to the prime $p=2$?

Similarly, the representations $Spin(n)\rightarrow SO_n$ give rise to a stable object.

$$\boldsymbol{M}(Spin)=(o,M(Spin(1)),M(Spin(2)),\cdots).$$

Again there is no odd torsion; but the case $p=2$ seems difficult. As a final question, consider the stable object $\boldsymbol{M}(SU)$ corresponding to the special unitary group.

Problem. What can be said about $\pi_*\boldsymbol{M}(SU)$?

REFERENCES.

[1] J. F. Adams, "On the structure and applications of the Steenrod algebra," *Commentarii Mathematici Helvetici*, vol. 32 (1958), pp. 180-214.

[2] ———, "Sur la théorie de l'homotopie stable," *Bulletin de la Société Mathématique de France*, vol. 87 (1959), pp. 277-280.

[3] J. Adem, "The relations on Steenrod powers of cohomology classes," *Algebraic geometry and topology*, Princeton, 1957, pp. 191-238.

[4] A. Borel and J.-P. Serre, "Groupes de Lie et puissances réduites de Steenrod," *American Journal of Mathematics*, vol. 75 (1953), pp. 409-448.

[5] H. Cartan, "Sur les groupes d'Eilenberg-MacLane II," *Proceedings of the National Academy of Sciences U. S. A.*, vol. 40 (1954), pp. 704-707.

[6] ———, "Sur l'iteration des opérations de Steenrod," *Commentarii Mathematici Helvetici*, vol. 29 (1955), pp. 40-58.

[7] C. Ehresmann, "Sur la topologie de certains espaces homogenes," *Annals of Mathematics*, vol. 35 (1934), pp. 396-443.

[8] F. Hirzebruch, *Neue topologische Methoden in der algebraischen Geometrie*, Springer, 1956.

[9] E. L. Lima, "Stable Postnikov invariants," *American Mathematical Society Notices*, vol. 5 (1958), abstract number 544-27, p. 215.

[10] J. Milnor, "The Steenrod algebra and its dual," *Annals of Mathematics*, vol. 67 (1958), pp. 150-171.

[11] F. Peterson, "Generalized cohomotopy groups," *American Journal of Mathematics*, vol. 78 (1956), pp. 259-281.

[12] J.-P. Serre, "Groupes d'homotopie et classes de groupes abéliens," *Annals of Mathematics*, vol. 58 (1953), pp. 258-294.

[13] E. Spanier, "Function spaces and duality," *Annals of Mathematics*, vol. 70 (1959), pp. 338-378.

[14] R. Thom, "Espaces fibrés en sphères et carrés de Steenrod," *Annales Scientifiques de l'Ecole Normale superieure*, vol. 69 (1952), pp. 109-181.

[15] ———, "Quelques propriétés globales de variétés différentiables," *Commentarii Mathematici Helvetici*, vol. 28 (1954), pp. 17-86.

Travaux de Milnor sur le Cobordisme

René Thom

L'article de J. Milnor [**3**] (à paraître) apporte sur la structure de l'anneau de cobordisme Ω de nouveaux renseignements importants; de plus, Milnor généralise la théorie du cobordisme aux variétés presque-complexes, ce qui conduit à une connaissance plus complète des propriétés de divisibilité des nombres caractéristiques de Chern des variétés presque-complexes (a fortiori des variétés complexes et algébriques). La démonstration fait appel à un nouvel instrument en théorie de l'homotopie: la suite spectrale de F. Adams dont on rappellera ici les principales propriétés. Elle met en oeuvre également, de façon essentielle, la structure algébrique de l'algèbre de Steenrod, élucidée pour une large part dans un article antérieur de Milnor [**2**].

1. Rappel sur l'anneau de cobordisme Ω.

Deux variétés différentiables (C^∞), V^n, V'^n, de même dimension n, compactes, orientées, son dites ***cobordantes***, s'il existe une variété à bord M^{n+1}, compacte, orientée, telle que le bord ∂M^{n+1} soit formé de V' et de $-V$, $-V$ désignant la variété V pourvue de l'orientation opposée à l'orientation donnée. Ceci définit, dans l'ensemble des variétés différentiables orientées compactes, une relation d'équivalence, dont les classes forment, par réunion des variétés représentantes, un groupe abélien Ω^n. De plus le produit topologique fait de la somme directe $\Omega = \sum_n \Omega^n$ un anneau à multiplication anticommutative, noté Ω.

Soit $\mathrm{A(SO(n))}$, le fibré universel en n-boules associé à $\mathrm{SO(n)}$, dont la base est la grassmannienne des n-plans dans un espace euclidien de grande dimension; $\mathrm{A(SO(n))}$ est une variété à bord compacte; si l'on identifie son bord en un point, on obtient un espace noté $\mathrm{M(SO(n))}$; la même construction peut s'effectuer à partir d'un sous-groupe fermé G de $\mathrm{SO(n)}$; elle conduit à construire un espace $\mathrm{M}(G)$, et l'injection $G \subset \mathrm{SO(n)}$ induit une application canonique de $\mathrm{M}(G)$ dans $\mathrm{M(SO(n))}$. En particulier, si l'on considère $\mathrm{SO(n)}$ comme sous-groupe de $\mathrm{SO(n+1)}$, on déduit une injection de $\mathrm{M(SO(n))}$ dans $\mathrm{M(SO(n+1))}$; il est aisé de voir que cette injection possède la propriété homotopique essentielle d'une suspension, à savoir qu'elle induit un isomorphisme des groupes d'homotopie

$$\pi_{n+k}(\mathrm{M(SO(n))}) \to \pi_{\mathrm{n+k+1}}(\mathrm{M(SO(n+1))})$$

pour $k < n$; on désignera par la notation $\pi_k(\mathrm{MSO})$ ces groupes stables. Cette construction est justifiée par le résultat suivant (Cf. Thom [**4**]): Le groupe de cobordisme Ω^k est isomorphe au groupe stable $\pi_k(\mathrm{MSO})$.

2. Les résultats de Milnor.

Désignons par T l'idéal de Ω formé des éléments d'ordre fini. On a les résultats suivants:

THÉORÈME 1. *L'ideal T n'a pas d'élément d'ordre impair.*

THÉORÈME 2. *Le quotient Ω/T est un anneau de polynômes (sur $\mathbb{Z}$), $S(Y_1, Y_2, \ldots, Y_k, \ldots)$ admettant un générateur Y_k de dimension $4k$.*

Ces générateurs Y_k sont réalisés par des variétés qu'on sait construire (au moins théoriquement).

3. La suite spectrale de F. Adams.

Soient X un complexe fini, $S^m(X)$ la m-ième suspension de X; les groupes d'homotopie $\pi_{n+m}(S^m(X))$ sont indépendants de m dès que $m > n$; on les désignera par $\Sigma_n(X)$. Soient q un nombre premier, A l'algèbre de Steenrod relative à l'entier q (i.e. l'algèbre engendrée par les puissances de Steenrod $\mathfrak{P}_q$, le produit étant défini par la composition de ces opérations). On désigne par $q^\infty\,\Sigma_n(X)$ le sous-groupe des éléments de $\Sigma_n(X)$ divisibles par toute puissance de q . On a alors le

THÉORÈME de F. Adams. *Il existe une suite spectrale $(E_r^{s,t}, d_r)$ associée à X, telle que:*

$$E_2^{s,t} = \mathrm{Ext}_A^{s,t}(H^*(X\,;\mathbb{Z}_q)\,;\mathbb{Z}_q)$$

et

$$E_\infty^{s,t} = B^{s,t}/B^{s+1,t+1},$$

où les groupes $B^{i,n+i}$ définissent une filtration du groupe stable, $\Sigma_n(X)$:

$$\Sigma_n(X) = B^{0,n} \supset B^{1,n+1} \supset \ldots \supset B^{s,n+s}$$

telle que:

$$\bigcap_s B^{s,n+s} = q^\infty\,\Sigma_n(X)\,.$$

Le foncteur $\mathrm{Ext}_A^{s,t}$ est défini comme suit: soient M , N deux A-modules à gauche, gradués; désignons par $\mathrm{Hom}_A^t(M,N)$ le groupe des A-homomorphismes de degré $-t$ de M dans N . Etant données une résolution de M par des A-modules projectifs:

$$\to P_s \to P_{s-1} \to \cdots \to P_2 \to P_1 \to P_0 \to M \to 0$$

on forme le complexe défini par:

$$\to \operatorname{Hom}_A^t(P_{s-1}, N) \to \operatorname{Hom}_A^t(P_s, N) \to \operatorname{Hom}_A^t(P_{s+1}, N) \to \quad ;$$

son homologie est alors données par les groupes $\operatorname{Ext}_A^{s,t}(M, N)$. Le principe de la démostration consite à définir une suite d'espaces

$$S^m(X) = Y_0 \supset Y_1 \supset Y_2 \supset \cdots \supset Y_s \supset Y_{s+1} \supset \cdots$$

telle que la suite de A-homomorphismes:

$$0 \leftarrow H^*(Y_0 ; \mathbb{Z}_q) \leftarrow H^*(Y_1, Y_0 ; \mathbb{Z}_q) \leftarrow H^*(Y_2, Y_1 ; \mathbb{Z}_q) \leftarrow$$

donne une résolution de $H^*(Y_0 ; \mathbb{Z}_q)$ par des modules libres sur l'algèbra de Steenrod A. De plus, l'image de $H_{m+n}(Y_s ; \mathbb{Z})$ dans $H_*(Y_0 ; \mathbb{Z})$ se compose d'éléments divisibles par q^s, ce qui permet d'interpréter, dans une certaine mesure, la filtration de $\Sigma_n(X)$ par les $B^{s,n+s}$.

4. La cohomologie de $M(\mathrm{SO})$ en tant que A-module.

Désignons par Q_0 l'opérateur de Bockstein $1/q$; si q est impair, la cohomologie de M(SO) n'a pas de torsion d'ordre impair (car, en vertu d'un résultat de Borel, la cohomologie de la grassmannienne n'a pas de torsion d'ordre impair). Il en résulte que le Bockstein Q_0 opère trivialement dans la cohomologie $H^*(\mathrm{MSO})$; il est donc raisonnable de considérer d'abord $H(\mathrm{MSO})$ comme un $A/(Q_0)$-module, où $A/(Q_0)$ est le quotient de A par l'idéal bilatère engendré par (Q_0). On alors le résultat suivant:

PROPOSITION 1. *La cohomologie $H(\mathrm{MSO})$ est un $A/(Q_0)$-module libre.*

Il est assez facile de donner une base de $A/(Q_0)$: à toute suite d'entiers $r_1, r_2, \ldots, r_k, \ldots$ tous nuls sauf un nombre fini, on associe une opération $\mathfrak{P}^{(r)}$ de A, de degré:

$$2r_1(q-1) + 2r_2(q^2-1) + \cdots + 2r_i(q^i-1) + \cdots$$

et ces opérations forment une base de $A/(Q_0)$. Par example, sur une classe t de dimension 2, on a:

$$\mathfrak{P}^1 t = t^q, \qquad \mathfrak{P}^{0\,1} t = t^{q^2} \quad \cdots \quad \mathfrak{P}^{0\,\cdots\,1} = t^{q^s}$$

et

$$\mathfrak{P}^{(r)} t = 0 \qquad \text{si} \qquad r_1 + r_2 + \cdots + r_s > 1 .$$

En définissant de manière évidente la somme de deux suites (r), on a la formule de produit:

$$\mathfrak{P}^{(r)}(X \cdot Y) = \sum_{(s)+(t)=(r)} \mathfrak{P}^{(s)}(X) \cdot \mathfrak{P}^{(t)}(Y) .$$

De là résulte sur un produit de n classes t_i de degré 2:

$$\mathfrak{P}^{(r)}(t_1 t_2 \ldots t_n) = \sum (t_1)^q (t_2)^q \cdots (t_{r_1})^q (t_{r_1+1})^{q^2} \cdots (t_{r_2+1})^{q^3} \cdots t_n$$

où l'exposant q apparait r_1 fois, l'exposant q^2 apparait r_2 fois, etc., et l'exposant 1 apparait $n - (r_1 + r_2 \cdots + r_s + \cdots)$ fois.

Or, la cohomologie $H^{4j}(\mathrm{MSO})$ peut s'identifier, suivant le procédé connu dû à Borel-Serre, aux fonctions symétriques en n variables t_i de degré 2, de la forme:

$$X_\omega = \sum t_1^{1+2i_1} \cdots t_s^{1+2i_s} \cdots t_n , \qquad i_1 + i_2 + \cdots + i_s = j.$$

On peut alors définir sur les partitions ω de j un ordre, par lequel on démontre que toutes les classes de la forme $\mathfrak{P}^{(r)} X_\omega$, $\deg(r) < n$, sont des éléments linéairement indépendants de $H^*(\mathrm{MSO})$, dès que la partition ω ne contient aucun entier de la forme $\frac{1}{2}(q^i - 1)$.

Il en résulte que $H^*(\mathrm{MSO})$ est un module libre sur $A/(Q_0)$; on obtient une base de $H^{4j}(\mathrm{MSO})$ en tant que A-module en formant toutes les fonctions symétriques X_ω telles que la partition ω ($\deg \omega = j$) ne contienne aucun entier de la forme $\frac{1}{2}(q^i - 1)$.

Pour passer de $A/(Q_0)$ à A elle-même, Milnor construit explicitement une résolution de $H(\mathrm{MSO})$:

$$0 \leftarrow H(\mathrm{MSO}\,; \mathbb{Z}_\mathrm{q}) \leftarrow \mathrm{F}_0 \leftarrow \mathrm{F}_1 \leftarrow \cdots \leftarrow \mathrm{F}_\mathrm{r} \leftarrow \cdots$$

où les F_r sont des A-modules libres.

A l'usage des spécialistes, précisons les détails suivants: soit Q_0 l'opérateur de Bockstein; on pose par récurrence:

$$Q_{i+1} = \mathfrak{P}^{q^i} Q_i - Q_i \mathfrak{P}^{q^i} ;$$

le degré de Q_i est $2q^i - 1$; les Q_i anticommutent et engendrent dans A une algèbre extérieure. On obtient une base de A en formant tous les produits:

$$\mathfrak{P}^{(r)} Q_{j_1} Q_{j_2} \cdots Q_{j_a} , \qquad \text{où} \qquad a \geq 0, \qquad 0 \leq j_1 \leq j_2 \cdots \leq j_a .$$

Cela étant, on donnera pour base au groupe F_s les symboles $y(\omega, i_1, i_2, \ldots, i_s)$ où ω est une partition ne contenant aucun entier de la forme $1/2(q^j - 1)$, où $i_1, i_2, \ldots, i_s$ est une suite non décroissante d'entiers positifs, et où le degré du symbole $y(\omega, i_1, \ldots, i_s)$ est égal à

$$\dim X_\omega + 2(q^{i_1} - 1) + 2(q^{i_2} - 1) + \cdots + 2(q^{i_s} - 1) + s .$$

On définit l'opérateur bord: $F_s \xrightarrow{\mathrm{d}} F_{s-1}$ par la formule:

$$d : y(\omega, i_1, i_2, \ldots, i_s) = \sum_i Q_i\, y(\omega, i_1, \ldots, \check{i}, \ldots, i_s)$$

où la sommation s'étend à tous les indices i apparaissant dans la suite $i_1, i_2, \ldots, i_s$. On vérifie alors $dd = 0$; pour montrer qu'on a affaire à une suite exacte, on construit ensuite un opérateur d'homotopie $D : F_s \to F_{s+1}$ tel que $dD + Dd =$ identité.

Il en résulte que les éléments des groupes F_s de la résolution sont tous de dimension $4k + s$; par suite, les différentielles de Leray d_r de la suite spectrale d'Adams son toutes nulles. Il en résulte immédiatement que les seuls groupes de cobordisme Ω^k qui peuvent présenter de la torsion d'ordre impair sont les groupes de dimension divisible par 4.

Pour démontrer que cette torsion n'existe pas, même en dimension $\equiv 0 \bmod 4$, Milnor exhibe un sous-groupe Ω_0^k libre, de Ω^k. Ce sous-groupe est engendré par les produits de certaines variétés Y_j de dimension $4j$. Ces variétés apparaissent en quelque sorte comme les "plus simples" de toutes les variétés de dimension $4j$. Rappelons que le nombre de Pontrjagin s_j défini en fonction symétrique par $\sum t^j$ possède la propriété suivante: s_j s'annule sur toutes les variétés M^{4j} qui sont "décomposables" au sens du cobordisme (M^{4j} équivalente à une somme de produits de variétés de dimension $< 4j$); seules peuvent être prises comme générateurs dans $\Omega \otimes Q$ les variétés M^{4j} telles que $s_j(M^{4j}) \neq 0$. Mais pour obtenir un vrai générateur, il est très important de savoir déterminer la plus petite valeur possible de ce nombre s_j. A l'aide de propriétés de divisibilité de coefficients binomiaux (trop longues pour être rapportées ici), Milnor démontre:

THÉORÈME 3. *Il existe une variété Y_j de dimension $4j$, telle que:*

$$s_j(Y_j) = \begin{cases} q, & si \quad 2j+1 \quad \text{est de la forme}\, q^s \\ 1, & si \quad 2j+1 \quad \text{n'est pas une puissance de nombre premier.} \end{cases}$$

Les espaces projectifs complexes (de dimension paire) ne suffisent pas pour construire ces variétés; il est nécessaire d'utiliser des variétés qui sont des hypersurfaces de degré $(1, 1)$ dans un produit d'espaces projectifs $\mathbb{PC}(m) \times \mathbb{PC}(n)$ avec $m, n > 1$, $m + n = 2j + 1$. On se sert alors du fait que le nombre s_j d'une telle [1] hypersurface est égal à $-(m+n)!/m!\,n!$.

On peut alors calculer l'indice de l'image $\phi : \Omega_0^k$ dans le groupe d'homologie $H_{4k}(\mathrm{MSO})$ (modulo la 2-torsion): il suffit dans ce but de calculer le déterminant $s_\omega(Y_{j_1} \times \cdots \times Y_{j_s})$, où ω et $(j_1 \ldots j_s)$ parcourent indépendamment toutes les partitions de k. En usant de la formule du produit pour s_ω, on peut mettre cette matrice sous forme triangulaire, et calculer ainsi son déterminant. En comparant ensuite avec l'indice de l'image ϕ, B_s, dans $H(\mathrm{MSO})/2$ torsion, qu'on sait calculer, puisqu'on connait une forme explicite de B_s, on montre, en comparant les indices, qu'il n'y a plus de place dans Ω^k pour de la torsion impaire dans Ω^k/Ω_0^k. Ceci montre que le groupe Ω^k est somme directe du groupe libre Ω_0^k et d'un groupe de 2-torsion.

Ceci démontre les théorèmes 1 et 2, et montre de plus que les valeurs pour $s_j(Y_j)$ trouvées au théorème 3 sont les plus petites possibles. Parmi les autres conclusions qu'en tire Milnor, citons la suivante: désignons par $\pi(k, u, q)$ le nombre des partitions de l'entier k qui contiennent au plus u entiers de la forme $1/2(q^i - 1)$, par $\pi(k)$ le nombre global de toutes les partitions de k. Soit dès lors un ensemble de $\pi(k)$ entiers. Pour que ces nombres soint les nombres de Pontrjagin d'une variété M^{4k}, il faut que ces nombres vérifient un certain nombre de congruences:

[1] This paragraph has been mildly corrected (JWM 2006).

le nombre de ces congruences modulo q^u qui sont linéairement indépendantes est égal à $\pi(k,u,q)$.

5. Généralisation "complexe" du cobordisme.

Deux variétés presque-complexes V^n, V'^n, de même dimension complexe n, son dites C-équivalentes, si leurs nombres caractéristiques de Chern sont égaux:

$$C_{i_1} \cdot C_{i_2} \cdots C_{i_k}(V^n) = C_{i_1} \cdot C_{i_2} \cdots C_{i_k}(V'_n).$$

L'ensemble des classes de C-équivalence forme un monoïde abélien Γ^k, l'addition étant définie par la réunion des variétés représentantes; le produit topologique permet de définir une multiplication commutative: $\Gamma^k \times \Gamma^{k'} \to \Gamma^{k+k'}$. On montre d'abord que le monoïde Γ^k est en fait un groupe abélien; il suffit pour cela de montrer que toute variété a un opposé; ceci se démontre par récurrence sur les dimensions; c'est vrai pour $n = 1$, car la sphère de Gauss a pour opposée la courbe algébrique de genre deux. On établit alors la propriété par récurrence sur k. Par ailleurs le rang de Γ^k est égal à $\pi(k)$, nombre des partitions de k. Il est facile, en effet, de montrer que les $\pi(k)$ produits d'espaces projectifs complexes

$$\mathbb{PC}(i_1) \times \mathbb{PC}(i_2) \times \cdots \times \mathbb{PC}(i_r), \qquad i_1 + i_2 + \cdots + i_r = k$$

sont linéairement indépendants au point de vue de la C-équivalence. Soit M_k une variété presque complexe de dimension k; si $s_k(M_k) = 0$, alors un multiple positif aM_k est C-équivalent à une combinaison linéaire de produits d'espaces projectifs complexes de dimension inférieure; par induction sur k, aM_k a un opposé, donc aussi M_k (car $aM_k = M_k + (a-1)M_k$). Si $s_k(M_k) \neq 0$, on peut trouver une variété M'_k (par exemple $\mathbb{PC}(k)$ ou une hypersurface régulière, suivant le signe de $s_k(M_k)$), telle qu'il existe une combinaison à coefficients entiers positifs a, b pour lesquels $s_k(a M_k + b M'_k) = 0$. On est alors ramené au cas précédent.

On remarquera que l'opposée de toute variété presque-complexe peut être réalisée par une variété algébrique sans singularité. De même, si on se donne "a priori" un ensemble de $\pi(k)$ entiers $m_{i_1 \cdots i_r}$, il existe une variété algébrique M^k telle que

$$C_{i_1} C_{i_2} \cdots C_{i_r}(M_k) = p\, m_{i_1 \ldots i_r},$$

p entier positif.

On peut alors construire, comme dans le cas réel, pour chaque k, une variété algébrique Y'_k telle que $s_k(Y'_k) = q$, si $k+1$ est de la forme q^s, ou $s_k(Y'_k) = 1$ dans tout autre cas. La théorème central est:

THÉORÈME 4. *L'anneau $\Gamma = \sum \Gamma^k$ est un anneau de polynômes sur $\mathbb{Z}$ engendré par les Y'_j.*

Donnons, avant la démonstration, quelques conséquences: comme les Y'_j sont des variétés algébriques, toute propriété de divisibilité démontrée pour les variétés algébriques s'étend "ipso facto" aux variétés presque-complexes. En particulier, ce qui répond à une question posée par Hirzebruch [**1**], *le genre de Todd d'une variété presque-complexe est toujours un entier.*

La démonstration du théorème 4 peut s'interpréter comme suit: disons qu'une variété X^n compacte orientée, (éventuellement avec bord) est *faiblement presque-complexe*, si elle peut être plongée dans un espace euclidien $\mathbb{R}^{n+2k}$ de telle façon que le fibré des vecteurs normaux à X^n admette une structure unitaire; deux variétés X, X' faiblement presque-complexes seront "faiblement" équivalentes si elles constituent le bord (algébrique) d'une variété à bord M^{n+1}, elle-même faiblement presque-complexe, de telle façon que la structure unitaire d'un plongement de M induise sur le bord les structures unitaires déjà données sur X et X'. Alors la construction générale de la théorie du cobordisme s'applique, et permet de définir sur l'ensemble des classes d'équivalence une structure de groupe; ces groupes sont alors isomorphes aux groupes stables $\pi_n(\mathrm{M(U)})$. Il n'est pas difficile de montrer qu'une variété presque-complexe est "faiblement-presque-complexe", à condition de la plonger dans un espace euclidien de dimension assez grande. De là résulte que les groupes Γ^k de Milnor s'identifient aux groupes d'homotopie stables $\pi_{2k}(\mathrm{M(U)})$. Or ces groupes peuvent se calculer, comme ceux de MSO, en usant de la suite spectrale de F. Adams, qui cette fois peut s'appliquer *pour tout nombre premier, y compris* 2. On obtient:

Les groupes stables $\pi_j(\mathrm{M(U)})$ *son nuls pour* j *impair; pour* $j = 2k$, *le groupe* $\pi_j(\mathrm{M(U)})$ *est isomorphe à la somme directe de* $\pi(k)$ *groupes* $\mathbb{Z}$.

La structure multiplicative de l'anneau Γ peut alors se déterminer aisément; elle admet les Y'_j pour générateurs.

Signalons, pour terminer, une conséquence purement topologique de ce résultat: *si une variété* X^n *peut être plongée dans l'espace euclidien de telle façon que le fibré de vecteurs normaux soit trivial, alors* X^n *est une variété-bord.*

Les complexes $\mathrm{M(U}(k))$ semblent d'ailleurs présenter un réel intérêt pour l'étude des propriétés topologiques des variétés algébriques, notamment dans les questions de désingularisation.

References

[1] HIRZEBRUCH, FRIEDRICH: *Some problems on differentiable and complex manifolds,* Annals of Math., Series 2 (**60**), 213–236, (1954).

[2] MILNOR, JOHN: *The Steenrod algebra and its dual,* Annals of Math., Series 2 (**67**), 150–171, (1958).

[3] MILNOR, JOHN: *On the cobordism ring, and a complex analogue,* (à paraître[2]).

[4] THOM, RENÉ: *Quelques propriétés globales des variétés différentiables,* Comment. Math. Helvet., (**28**), 17–86, (1954).

[2]Part I of this projected paper appears on pages 257–273 of this volume, but Part II was never written.

A SURVEY OF COBORDISM THEORY[1]

by J. MILNOR

This paper will start out with a discussion of known results and then will taper off into a discussion of unsolved problems.

The theory of cobordism was initiated by L. Pontrjagin and V. A. Rohlin [10, 12]. It came of age with the work of R. Thom [17]. The basic question in this theory is the following. Let $\mathscr{M}$ be some class of compact manifolds. Given $V \in \mathscr{M}$ how can one decide whether or not V is the boundary of some other manifold in $\mathscr{M}$? Of course a necessary condition is that V itself must be a *closed* manifold: that is the boundary ∂V must be vacuous.

1. THE CLASSICAL COBORDISM GROUPS N_k AND Ω_k.

As a first illustration of this problem let $\mathscr{D}$ denote the class of all compact differentiable manifolds. The manifolds $V \in \mathscr{D}$ need not be connected or orientable, and are allowed to have boundaries.

THEOREM 1 (Pontrjagin, Thom). — *A closed* k*-dimensional manifold* V $\in \mathscr{D}$ *is the boundary of some* (k + 1) *-dimensional manifold in* $\mathscr{D}$ *if and only if the Stiefel-Whitney numbers* $w_{i_1} \dots w_{i_n}$ [V] *are all zero.*

(Explanation: The Stiefel-Whitney cohomology classes[2]) $w_i \in H^i(V; J_2)$ are defined for example in Steenrod [15]. If $i_{i_1} + \dots + i_n = k$ is any partition of k then the cup product $w_{i_1} \dots w_{i_n}$ is a top dimensional cohomology class. Applying the canonical " integration " homomorphism

$$[V]: H^k(V; J_2) \rightarrow J_2$$

we obtain a " Stiefel-Whitney number " $w_{i_1} \dots w_{i_n} [V] \in J_2$.)

---

1) Talk delivered at the Zurich Colloquium on Differential Geometry and Topology, June 1960.

2) The notation J will be used for the integers and J_2 for the integers modulo 2.

The *non-oriented cobordism group* $\mathrm{N}_k = H_k(\mathcal{D})$ is constructed as follows. Given two k-manifolds V, $V' \in \mathcal{D}$ the *sum* $V + V'$ will mean the (disjoint) topological sum, provided with a differentiable structure in the obvious way.

Definition. Two closed manifolds V, $V' \in \mathcal{D}$ are *congruent modulo* $\partial\mathcal{D}$ if $V + V'$ is the boundary of some manifold in $\mathcal{D}$. The set of all congruence classes of closed k-manifolds, under the composition operation $+$, forms the required group N_k. We will also use the notation $H_k(\mathcal{D})$ for this group since it is something like a homology group. (The Russian term for " cobordism " is " intrinsic homology ".)

It follows from Theorem 1 that each N_k is a finite abelian group of the form $J_2 \oplus \ldots \oplus J_2$.

The cartesian product operation between differentiable manifolds gives rise to a bilinear pairing

$$N_k \oplus N_l \to N_{k+l} .$$

Thus the graded group $N_* = (N_0, N_1, \ldots)$ has the structure of a graded ring.

Theorem 2 (Thom). — *The non-oriented cobordism ring* N_* *has the structure of a polynomial algebra*

$$J_2[X_2, X_4, X_5, X_6, X_8, X_9, \ldots]$$

with one generator $\mathrm{X}_k \in \mathrm{N}_k$ *for each dimension which is not of the form* $2^m - 1$.

If k is even then the real projective k-space can be taken as generator. For k odd generators have been constructed by Dold [4].

Thom's proof of Theorems 1 and 2 involves a brilliant mixture of algebra and geometry. A key step in the argument is his proof that N_k is isomorphic to a certain homotopy group. I will not try to give details.

Next consider the class $\mathcal{D}_0$ consisting of all *oriented* compact differentiable manifolds.

Theorem 1'. — *A closed manifold in* $\mathcal{D}_0$ *is the boundary of a manifold in* $\mathcal{D}_0$ *if and only if both its Stiefel-Whitney numbers and its Pontrjagin numbers are zero.*

This result is due to Pontrjagin, Thom, Milnor, Averbuh, and Wall. (See [2, 9, 19].) For the definition of the Pontrjagin numbers $p_{i_1} \dots p_{i_n} [V] \in J$ the reader is refered to Hirzebruch [6]. These numbers are defined only if the dimension k is a multiple of 4.

The *oriented cobordism ring* $\Omega_* = H_* (\mathscr{D}_o)$ is defined as follows. For $V \in \mathscr{D}_o$ let $-V$ denote the same manifold V with the opposite orientation. We will say that

$$V \equiv V' \quad (mod\ \partial \mathscr{D}_o)$$

if $(-V) + V'$ is the boundary of some manifold in $\mathscr{D}_o$. As an example, for any closed manifold V we have $V \equiv V$ (mod $\partial \mathscr{D}_o$) since

$$(-V) + V \approx \partial (V \times I)$$

where I denotes the unit interval. The set of all such congruence classes form the required group Ω_k. Again the cartesian product operation makes $\Omega_* = (\Omega_0, \Omega_1, \dots)$ into a graded ring.

It follows from Theorem 1′ that Ω_k is a finitely generated group of the form

$$J \oplus \dots \oplus J \oplus J_2 \oplus \dots \oplus J_2$$

where infinite cyclic summands can occur only if $k \equiv 0$ (mod 4).

THEOREM 2′. — *The ring Ω_*, modulo the ideal consisting of 2-torsion elements, is a polynomial ring* $\mathrm{J}[\mathrm{Y}_4, \mathrm{Y}_8, \mathrm{Y}_{12}, \dots]$ *with one generator in each dimension divisible by* 4.

The complex projective space of real dimension $4m$ can be taken as generator for $m = 1, 2, 3$. However a different generator is needed in dimension 16.

For a description of the 2-torsion in Ω_* the reader is referred to Wall's paper.

2. MANIFOLDS WITH X-STRUCTURE.

In this section we will define the concept of an " X-structure " on the tangent bundle of a differentiable manifold; and study the corresponding cobordism theory.

First recall Steenrod's definition of a tensor field [15, § 6.4 and § 9.1 with mild alterations]. Every differentiable k-manifold V can be made Riemannian and hence has a tangent bundle with structural group O_k. Let X be any topological space on which the group O_k acts. Then we can form the weakly associated bundle with base space V and fibre X. This may be called the " tensor bundle of type X " and its cross-sections are " tensor fields ". As an example, if $k = 2m$, then O_{2m} acts on the coset space O_{2m}/U_m.

A cross-section of the corresponding bundle is called a *quasi-* (or almost) *complex structure* on V. (See [15, § 41.10].)

We will modify this definition as follows, so that it makes sense for all dimensions simultaneously. Let O denote the union of the orthogonal groups $O_1 \subset O_2 \subset O_3 \subset \ldots$ in the fine topology. Then we require that this infinite orthogonal group O act on the space X. It follows that each O_k acts on X. Hence there is a tensor bundle of type X over any manifold $V \in \mathscr{D}$.

Definition: A homotopy class of cross-sections of the tensor bundle with fibre X over V is called an *X-structure* on V. A manifold $V \in \mathscr{D}$ together with an X-structure on V is called an *X-manifold.* We will still use the single symbol V to denote this pair.

Now if V is an X-manifold then ∂V is also. Given any closed X-manifold V one can define a second X-manifold $-$ V so that

$$\partial(V \times I) \approx V + (-V) .$$

Thus one can define a cobordism group for the class of X-manifolds. The resulting group will be denoted by $N_k(X)$ and called the X-*cobordism group.* (Following Atiyah [1] this could also be called the k-th " bordism group " of the O-space X.)

Example 1. Let O/U denote the union of the spaces

$$O_2/U_1 \subset O_4/U_2 \subset O_6/U_3 \subset \ldots$$

in the fine topology with O acting on O/U in the usual way. Then a manifold with an O/U-structure will be called a *weakly complex manifold.* (Compare Hirzebruch [7].) For example any complex manifold is quasi-complex and hence weakly complex. Any sphere can be given an O/U-structure although only S^2 and S^6 possess quasi-complex structures.

The following results are due to Milnor and Novikov.

THEOREM 1″. — *A closed weakly complex manifold* V *is the boundary of a weakly complex manifold if and only if its Chern numbers* $c_{i_1} \dots c_{i_n}[V]$ *are all zero.*

(Explanation: an O/U-structure on V determines a preferred U-bundle over V. Hence Chern classes are defined.) It follows that $N_k(O/U)$ is zero for k odd and is free abelian for k even.

THEOREM 2″. — *The graded group* $N_*(O/U)$ *has a natural ring structure, making it into a polynomial ring* $J[Y_2, Y_4, Y_6, \dots]$ *with one generator in each even dimension.*

As generators one can take certain algebraic varieties with their natural complex structures. (Compare [7]. It is not known whether connected varieties will suffice.)

Example 2. More generally one could use any subgroup G of the infinite orthogonal group in place of U. For example using the infinite symplectic group Sp we would obtain a cobordism ring $N_*(O/Sp)$ which is appropriate for the study of " weakly quaternionic manifolds ". The following six groups seem particularly interesting:

$$1 \subset Sp \subset SU \subset U \subset SO \subset O .$$

Starting from the right, the ring $N_*(O/O)$ is just the non-oriented cobordism ring N_* and $N_*(O/SO)$ is the oriented cobordism ring Ω_*. The rings $N_*(O/SU)$ and $N_*(O/Sp)$ are more or less unknown. (Compare the concluding remarks in [9].)

The ring $N_*(O/1) = N_*(O)$ has essentially been studied by Pontrjagin [11]. An O-structure on V is a trivialization of the tangent O-bundle of V (the " stable " tangent bundle). Manifolds which admit such a structure are called " π-manifolds ". It turns out that $N_k(O)$ is isomorphic to the stable homotopy groups $\pi_{k+n}(S^n)$ of the n-sphere, with n large. This fact is the basis for Pontrjagin's method of studying homotopy groups.

Example 3. Let X be a space on which O operates trivially. Then an X-structure on V is just a prefered homotopy class of maps $V \to X$. As cases of particular interest X might be an Eilenberg-MacLane space or the classifying space of a group. How does one compute he groups $N_k(X)$?

The above definitions can be modified slighlly by admitting only oriented manifolds. Thus one obtains groups $\Omega_k(X)$ where X is any space on which the rotation group SO acts. Again I do not know how to compute these groups. (Added in proof: See Conner and Floyd [21].)

Example 4. Let P denote the infinite real projective space, with the infinite rotation group SO acting in the natural way. The cobordism groups $\Omega_k(P)$ for oriented manifolds with P-structure can be called the *spinor cobordism groups.* This name is appropriate since a P-structure is roughly a " lifting " of the structural group of the tangent bundle to the infinite spinor group. A manifold admits a P-structure if and only if its Stiefel-Whitney class w_2 is zero. The groups $\Omega_k(P)$ have no odd torsion, but otherwise I do not know much about them.

3. Miscellaneous cobordism theories.

So far we have concentrated on differentiable manifolds. However one could equally well define a cobordism group based on the class $\mathscr{T}$ of all compact topological manifolds. (Compare Brown [3, Theorem 3].) The natural correspondence $\mathscr{D} \to \mathscr{T}$ induces a homomorphism from the differentiable cobordism group $N_k = H_k(\mathscr{D})$ to the topological cobordism group $H_k(\mathscr{T})$.

Since Thom [16] has shown that Stiefel-Whitney classes can be defined topologically, we have:

Theorem 3 (Thom). — *The homomorphism* $\mathrm{N}_k \to H_k(\mathscr{T})$ *has kernel zero.*

Problem: Is this homomorphism onto ?

Another possibility would be to consider the class $\mathscr{C}_o$ of all compact, oriented, combinatorial manifolds. Whitehead [20] has shown that each differentiable manifold has a preferred class of triangulations. Hence there is a natural homomorphism from

$\Omega_k = H_k(\mathscr{D}_o)$ to $H_k(\mathscr{C}_o)$. Thom, Rohlin and Švarč have shown that Pontrjagin classes can be defined for combinatorial manifolds. Therefore we have:

THEOREM 3'. — *The homomorphism* $\Omega_k \to H_k(\mathscr{C}_o)$ *has kernel zero.*

However examples show that this homomorphism is not onto. The reader is referred to [13, 18].

Another interesting possibility would be to look at the class of compact homology manifolds.

Returning to the differentiable case, interesting cobordism groups can be obtained by restricting the connectivities of the manifolds involved. As an extreme case we can consider only differentiable manifolds which are either homotopy spheres or homotopy cells. The resulting cobordism groups are closely related to the problem of classifying differentiable structures on spheres. The reader is referred to Milnor [8] and Smale [14].

As a final, quite different, example consider differentiable imbeddings of the circle S^1 in the 3-sphere S^3. Such an object (a knot) is said to *bound* if it can be extended to a differentiable imbedding of the disk D^2 in the disk D^4. The resulting cobordism group has been studied by Fox and Milnor [5]. This group is not finitely generated.

REFERENCES

[1] ATIYAH, M., Bordism and cobordism, Proc, Cambridge Phil. Soc., 57 (1961), 200-208.

[2] AVERBUH, B. G., Algebraic structure of internal homology groups. Doklady Akad. Nauk S.S.S.R., 125 (1959), 11-14.

[3] BROWN, M., Locally flat embeddings of topological manifolds. *A.M.S. Notices*, 7 (1960), 939-940.

[4] DOLD, A., Erzeugende der Thomschen Algebra *N. Math. Zeits.*, 65 (1956), 25-35.

[5] FOX, R. H. and J. MILNOR, Singularities of 2-spheres in 4-space and equivalence of knots. *Bull. Amer. Math. Soc.*, 63 (1957), 406.

[6] HIRZEBRUCH, F., *Neue topologische Methoden in der algebraischen Geometrie.* Springer (Berlin), 1956.

[7] —— Komplexe Mannigfaltigkeiten. *Proceedings Int. Congr. Math.*, 1958 (1960), 119-136.

[8] MILNOR, J., Sommes de variétés différentiables et structures différentiables des sphères. *Bull. Soc. Math. France*, 87 (1959), 439-444.

[9] —— On the cobordism ring Ω^* and a complex analogue, I. *American Journ. of Math.*, 82 (1960), 505-521. [See also *A.M.S. Notices*, 5 (1958), 457.]

[10] PONTRJAGIN, L. S., Characteristic cycles on differentiable manifolds. *Mat. Sbornik*, 21 (1947), 233-284.

[11] —— Smooth manifolds and their applications in homotopy theory. *A.M.S. Translations*, 11 (1959), 1-114.

[12] ROHLIN, V. A., A 3-dimensional manifold is the boundary of a 4-dimensional manifold. (Russian.) *Doklady Akad. Nauk S.S.S.R.*, 81 (1951), 355. [See also *Doklady*, v. 84, p. 221,, v. 89, p. 789 and v. 119, p. 876.]

[13] —— and A. S. ŠVARČ, Combinatorial invariance of the Pontrjagin classes. (Russian.) *Doklady Akad. Nauk S.S.S.R.*, 114 (1957), 490-493.

[14] SMALE, S., Generalized Poincaré's conjecture in dimensions greater than four. Annals of Math., 74 (1961), 391-406.

[15] STEENROD, N., *The topology of fibre bundles.* Princeton, 1951.

[16] THOM, R., Espaces fibrés en sphères et carrés de Steenrod. *Ann. Ecole Norm. Sup.*, 69 (1952), 109-181.

[17] —— Quelques propriétés globales des variétés différentiables. *Comment. Math. Helv.*, 28 (1954), 17-86.

[18] —— Les classes caractéristiques de Pontrjagin des variétés triangulées. *Symposium Internacional de Topologia Algebraica, Mexico* (1958), pp. 259-272.

[19] WALL, C. T. C., Determination of the cobordism ring. *Annals of Math.*, 72 (1960), 292-311.

[20] WHITEHEAD, J. H. C., On C^1-complexes. *Annals of Math.*, 41 (1940), 809-824.

[21] CONNER, P. E. and E. E. FLOYD, Differentiable periodic maps, Technical Note 13, Dept. of Math., Univ. of Virginia, 1961.

[22] NOVIKOV, S. P., Some problems in the topology of manifolds connected with the theory of Thom spaces. Dokl, Akad. Nauk SSSR 132 (1960), 1031-1034 (Russian); translated as Soviet Math. Dokl. 1, 717-720.

Princeton University.

Erratum
to "A Survey of Cobordism Theory"

Example 4, in §2 of "A Survey of Cobordism Theory," attempts to define the concept of "spin structure" on an oriented manifold; but there is a subtle mistake in the exposition. Let me explain the problem. If G is a subgroup of the Lie group H, then H acts on the left of the coset space H/G. If $P \twoheadrightarrow B$ is a principal H-bundle (with H acting freely on the right of P), then we can form the associated H/G-bundle $P \times_H H/G \twoheadrightarrow B$. A section of this associated bundle gives rise to an associated principal G-bundle $P \supset P' \twoheadrightarrow B$, or in other words to a ***reduction***[1] of the structural group of the bundle from H to G.

However, the group $\mathrm{Spin}(n)$ is not a subgroup of the rotation group but rather a 2-fold covering group

$$1 \to \{\pm 1\} \to \mathrm{Spin}(n) \to \mathrm{SO(n)} \to 1 .$$

Thus, in order to specify a ***spin structure*** on an oriented n-manifold, instead of reducing the structural group for the tangent bundle to a subgroup of SO(n) we must lift it to a covering group. More generally, consider any homomorphism $\varphi : G \to H$ between Lie groups (or between well behaved infinite limits of Lie groups). The quotient space of H modulo G no longer makes sense in this context, but we can still form the ***homotopy quotient*** $H/\!\!/G$, as follows. Let E_G be a contractible space on which G acts freely. Then we can "thicken" H and P to spaces

$$\widehat{H} = H \times E_G \qquad \text{and} \qquad \widehat{P} = P \times E_G$$

of the same homotopy type on which G operates freely by the diagonal actions

$$(h,\, e) \cdot g = \Big(h \cdot \varphi(g)\,,\; e \cdot g\Big) \qquad \text{and} \qquad (p,\, e) \cdot g = \Big(p \cdot \varphi(g)\,,\; e \cdot g\Big)\,.$$

Now set $H/\!\!/G$ equal to the quotient $\widehat{H}/G$. Then the quotient space $\widehat{P}/G$ fibers over B with fiber $\widehat{H}/G = H/\!\!/G$, Any section of this fibration $\widehat{P}/G \twoheadrightarrow B$ picks out a copy of G in $\widehat{P}$ for each point of the base space, and hence specifies a principal G -bundle over B.

[1]Note: The examples which are cited in the paper all have the important property that such a reduction of the structure on the stable tangent bundle determines a similar reduction on the stable normal bundle up to homotopy, and conversely. (Compare Lashof [1963].) However, examples such as $G = \mathrm{O}(n) \subset \mathrm{O}$ show that a general subgroup will not have this property.

Note that this homotopy quotient $H/\!\!/G$ fits into a one-sided infinite sequence

$$\cdots \to \Omega(H) \to \Omega(H/\!\!/G) \to G \to H \to H/\!\!/G \to B_G \to B_H\ , \qquad (*)$$

which is *'homotopy-exact* in the sense that any two successive maps form the inclusion and projection, (fiber) $\hookrightarrow$ (total space) $\twoheadrightarrow$ (base), of a fibration, up to homotopy type. Any H-bundle is classified by a map into B_H, and we can lift to a map into B_G if and only if there exists a section of the associated $H/\!\!/G$-bundle.

In particular, taking H to be the infinite rotation group SO and taking G to be the infinite Spin group, defined as the non-trivial central extension

$$1 \to \{\pm 1\} \to \mathrm{Spin} \to \mathrm{SO} \to 1\ ,$$

it is not hard to check that SO$/\!\!/$Spin has the homotopy type of the infinite real projective space $\mathbb{RP}^\infty$, and that the loop space $\Omega(\mathrm{SO}/\!\!/\mathrm{Spin})$ has the homotopy type of the finite set $\{\pm 1\}$. Then the homotopy-exact sequence $(*)$ takes the form

$$1 \to \Omega(\mathrm{Spin}) \to \Omega(\mathrm{SO}) \to \{\pm 1\} \to \mathrm{Spin} \to \mathrm{SO} \to \mathbb{RP}^\infty \to \mathrm{B}_{\mathrm{Spin}} \to \mathrm{B}_{\mathrm{SO}}\ ,$$

where the fundamental group $\pi_1(\mathrm{SO}) \cong \mathbb{Z}/2$ maps isomorphically to $\pi_1(\mathbb{RP}^\infty)$. (Here we can equally well substitue Spin(n) and SO(n) for any $n \geq 3$.)

So far, the exposition in "A Survey of Cobordism Theory" was correct. However, it described the map $\mathrm{SO} \to \mathbb{RP}^\infty$ as coming from the natural action of SO on $\mathbb{RP}^\infty$ (corresponding to the action of the rotation group SO(n) on the projective space of lines through the origin in $\mathbb{R}^n$). Since this lifts to an action on the sphere S^{n-1}, it would induce the zero homomorphism on the fundamental group.

Not surprisingly, the correct construction involves a spin representation. In fact, the group Spin(n) can be defined as a subgroup of the units in the 2^n-dimensional Clifford algebra $C(n)$. It acts by left translation on $C(n)$. Since the central element $-1 \in \mathrm{Spin}(n)$ takes each point of the unit sphere S^{2^n-1} to its antipode, there is an induced action of the quotient group SO(n) on the projective space $\mathbb{RP}^{2^n-1}$. Passing to the limit as $n \to \infty$, we obtain the correct action of SO on $\mathbb{RP}^\infty$.

(An interesting variant of this example is provided by the group Spin^c, defined as the non-trivial central extension

$$1 \to S^1 \to \mathrm{Spin}^c \to \mathrm{SO} \to 1\ .$$

In this case, the homotopy quotient SO$/\!\!/$Spin^c has the homotopy type of the infinite complex projective space $\mathbb{CP}^\infty$. Since the associated map $\mathrm{SO} \to \mathrm{SO}/\!\!/\mathrm{Spin}^c \simeq \mathbb{CP}^\infty$ factors through $\mathrm{SO}/\!\!/\mathrm{Spin} \simeq \mathbb{RP}^\infty$, it follows easily that it induces a non-trivial homomorphism of two-dimensional cohomology.)

SPIN STRUCTURES ON MANIFOLDS

by J. MILNOR

Let M be an oriented, Riemannian manifold. Then the tangent bundle of M has the rotation group $SO(n)$ as structural group. If it is possible to replace $SO(n)$ by the 2-fold covering group Spin (n) as structural group, then one says that M can be given a " spin structure ". The object of this note will be to make this concept precise, and to discuss the related concept of " spin cobordism ".

Let me take this opportunity to point out an error in a previous paper. The definition of spinor cobordism group which was proposed in my paper " A survey of cobordism theory " [7, §2, Example 4] is erroneous. A corrected version of this definition will be given at the end of the present paper.

Let ξ denote a principle fibre bundle with structural group $SO(n)$. Here n can be any positive integer. The value $n = +\infty$ is also acceptable. The total space of ξ will be denoted by $E(\xi)$ and the base space by B. We will always assume that B is a CW-complex, or a manifold.

Definition: A *spin structure* on ξ is a pair (η, f) consisting of

(1) A principal bundle η over B with the spinor group Spin (n) as structural group; and

(2) A map $f\colon E(\eta) \to E(\xi)$ such that the following diagram is commutative.

$$\begin{array}{ccccc}
E(\eta) \times \mathrm{Spin}\,(n) & \xrightarrow{\text{right transl.}} & E(\eta) & \searrow & \\
\downarrow f \times \lambda & & \downarrow & & B \\
E(\xi) \times SO(n) & \xrightarrow{\text{right transl.}} & E(\xi) & \nearrow &
\end{array}$$

Here λ denotes the standard homomorphism from Spin (n) to $SO(n)$.

This definition must be qualified as follows. *A second spin structure (η', f') on ξ should be identified with (η, f) if there exists an isomorphism g from η' to η so that $f \circ g = f'$.*

Note that the definition makes sense even in the special cases $n = 2$ and $n = 1$. It is to be understood that Spin (2) is the 2-fold covering group of the circle $SO(2)$; and that Spin (1) is a cyclic group of order 2.

As an example, the tangent bundle of the 2-sphere S^2 has a unique spin structure (η, f) where $E(\eta)$ is a 3-sphere. The tangent bundle of the circle S^1 has two distinct spin structures.

The above definition is straightforward, but is rather cumbersome. An elegant variant was suggested to the author by M. Hirsch:

Alternative definition 1: A *spin structure* on ξ is a cohomology class[1] $\sigma \in H^1(E(\xi); J_2)$ whose restriction to each fibre is a generator of the cyclic group H^1 (Fibre; J_2). (Note: This last clause disappears in the special case of an $SO(1)$-bundle.)

The idea is the following: Any such cohomology class determines a 2-fold covering of $E(\xi)$. This 2-fold covering space is to be taken as the total space $E(\eta)$. The condition on $\sigma \mid$ Fibre guarantees that each fibre is covered by a copy of Spin(n), the unique 2-fold covering of $SO(n)$. With this interpretation it is not difficult to show that the alternative definition is completely equivalent to the original. Henceforth we will use the two definitions interchangeably.

It is known that an $SO(n)$-bundle can be given a spin structure if and only if its Stiefel-Whitney class w_2 is zero. (Compare Borel and Hirzebruch [2, pg. 350].)

Lemma: *If $w_2(\xi) = 0$ then the number of distinct spin structures on ξ is equal to the number of elements in $H^1(B; J_2)$.*

Proof: If B is connected, then this follows from the exact sequence

$$0 \to H^1(B; J_2) \to H^1(E(\xi); J_2) \to H^1(SO(n); J_2) \to H^2(B; J_2)$$

which can be extracted from the spectral sequence of the fibration ξ. The general case follows easily.

[These facts suggest an analogy between the concept of " spin structure " for $SO(n)$-bundles and the concept of " orientation " for $O(n)$-bundles. Thus an $O(n)$-bundle can be oriented if and

1) The notation J_2 is used for the integers modulo 2.

only if $w_1 = 0$. If $w_1 = 0$ then the number of distinct orientations is equal to the number of elements in $H^0(B; J_2)$.]

Now a word of warning. It may happen that two spin structures (η, f) and (η', f') on ξ are distinct, even though the corresponding spinor group bundles η and η' are isomorphic. As an illustration, consider the following.

Example: Let ε^n denote the trivial $SO(n)$-bundle over the real projective plane P^2. Since $H^1(P^2; J_2) \cong J_2$ this bundle can be given two distinct spin structures. For $n = 1$ or for $n = 2$ the two corresponding Spin(n)-bundles are distinct from each other. However for $n > 2$ it can be shown that the two Spin(n)-bundles are isomorphic: in fact both are trivial.

[This example suggests the conjecture that if (η, f) and (η', f') are two spin structures on the same $SO(n)$-bundle, with $n >$ dim B, then η is necessarily isomorphic to η'. The analagous statement for orientations of an $O(n)$-bundle is known to be true.]

Now we will restrict attention to tangent bundles.

Definition: A *spin manifold* will mean an oriented Riemannian manifold M, together with a spin structure on the tangent bundle of M.

To be more explicit let FM denote the space of oriented orthonormal n-frames on M. Then we will think of the spin structure as being a cohomology class $\sigma \in H^1(FM; J_2)$ whose restriction to each fibre is non-trivial (if $n > 1$).

The notation (M, σ) will be used for such a spin manifold. However if M happens to be simply connected, so that σ is uniquely determined, then we will simply say that M is a spin manifold.

Suppose that V is a k-dimensional submanifold of M with a specified field of normal $(n - k)$-frames. Then $FV \subset FM$; hence any spin structure σ on M determines a spin structure $\sigma \mid FV$ on V. In particular this is true if V is the boundary ∂M.

Definition: A closed spin manifold (V, σ_1) will be called a *spin boundary* if there exists a compact spin manifold (M, σ) with $\partial M = V$ and $\sigma \mid FV = \sigma_1$.

As an illustration consider the following.

Example: The 2-dimensional disk D^2 has a unique spin structure σ. Restricting to the boundary $\partial D^2 = S^1$ we obtain the non-zero cohomology class

$$\sigma_1 \in H^1 (FS^1; J_2) \cong J_2.$$

Thus (S^1, σ_1) is a spin boundary. On the other hand if we take the zero cohomology class in $H^1(FS^1; J_2)$ (this is permissible since $n = 1$) we obtain a different spin manifold $(S^1, 0)$. It can be shown that $(S^1, 0)$ is not a spin boundary.

Similarly one can define the relation of cobordism between closed n-dimensional spin manifolds. The corresponding cobordism group will be denoted by Ω_n^{spin}. Here is a list of the first eight spin cobordism groups.

$\Omega_0^{\text{spin}} \cong J$ (infinite cyclic) by definition.

$\Omega_1^{\text{spin}} \cong J_2$ generated by $(S^1, 0)$.

$\Omega_2^{\text{spin}} \cong J_2$ generated by the torus with a suitable spin structure.

$\Omega_3^{\text{spin}} = 0$.

$\Omega_4^{\text{spin}} \cong J$ generated by a Kummer surface K^4. (Compare [5 pg. 127].)

$\Omega_5^{\text{spin}} = 0$.

$\Omega_6^{\text{spin}} = 0$. [1]

$\Omega_7^{\text{spin}} = 0$.

$\Omega_8^{\text{spin}} \cong J \oplus J$ generated by the quaternion projective plane and by a manifold L^8 such that $L^8+L^8+L^8+L^8$ is spin cobordant to $K^4 \times K^4$. Alternatively, as second generator, one could take the almost parallelizable manifold M_0^8 of reference [4].

It follows that in dimensions 4 and 8 the spin cobordism class of a manifold (V, σ) is completely determined by the Pontrjagin numbers of V. In dimension 4 the Pontrjagin number $p_1 [V^4]$ is subject only to Rohlin's relation

$$p_1 [V^4] \equiv O \quad (\text{mod } 48).$$

[1]) Compare Wall [8, p. 428].

In dimension 8 the two Pontrjagin numbers are subject only to the Borel-Hirzebruch relation

$$7p_1^2[V^8] \equiv 4p_2[V^8] \pmod{5760}.$$

(Compare references [1, Corollary 2], [3, §3.1] and [4].)

The computation of these eight groups is similar to the usual computations in cobordism theory. Thus one first shows that Ω_n^{spin} is isomorphic to the stable homotopy group $\pi_{n+k}\ M(\mathrm{Spin}(k))$ of a suitable Thom complex; and then determines these homotopy groups by a formidable computation. No details will be given.

For $n \geqq 6$ the spin cobordism group can also be interpreted as the cobordism group for the class of 2-connected oriented Riemannian manifolds. In fact if $n \geqq 6$ then:

Assertion 1: Any closed n-dimensional spin manifold is spin cobordant to a 2-connected manifold.

Assertion 2: If a 2-connected n-manifold is a spin boundary, then it bounds a 2-connected manifold. Proofs are easily given using the technique of surgery (=spherical modification) which is described in references [6], [9].

In conclusion let me mention two other variant definitions for the concept of spin structure, which may prove useful for special purposes. We will assume that $n \geqq 2$.

Let ξ be an $SO(n)$-bundle over a CW-complex B. The k-skeleton of B will be denoted by B^k.

Alternative definitions 2: A *spin structure* on ξ is a homotopy class of cross-sections of $\xi \mid B^1$ which can be extended to cross-sections of $\xi \mid B^2$.

It can be shown that every " spin structure " in this sense determines a spin structure in the original sense, and conversely. No details will be given.

Now let the group Spin(n) act on a high dimensional sphere S^N in such a way that the cyclic subgroup $J_2 \subset \mathrm{Spin}(n)$ acts freely on S^N. [Such an action can be obtained by using a spinor representation of the group Spin(n).] We will assume that $N > \dim B$.

Then the quotient group $SO(n) = \mathrm{Spin}(n)/J_2$ acts on the quotient space $P^N = S^N/J_2$. Hence to every $SO(n)$-bundle ξ over B there corresponds an associated bundle ξ' having the projective space P^N as fibre.

Alternative definition 3: A *spin structure* on ξ is a homotopy class of cross-sections of the associated bundle ξ'.

Again it can be seen that this definition is equivalent to the original definition.

REFERENCES

[1] Atiyah, M. and F. Hirzebruch, Riemann-Roch theorems for differentiable manifolds, *Bull. Amer. Math. Soc.*, 65 (1959), 276-281.

[2] Borel, A. and F. Hirzebruch, Characteristic classes and homogeneous spaces II, *Amer. Journ. Math.*, 81 (1959), 315-382.

[3] —— Characteristic classes and homogeneous spaces III, *Amer. Journ. Math.*, 82 (1960), 491-504.

[4] Kervaire, M. and J. Milnor, Bernoulli numbers, homotopy groups, and a theorem of Rohlin, *Proc. Intern. Congress Math.*, Edinburgh 1958 (1960), 454-458.

[5] Milnor, J., On simply connected 4-manifolds, *Symposium Internacional de Topologia Algebraica, Mexico* (1958), 122-128.

[6] —— A procedure for killing homotopy groups of differentiable manifolds, *A.M.S. Symposia in Pure Math. III Differential geometry, Providence* (1961), 39-55.

[7] —— A survey of cobordism theory, *L'Enseignement Math.*, 8 (1962), 16-23.

[8] Wall, C. T. C., Killing the middle homotopy groups of odd dimensional manifolds, *Trans. Amer. Math. Soc.*, 103 (1962), 421-433.

[9] Wallace, A. H., Modifications and cobounding manifolds, *Canadian J. Math.*, 12 (1960), 503-528.

Princeton University.

Remarks Concerning Spin Manifolds

JOHN W. MILNOR

This paper will consist of several unrelated remarks. In §2, as an application of the concept of "spin manifold," a non-standard involution of the 7-sphere is constructed. In §3 the spin cobordism ring is discussed.

1. Spin structures

Let ξ denote an oriented vector bundle with base space B. The concept of a "spin structure" on ξ can be defined in various ways. (See [4] §6.5, and [11].) The following definition is perhaps easiest to use. Let $P\xi$ denote the total space of the associated principal bundle. (One may use either the principal bundle with the rotation group $SO(n)$ as fibre, or with the linear group $GL^+(n, R)$ as fibre. It will not matter which is used, since the first is a deformation retract of the second.) First suppose that $n \geq 2$.

Definition. A *spin structure* on ξ is a cohomology class $s \in H^1(P\xi, Z_2)$ whose restriction to each fibre is non-zero.

For the special case $n = 1$ we will allow s to be an arbitrary element of $H^1(P\xi, Z_2) \cong H^1(B; Z_2)$.

Such a class s determines a 2-fold covering of $P\xi$. This covering can be taken as the total space of a new principal bundle whose structural group is the 2-fold covering group of $SO(n)$ or $GL^+(n, R)$.

The following three lemmas give the basic properties of this concept.

Lemma 1. *ξ admits a spin structure if and only if the Stiefel–Whitney class $w_2(\xi)$ is zero. This spin structure is unique if and only if $H^1(B; Z_2) = 0$.*

Lemma 2. *Let U_1 and U_2 be open sets whose union is the base space B. Given spin structures s_1 on $\xi | U_1$ and s_2 on $\xi | U_2$ which coincide on $\xi | U_1 \cap U_2$, there exists a spin structure s on ξ which extends both s_1 and s_2.*

(Caution: In general s will not be uniquely determined by s_1 and s_2.)

Lemma 3. *Given spin structures on two of the three bundles ξ, η, $\xi \oplus \eta$, there is a uniquely determined spin structure on the third.*

Proofs. Lemma 1 follows (as in [11]) from the exact sequence

$$0 \to H^1(B) \to H^1(P\xi) \to H^1(\text{Fibre}) \to H^2(B),$$

where the coefficient group Z_2 is understood, and B is assumed to be connected. Lemma 2 follows from the Meyer–Vietoris sequence

$$\tilde{H}^0(P_1 \cap P_2) \to H^1(P\xi) \to H^1(P_1) \oplus H^1(P_2) \to H^1(P_1 \cap P_2),$$

where P_i stands for $P(\xi|U_i)$.

To prove Lemma 3, first consider the Cartesian product $\xi \times \eta$, and note that $P\xi \times P\eta \subset P(\xi \times \eta)$. Given spin structures $s \in H^1(P\xi)$ and $t \in H^1(P\eta)$, consider the class $s \times 1 + 1 \times t \in H^1(P\xi \times P\eta)$.

Assertion. There exists one and only one cohomology class $u \in H^1(P(\xi \times \eta))$ which is a spin structure, and which extends the class $s \times 1 + 1 \times t$.

If B is connected then this fact is proved by chasing around the following diagram.

$$\begin{array}{ccccccc} & H^1(B \times B) & \to & H^1P(\xi \times \eta) & \to & H^1(\text{Fibre}) & \to H^2(B \times B) \\ & \downarrow & & \downarrow & & \downarrow & \downarrow \\ 0 \to & H^1(B \times B) & \to & H^1(P\xi \times P\eta) & \to & H^1(\text{Fibre} \times \text{Fibre}) & \to H^2(B \times B) \end{array}$$

The general case follows easily.

Now let $\Delta\colon P(\xi \oplus \eta) \to P(\xi \times \eta)$ denote the diagonal map. Then we can form the cohomology class $\Delta^*u \in H^1P(\xi \oplus \eta)$.

Definition. Δ^*u is the spin structure on $\xi \oplus \eta$ which is determined by s and t.

Finally we must verify that each of the cohomology classes s, t, Δ^*u is uniquely determined by the other two. The most general spin structure on ξ is $s + \pi_\xi^* a$, where $\pi_\xi\colon P\xi \to B$ denotes the projection map and a denotes an arbitrary element of H^1B. If we replace s by $s + \pi_\xi^* a$ and similarly replace t by $t + \pi_\eta^* b$, then Δ^*u will be replaced by $\Delta^*u + \pi_{\xi\oplus\eta}^*(a + b)$. It is now clear that any two determine the third uniquely. This completes the proof of Lemma 3.

As an application of Lemma 3, consider a smoothly imbedded, oriented manifold $M^n \subset R^{n+k}$, with tangent bundle τ and normal bundle ν. Then $\tau \oplus \nu \simeq \tau'|M^n$, where the tangent bundle τ' of R^{n+k} has a unique spin structure. Thus a spin structure on τ determines a spin structure on ν and conversely.

Definition. The pair consisting of a manifold M^n together with a spin structure on its tangent bundle is called a *spin manifold*.

As another application of Lemma 3 one sees that the boundary of any spin manifold is again a spin manifold. Similarly the product of two spin manifolds is a spin manifold.

2. An exotic involution of S^7

In order to illustrate the utility of the concept of spin structure, we will use it to prove the following.

THEOREM 1. *There exists a smooth involution $f: S^7 \to S^7$ without fixed points so that the orbit manifold S^7/f is not diffeomorphic to P^7.*

The proof will depend on an invariant which Eells and Kuiper [6] have defined for certain spin manifolds of dimension $4k - 1$. In order to simplify the exposition we will define this invariant only for 7-manifolds.

First consider a closed 8-dimensional spin manifold M with signature σ. According to Thom and Hirzebruch [7] one has the formula

$$\sigma = \tfrac{1}{45}(7p_2 - p_1^2)[M].$$

On the other hand Atiyah and Hirzebruch [3] define the $\hat{A}$-genus

$$\hat{A}[M] = \tfrac{1}{5760}(7p_1^2 - 4p_2)[M]$$

and prove that this is also an integer (for a spin manifold). Eliminating $p_2[M]$ between these two linear equations, we obtain the formula

$$\hat{A}[M] = (p_1^2[M] - 4\sigma)/896.$$

Now consider a compact 8-dimensional spin manifold (W, s') with boundary $(\partial W, s)$. Following Eells and Kuiper, one must assume:

HYPOTHESIS. The natural homomorphism $H^4(W, \partial W; Q) \to H^4(W; Q)$ between rational cohomology groups is an isomorphism.

The signature σ of W is easily defined. Furthermore, after lifting the Pontrjagin class p_1 of W back to $H^4(W, \partial W; Q)$, the Pontrjagin number $p_1^2[W]$ is well defined. Now define $\mu(\partial W, s)$ to be the residue class, modulo 1, of the rational number $(p_1^2[W] - 4\sigma)/896$. It is not difficult to show that this is an invariant of the spin manifold $(\partial W, s)$: that is, it does not depend on the choice of (W, s').

Example 1. Let M^7 be a homotopy sphere which bounds a parallelizable manifold W having signature $\sigma = +8$. (Compare Kervaire and Milnor [9].) Then

$$\mu(M^7) \equiv (0 - 4.8)/896 \equiv -\tfrac{1}{28} \pmod 1.$$

But according to [9] the group Θ_7 consisting of all homotopy 7-spheres is cyclic of order 28, with generator (M^7). It follows easily that μ provides a complete invariant for homotopy 7-spheres.

Example 2. The real projective space P^{2n+1} can be thought of as the total space of a circle bundle over the complex projective space $P_n(C)$. (An explicit projection map is given by the formula,

$$(x_0 : y_0 : x_1 : y_1 : \cdots : x_n : y_n) \to (x_0 + iy_0 : x_1 + iy_1 : \cdots : x_n + iy_n).)$$

Let ξ denote the associated 2-disk bundle, and let W denote its total space, so that $\partial W = P^{2n+1}$. If n is odd then $P_n(C)$ is a spin manifold; and it follows that W is a spin manifold. The Euler class $e(\xi)$ is equal to twice a generator:

$$e(\xi) = 2a \in H^2(P_n(C); Z),$$

hence the Pontrjagin class $p_1(\xi) = e^2$ is equal to $4a^2$. The first Pontrjagin class for the tangent bundle of $P_n(C)$ is equal to $(n + 1)a^2$. Therefore the class

$$p_1(W) \in H^4(W; Z) \cong H^4(P_n(C); Z)$$

corresponds to $(n + 5)a^2$.

Now let us specialize to the case $n = 3$, and compute $p_1^2[W]$. From the sequence

$$0 \to H^4(W, P^7; Z) \to H^4(W: Z) \to H^4(P^7; Z) \to 0,$$

where the first two groups are infinite cyclic and the third is Z_2, one sees that the class $p_1(W) = 8a^2$ lifts back to 4 times a generator. By Poincaré duality the groups $H^4(W, P^7; Z)$ and $H^4(W; Z)$ are dually paired to $H^8(W, P^7; Z)$ by the cup product. Therefore $p^2[W] = \pm 4.8$. Choosing the orientation of W correctly, and letting s_1 denote the induced spin structure on $\partial W = P^7$, this yields

$$\mu(P^7, s_1) \equiv (p_1^2[W] - 4\sigma)/896 = (32 - 4)/896 \equiv \tfrac{1}{32} \pmod 1.$$

Thus we have computed the invariant μ for P^7 with one of its two possible spin structures.

Now consider the manifold $-W$ with reversed orientation. Its boundary is a spin manifold $(-P^7, s')$ with $\mu(-P^7, s') \equiv -\frac{1}{32}$. But the projective space P^7 possesses an orientation reversing diffeomorphism. This proves that there exists a spin structure s_2 on P^7 so that $\mu(P^7, s_2) \equiv -\frac{1}{32}$. In other words:

LEMMA 4. *The invariant μ of P^7 is equal to $\pm\frac{1}{32}$, where the sign depends on which of the two possible spin structures is chosen.*

Now consider connected sums of the form $P^7 \# M^7 \# \cdots \# M^7$; where M^7 is the homotopy sphere considered earlier. It can be shown that each such sum is homeomorphic to P^7. Clearly

$$\mu(P^7 \# M^7 \# \cdots \# M^7) \equiv \pm\tfrac{1}{32} - \tfrac{1}{28} - \cdots - \tfrac{1}{28}.$$

Since one can obtain 28 distinct invariants in this way, we have proved that the oriented manifold P^7 admits 28 distinct differentiable structures.

Let $r(M^7)$ denote the sum of r copies of M^7. Next note that the 2-fold covering space of $P^7 \# r(M^7)$ is diffeomorphic to $S^7 \# 2r(M^7)$. Taking

$r = 14$ it follows that the 2-fold covering space of $P^7 \# 14(M^7)$ is diffeomorphic to the standard sphere S^7. In other words there exists an involution of S^7 whose orbit space is diffeomorphic to $P^7 \# 14(M^7)$. Since $P^7 \# 14(M^7)$ is distinct from P^7; this proves Theorem 1.

3. Spin cobordism groups

Using Lemma 2 one sees that the relation of "spin cobordism" between closed n-dimensional spin manifolds is an equivalence relation. Let Ω_n^{spin} denote the resulting cobordism group. The cartesian product operation between spin manifolds (compare Lemma 3) gives rise to a product operation

$$\Omega_m^{\text{spin}} \otimes \Omega_n^{\text{spin}} \to \Omega_{m+n}^{\text{spin}}.$$

Thus one obtains a graded ring.

Our ignorance concerning these groups is quite extensive. For example, to the best of my knowledge, it is not known whether any Ω_n^{spin} contains an element of order 4.

In studying spin cobordism it is natural to look at the homomorphism $\Omega_n^{\text{spin}} \to \Omega_n$ into the ordinary (oriented) cobordism group. Here is a description of this homomorphism for three particular cases.

For $n = 8$ the group $\Omega_8^{\text{spin}} \cong Z + Z$ maps monomorphically into Ω_8, which is also isomorphic to $Z + Z$. The quotient group $\Omega_8/\Omega_8^{\text{spin}}$ is cyclic of order 2^7. (Compare [11].)

Now let $n = 9$. According to Adachi [1] an oriented 9-manifold bounds if and only if the Stiefel–Whitney numbers $w_2^3 w_3[M]$ and $w_2 w_7[M]$ are zero. (Compare Wall [12].) But these clearly vanish for a spin manifold. Therefore the homomorphism

$$\Omega_9^{\text{spin}} \to \Omega_9 \cong Z_2 + Z_2$$

is zero. (Similarly this homomorphism is zero for $n = 5$, 11, and presumably for many higher dimensions.) On the other hand we will see presently that $\Omega_9^{\text{spin}} \neq 0$.

Finally, let $n = 10$. Let $H_{2,4}$ denote a non-singular hypersurface of degree (1, 1) in the product $P_2(C) \times P_4(C)$ of complex projective spaces. Then one can verify that

$$w_2(H_{2,4}) = 0, \qquad w_4 w_6[H_{2,4}] \neq 0.$$

Thus $H_{2,4}$ is a spin manifold, but is not an oriented boundary. It follows that the homomorphism

$$\Omega_{10}^{\text{spin}} \to \Omega_{10} \cong Z_2$$

is non-zero.

An extremely interesting tool for studying the spin cobordism ring was suggested to the author by Atiyah. Namely there is a homomorphism

$$\alpha\colon \Omega_n^{\text{spin}} \to \tilde{K}_O S^n$$

which is defined as follows. (Here $\tilde{K}_O S^n$ denotes the group of s-classes of vector bundles over S^n.) Consider a disk-bundle with spin structure, the fibre dimension being $8k$. Let E denote the total space, $\dot{E}$ the total space of the associated sphere bundle, and $E/\dot{E}$ the Thom space. The two half-spin representations of Spin $(8k)$ give rise to two vector bundles over E. Since these representations coincide over Spin $(8k-1)$, the two vector bundles coincide over $\dot{E}$. Thus their difference determines an element of $K_O(E, \dot{E}) \cong \tilde{K}_O(E/\dot{E})$.

Applying this construction to the universal Spin $(8k)$-bundle, with k large we obtain a canonical class $u \in \tilde{K}_O M$ (Spin $8k$). Now for any map $f\colon S^{n+8k} \to M(\text{Spin } 8k)$ the class $f^*u \in \tilde{K}_O S^{n+8k} \cong \tilde{K}_O S^n$ is defined. Thus we obtain a homomorphism from

$$\pi_{n+8k} M(\text{Spin } 8k) \cong \Omega_n^{\text{spin}}$$

to $\tilde{K}_O S^n$.

For $n \leq 7$ one can verify that this construction defines an isomorphism between Ω_n^{spin} and $\tilde{K}_O S^n$. (Compare [11].)

If $n \equiv 0 \pmod 4$ then the composition

$$\Omega_n^{\text{spin}} \xrightarrow{\alpha} \tilde{K}_O S^n \subset \tilde{K}_U S^n \cong Z$$

turns out to be precisely the $\hat{A}$-genus. In fact this is essentially the method which Atiyah and Hirzebruch [3] used to prove that the $\hat{A}$-genus of a spin manifold is an integer (or an even integer for $n \equiv 4 \pmod 8$).

Now recall that the tensor product operation between vector bundles gives rise to a bilinear product operation $\tilde{K}_O S^m \otimes \tilde{K}_O S^n \to \tilde{K}_O S^{m+n}$. (See [5].) The following result will be proved in a forthcoming paper by Atiyah, Bott, and Shapiro [2].

THEOREM. *The function* $\alpha\colon \Omega_n^{\text{spin}} \to \tilde{K}_O S^n$ *is a ring homomorphism: that is* $\alpha(M \times N) = \alpha(M)\alpha(N)$.

One interesting consequence of this theorem is the following. Let $n = 9$, 10, 17, or 18.

THEOREM 2. *For these n there exists a smooth manifold Σ^n which is a homotopy n-sphere, but is not a spin boundary.*

Thus the homomorphism $\Omega_n^{\text{spin}} \to \Omega_n$ has a non-trivial kernel for $n = 9$, 10, 17, 18.

PROOF. First consider the case $n \equiv 1 \pmod 8$. Let $k = (n-1)/4$.

According to Kervaire and Milnor [8] there exists an almost parallelizable manifold M^{4k} with

$$\hat{A}[M^{4k}] = B_k a_k j_k / 4k,$$

where B_k is the kth Bernoulli number, $a_k = 1$ for k even, and where $j_2 = 240$, $j_4 = 480$. Thus for $k = 2, 4$ there exists such a manifold with $\hat{A}[M^{4k}] = 1$. We may assume that M^{4k} is $(2k - 1)$-connected, hence has a unique spin structure.

Let (S^1, s) denote the circle with the "bad" spin structure, so that $\alpha(S^1, s) \neq 0$. Since the product

$$\tilde{K}_O S^{4k} \otimes \tilde{K}_O S^1 \to \tilde{K}_O S^{4k+1}$$

is non-trivial for k even, it follows that $\alpha(M^{4k} \times (S^1, s)) \neq 0$. Now apply a surgery to $M^{4k} \times S^1$ so as to kill the infinite cyclic fundamental group. care must be taken that the surgery is compatible with the spin structure. (Compare [10, Theorem 2] or [9, §6].) The resulting manifold will be a $(2k - 1)$-connected spin manifold which is parallelizable except along an imbedded circle. Now apply a sequence of surgeries to kill the free abelian group π_{2k}. The resulting manifold Σ^{4k+1} will be a homotopy sphere which is spin-cobordant to $M^{4k} \times (S^1, s)$; so that $\alpha(\Sigma^{4k+1}) \neq 0$.

To obtain a corresponding example in dimension $4k + 2$ it is only necessary to apply a single surgery to $\Sigma^{4k+1} \times (S^1, s)$. Again the surgery must be compatible with the spin structure. This completes the proof of Theorem 2.

It may be conjectured that similar manifolds Σ^n exist whenever $n \equiv 1$ or 2 (mod 8). The above proof works for $n = 4k + 1$ whenever $B_k j_k / 4k \equiv 1$ (mod 2). An alternative attack would be based on the following.

CONJECTURE. *For $n \equiv 1$ (mod 8) there exists a map $f: S^{8r+n} \to S^{8r}$ so that the induced $f^*: \tilde{K}_O S^{8r} \to \tilde{K}_O S^{8r+n} \cong Z_2$ is non-zero.*

(The corresponding statement for $n \equiv 2$ (mod 8) would be a consequence.)

This is related to Theorem 2 as follows. Consider the composition

$$S^{8r} \subset M(\text{Spin } 8r) \xrightarrow{u'} BO$$

where u' corresponds to the Atiyah–Hirzebruch class $u \in K_O M(\text{Spin } 8r)$ which was described above. Thus any element of $\pi_{8r+n}(S^{8r})$ determines a spin cobordism class. If $n \not\equiv 2$ (mod 4) then it follows from [9] that this cobordism class has a representative which is a homotopy sphere. Now, assuming the Conjecture, it is not difficult to prove the generalized Theorem 2.

Conversely, using Theorem 2, it is easy to verify that the Conjecture is true for $n = 9, 17$.

PRINCETON UNIVERSITY

REFERENCES

[1] M. ADACHI, On the groups of cobordism Ω^k, *Nagoya Math. J.*, *13* (1958); p. 135–156.

[2] M. ATIYAH, R. BOTT and A. SHAPIRO, Clifford modules, *Topology*, *3*, *Suppl. 1* (1964), p. 3–38.

[3] M. ATIYAH and F. HIRZEBRUCH, Riemann–Roch theorems for differentiable manifolds, *Bull. Amer. Math. Soc.*, *65* (1959), p. 276–281.

[4] A. BOREL and F. HIRZEBRUCH, Characteristic classes and homogeneous spaces, *Amer. J. Math.*, *80* (1958), p. 458–538 and *81* (1959), p. 315–382.

[5] R. BOTT, Some remarks on the periodicity theorems, *Bull. Soc. Math. France*, *89* (1959), p. 293–310.

[6] J. EELLS and N. KUIPER, An invariant for certain smooth manifolds, *Annali di Mat.*, *60* (1962), p. 93–110.

[7] F. HIRZEBRUCH, *Neue topologische Methoden in der algebraischen Geometrie*, Springer, Berlin, 1956.

[8] M. KERVAIRE and J. MILNOR, Bernoulli numbers, homotopy groups and a theorem of Rohlin, *Proc. Int. Congress of Math.*, Edinburgh (1960), p. 454–458.

[9] M. KERVAIRE and J. MILNOR, Groups of homotopy spheres I, *Annals of Math.*, *77* (1963), p. 504–537. Part II is in preparation.*

[10] J. MILNOR, A procedure for killing homotopy groups of differentiable manifolds, *A.M.S. Symposia in Pure Math.*, vol. III (1961), p. 39–55.

[11] J. MILNOR, Spin structures on manifolds, *L'Enseignement Math.*, *9* (1963), p. 198–203.

[12] C. T. C. WALL, Determination of the cobordism ring, *Annals of Math.*, *72* (1960), p. 292–311.

* Never published. (See this volume p. 7.)

Topology Vol. 3, pp. 223–230. Pergamon Press, 1965. Printed in Great Britain

ON THE STIEFEL–WHITNEY NUMBERS OF COMPLEX MANIFOLDS AND OF SPIN MANIFOLDS

J. MILNOR

(*Received* 8 *October* 1963)

THE FIRST section of this paper will characterize those cobordism classes in the Thom cobordism rings $\mathfrak{N}_*$ and Ω_* which contain complex manifolds†. The second section attempts to characterize those classes in $\mathfrak{N}_*$ which contain spin manifolds†. The attempt succeeds only through dimension 23.

§1. COMPLEX MANIFOLDS

Conner and Floyd have proved the following remarkable theorem [1]. *Let V_R be a 'real form' of the complex algebraic variety V_C, both being non-singular. Then V_C is non-oriented cobordant to $V_R \times V_R$.*

For example the complex projective space $P_k(C)$ is cobordant to the product $P_k(R) \times P_k(R)$.

An interesting consequence of this result is the following.

THEOREM (1). *A non-oriented cobordism class contains a complex manifold if and only if it contains a square $N \times N$.*

Proof. First consider the following example of the Conner–Floyd theorem. Let $H_{mn}(C)$ denote a non-singular hypersurface of degree (1,1) in the product $P_m(C) \times P_n(C)$. [In terms of homogeneous co-ordinates $(w_0, \ldots, w_m)$ and $(z_0, \ldots, z_n)$ with $m \leq n$ this hypersurface can be defined as the locus $w_0 z_0 + w_1 z_1 + \ldots + w_m z_m = 0$. It can also be thought of as a $P_{n-1}(C)$-bundle over $P_m(C)$.] Then $H_{mn}(C)$ is non-oriented cobordant to the square $H_{mn}(R) \times H_{mn}(R)$ of the corresponding real variety.

These manifolds are of interest since it is known that the weakly complex cobordism ring Ω_*^U is generated (redundantly) by the complex cobordism classes of the manifolds $P_k(C)$ and $H_{mn}(C)$. (See [4, 6].)

This proves that the image of the natural ring homomorphism $j: \Omega_*^U \to \mathfrak{N}_*$ is generated by cobordism classes $(P_k(R))^2$ and $(H_{mn}(R))^2$ which are squares. But the collection of all squares $(N \times N) \in \mathfrak{N}_*$ forms a sub-ring which will be denoted briefly by $(\mathfrak{N}_*)^2$. Thus we have proved that $j(\Omega_*^U) \subset (\mathfrak{N}_*)^2$.

† All manifolds are to be smooth, compact, and without boundary. Connectedness is not required.

Now note that the cobordism ring $\mathfrak{N}_*$ is generated by the cobordism classes $(P_k(R))$ and $(H_{mn}(R))$. In fact:

LEMMA (1). *$\mathfrak{N}_*$ is a polynomial ring over Z_2 with independent generators $(P_{2t}(R))$ and $(H_{2^k, 2t2^k}(R))$ where $t, k \geq 1$.*

Proof. According to Thom [10] we must verify (1) that the manifolds listed include precisely one manifold M^p for each dimension p which is not of the form $2^i - 1$, and (2) that $s_p(w_1, \ldots, w_p)\,[M^p] \neq 0$, where s_p denotes the polynomial which expresses the symmetric function $\sum t_i^p$ in terms of the elementary symmetric functions. A straightforward computation shows that

$$s_{m+n-1}(w_1, \ldots, w_{m+n-1})[H_{mn}(R)] = (m+n)!/m!\,n!$$

(reduced modulo 2) for $m, n \geq 2$. (Compare [6] or the proof of Lemma (4)). The rest of the argument can easily be supplied by the reader.

Now let us return to the proof of Theorem [1]. Since $\mathfrak{N}_*$ is generated by the $(P_k(R))$ and $(H_{mn}(R))$, it follows that $(\mathfrak{N}_*)^2$ is generated by $(P_k(R))^2 = (P_k(C))$ and $(H_{mn}(R))^2 = H_{mn}(C)$. This shows that $j(\Omega_*^U) \supset (\mathfrak{N}_*)^2$; which completes the proof.

Remark. This argument suggests the conjecture that every smooth manifold is diffeomorphic to a real form of some complex algebraic variety. (Compare Nash [8].)

Now consider the Stiefel–Whitney numbers of a complex manifold V. If i is odd, then the Stiefel–Whitney class $w_i \in H^i(V; Z_2)$ is zero. (Compare [9, §41.8].) Hence any Stiefel–Whitney number $w_{i_1} \ldots w_{i_k}[V]$ which involves an odd w_i will be zero. Conversely:

THEOREM (2). *Let M be a manifold such that every Stiefel–Whitney number involving an odd w_i is zero. Then M is cobordant to a complex manifold. In fact if M can be oriented, then M is oriented cobordant to a complex manifold.*

The proof in the non-oriented case will be based on the following.

LEMMA (2). *The Stiefel–Whitney numbers of a product $N \times N$ are given by*

$$w_{2i_1}w_{2i_2} \ldots w_{2i_r}[N \times N] = w_{i_1}w_{i_2} \ldots w_{i_r}[N],$$

while the number $w_{j_1} \ldots w_{j_s}[N \times N]$ is zero if some j_h is odd.

The proof will be left to the reader.

Now let M be a manifold such that every 'odd' Stiefel–Whitney number of M is zero. We will construct a manifold N of half the dimension so that

$$w_{i_1} \ldots w_{i_r}[N] = w_{2i_1} \ldots w_{2i_r}[M]$$

for all $i_1 \ldots i_r$. This will imply that M is cobordant to $N \times N$ and hence, by Theorem [1], to a complex manifold.

Dold [2] has shown that a given collection of Stiefel–Whitney numbers actually corresponds to a manifold N^n if and only if the following *Wu relations* are satisfied. For each $i_1 + \ldots + i_k + p = n$ we must have

$$(Sq^p(w_{i_1} \ldots w_{i_k}))[N^n] = u_p w_{i_1} \ldots w_{i_k}[N^n].$$

[This is to be understood as follows. The expression $Sq^p(w_{i_1} \ldots w_{i_k})$ is to be expressed as a polynomial in $w_1, w_2, \ldots$, using the Wu formula

$$Sq^p w_n = w_p w_n + \binom{p-n}{1} w_{p-1} w_{n+1} + \ldots + \binom{p-n}{p} w_0 w_{n+p};$$

where $\binom{a}{t} = a(a-1)\ldots(a-t+1)/t!$. The Wu classes $u_p = u_p(w_1, w_2, \ldots)$ are defined inductively by the conditions $u_1 = w_1$ and

$$u_n + Sq^1 u_{n-1} + Sq^2 u_{n-2} + \ldots = w_n.]$$

Formulating this more invariantly, let $\mathscr{R}^n \subset H^n(B_O; Z_2)$ be the vector space generated by all elements of the form $Sq^p x + u_p x$. The Stiefel–Whitney numbers of each manifold N determine a homomorphism

$$h_N : H^n(B_O; Z_2) \longrightarrow Z_2,$$

where $h_N(x) = x[N]$. Dold's theorem says that a given homomorphism $H^n(B_O; Z_2) \to Z_2$ corresponds to a manifold if and only if it annihilates $\mathscr{R}^n$.

Define the 'doubling homomorphism' $d : H^*(B_O; Z_2) \to H^*(B_O; Z_2)$ by $d(w_i) = w_{2i}$. Let M, of dimension $2n$, satisfy the hypothesis of Theorem (2). Then we will show that $h_M d : H^n(B_O; Z_2) \to Z_2$ annihilates $\mathscr{R}^n$. This will prove the existence of the required manifold N.

Let $\mathscr{I} = (w_1, w_3, w_5, \ldots) \subset H^*(B_O; Z_2)$ denote the ideal generated by the odd w_i.

LEMMA (3). *The doubling homomorphism d satisfies the congruence*

(1) $Sq^{2i} d(x) \equiv d(Sq^i x) \qquad \bmod \mathscr{I}$.

Furthermore

(2) $d(u_p) \equiv u_{2p} \qquad \bmod \mathscr{I}$.

Proof. In the special case $x = w_n$ we have

$$Sq^{2p} d(w_n) = Sq^{2p} w_{2n} = \sum_j \binom{2p-2n}{j} w_{2p-j} w_{2n+j}.$$

Deleting the odd terms, and using the congruence

$$\binom{2p-2n}{2s} \equiv \binom{p-n}{s} \qquad \text{(mod 2)},$$

this expression takes the required form

$$\sum_s \binom{p-n}{s} w_{2p-2s} w_{2n+2s} = d(Sq^p w_n).$$

Now assume that the congruence (1) has already been estalished for x and y. Then

$$Sq^{2p} d(xy) \equiv \sum_i (Sq^{2i} dx)(Sq^{2p-2i} dy)$$

$$\equiv \sum_i (dSq^i x)(dSq^{p-i} y) = dSq^p(xy) \qquad (\bmod \mathscr{I}).$$

Proceeding inductively, it follows that (1) is true for all elements of $H^*(B_O; Z_2)$.

C

The congruence $d(u_p) \equiv u_{2p}$ is proved by a straightforward induction on p. In order to carry out the induction, it is first necessary to verify that the ideal $\mathscr{I}$ is closed under the action of the squaring operations. Details will be left to the reader.

Now consider a manifold $M = M^{2n}$ which satisfies the condition $h_M(\mathscr{I}^{2n}) = 0$. For any generator $Sq^p x + u_p x$ of $\mathscr{R}^n$ we have

$$h_M d(Sq^p x + u_p x) = h_M(Sq^{2p}(dx) + u_{2p} dx + (\text{terms in } \mathscr{I}^{2n})) = 0.$$

Thus $h_M d$ satisfies the Wu relations, and hence is equal to h_N for some manifold N. This proves that M is cobordant to $N \times N$, and hence is cobordant to some complex manifold V.

Now suppose that M is an oriented manifold. Then the difference $M - V$ is an oriented manifold with all Stiefel–Whitney numbers zero. From the Rohlin exact sequence

$$\Omega_* \xrightarrow{2} \Omega_* \longrightarrow \mathfrak{N}_*$$

(compare [12]) it follows that $M - V$ is oriented cobordant to $2M_1$ for some oriented manifold M_1.

Recall that the ring Ω_* can be described as the direct sum of a polynomial ring $Z[Y^4, Y^8, Y^{12}, \ldots]$ and an ideal consisting of elements of order 2. The Y^{4i} are all complex manifolds. (For example we can take Y^{4i} equal to $P_{2i}(C)$ for $i = 1, 2, 3, 5, 6$; and equal to $9P_8(C) + H_{3,6}(C)$ for $i = 4$. Compare [5], [6].) Thus every oriented manifold is cobordant to a sum $V_1 - V_2 + T$ where V_1, V_2 are complex manifolds, and $2T \sim 0$. Replacing M_1 by such an expression, we see that $M - V \sim 2M_1 \sim 2V_1 - 2V_2 + 0$.

Thus M is cobordant to a difference of algebraic varieties, and hence (compare [4], [6]) is cobordant to an algebraic variety. This completes the proof of Theorem (2).

§2. SPIN MANIFOLDS

Let us start by considering some examples of spin manifolds.

LEMMA (4). *If* $m \equiv n \equiv 0 \pmod 2$ *then the complex hypersurface* $H_{mn}(C)$ *is a spin manifold.*

Proof. Let $a, b \in H^2(P_m(C) \times P_n(C); Z_2)$ be the standard generators. Then the Stiefel–Whitney class $w_2(P_m(C) \times P_n(C))$ is equal to $(m+1)a + (n+1)b$. The class $w_2(v)$ of the normal bundle of $H_{mn}(C)$ is equal to $(a+b) \mid H_{mn}(C)$, since $a + b$ is the cohomology class dual to this submanifold. Subtracting these two we obtain

$$w_2 H_{mn}(C) = (ma + nb) | H_{mn}(C)$$

which completes the proof.

A similar argument shows that the corresponding real variety $H_{mn}(R)$ is orientable, if m and n are even. Remembering that $H_{mn}(C)$ is cobordant to $H_{mn}(R) \times H_{mn}(R)$, we are tempted towards the following:

CONJECTURE. *If M is an orientable manifold then $M \times M$ is non-oriented cobordant to a spin manifold.* [Added in proof. This has been proved by P. G. Anderson: *Bull. Amer. Math. Soc.* **70** (1964), 818–819.]

As an example, consider the complex projective space $P_n(C)$.

LEMMA (5). *The product $P_n(C) \times P_n(C)$ is non-oriented cobordant to the quaternion projective space $P_n(H)$.*

Since $P_n(C)$ is orientable, and $P_n(H)$ is clearly a spin manifold, this tends to support the conjecture.

Proof. Both $P_n(C)$ and $P_n(H)$ have a mod 2 cohomology ring which has one generator a (of dimension 2 or 4 respectively) and one relation $a^{n+1} = 0$. In each case the total Stiefel–Whitney class is given by $w = (1 + a)^{n+1}$. (Compare Hirzebruch [3].) It follows that the Stiefel–Whitney numbers of these manifolds are given by the formula

$$w_{2i_1} \cdots w_{2i_k}[P_n(C)] = w_{4i_1} \cdots w_{4i_k}[P_n(H)] = \binom{n+1}{i_1} \cdots \binom{n+1}{i_k}.$$

Together with Lemma (2), this completes the proof.

The cobordism ring Ω_* has been computed by Wall [12]. In dimensions less than 15 it follows from Wall's work that Ω_* is generated by the classes of the manifolds $P_2(C)$, $H_{2,4}(R)$, $P_4(C)$, $H_{2,8}(R)$, $H_{4,8}(R)$, $P_6(C)$, and $H_{2,12}(R)$ (of dimensions 4, 5, 8, 9, 11, 12, 13 respectively). But for each of these manifolds we have verified that the square is cobordant to a spin manifold. Thus:

THEOREM (3). *If M is orientable of dimension less than 15 then $M \times M$ is non-oriented cobordant to a spin manifold.*

I do not know what happens in dimensions ≥ 15.

Conversely we may ask whether every spin manifold is non-oriented cobordant to the square of an orientable manifold. It will turn out that this is true for spin manifolds of dimension ≤ 23. Again I do not know what happens in higher dimensions.

Let us look at Stiefel–Whitney numbers. Consider manifolds M which satisfy the following:

HYPOTHESIS (1). *Every Stiefel–Whitney number $w_{i_1} \cdots w_{i_k}[M]$ which involves either w_1 or w_2 is zero.*

Clearly every spin manifold satisfies this hypothesis. Conversely we must ask:

PROBLEM. Does every non-oriented cobordism class which satisfies the hypothesis (1) contain a spin manifold?

We will verify that this is true in dimensions ≤ 23.

LEMMA (6). *If M^n satisfies (1) with $n \leq 23$, then every $w_{i_1} \cdots w_{i_k}[M^n]$ involving an odd w_i is zero.*

The proof, which will be outlined presently, involves a tedious case by case application of the Wu relations. This lemma is definitely false for $n = 24$. (In slightly higher dimensions, the lemma is probably true for $24 < n < 29$; but false for $n = 29$.)

Assuming Lemma (6), it follows from Theorem (2) that M is non-oriented cobordant to a product $N \times N$. The identity

$$w_{i_1} \dots w_{i_k}[N] = w_{2i_1} \dots w_{2i_k}[M]$$

now implies that every Stiefel–Whitney number of N which involves w_1 is zero. But according to Wall [12, §9] this means that N is cobordant to an orientable manifold. Finally, using Theorem (3), it follows that M is cobordant to a spin manifold. Thus we have proved:

THEOREM (4). *For a non-oriented cobordism class* (M) *of dimension* ≤ 23 *the following three conditions are equivalent*

(1) *each* $w_{i_1} \dots w_{i_k}[M]$ *involving* w_1 or w_2 *is zero*;

(2) (M) *contains a spin manifold;*

(3) (M) *contains the square* $N \times N$ *of an orientable manifold.*

It follows that the natural homomorphism $h: \Omega_n^{\text{Spin}} \to \mathfrak{N}_n$ is zero for odd values of n up to $n = 23$. The rank (over Z_2) of h can be tabulated as follows for even values of n. (Compare [7].)

n	2	4	6	8	10	12	14	16	18	20	22	24
rank $h(\Omega_n^{\text{Spin}})$	0	0	0	1	1	0	0	2	2	1	1	(3 or 4)

The ambiguity in dimension 24 can be described as follows. There exists an orientable manifold X of dimension 24 such that every Stiefel–Whitney number involving w_2 is zero, but such that

$$w_4 w_6 w_7 w_7[X] \neq 0.$$

(This is proved by an exhaustive examination of the Wu relations: to be more precise X can be chosen so that $w_4 w_6 w_7^2 = w_6^4 = w_4^6 = w_4^3 w_6^2 = (w_4 w_8)^2 \neq 0$, but so that all other Stiefel–Whitney numbers are zero.) It is not known whether or not this X is cobordant to a spin manifold.

This description can be transformed into one involving Pontrjagin numbers as follows. Using the Wu relation $(Sq^2 + u_2)(w_4 w_6^3)[X] = 0$ one finds that

$$w_6^4[X] = w_4 w_6 w_7^2[X] \neq 0.$$

But w_6^2 is the mod 2 reduction of the Pontrjagin class p_3. Therefore $p_3^2[X] \equiv 1 \pmod 2$.

The description can be further transformed by considering the polynomial

$$s_6 = s_6(p_1, \dots, p_6) = p_1^6 - 6p_1^4 p_2 \pm \dots - 6p_6,$$

which expresses the symmetric function $\sum t_i^6$ in terms of elementary symmetric functions. Since

$$s_6 \equiv s_3^2 \equiv p_1^6 + p_1^2 p_2^2 + p_3^2 \qquad \pmod 2,$$

and since $p_1^6[X] \equiv w_2^{12}[X] = 0$, and similarly $p_1^2 p_2^2[X] \equiv 0$, we see that

$$s_6(p_1, \dots, p_6)[X] \equiv p_3^2[X] \equiv 1 \qquad \pmod 2.$$

Thus we are left with the following:

PROBLEM. Does there exist a spin manifold Σ of dimension 24 so that $s_6(p_1, \ldots, p_6)[\Sigma] \equiv 1 \pmod 2$?

The rest of this paper will be concerned with the proof of Lemma (6). We first give two preliminary statements which are true in arbitrary dimensions.

LEMMA (7). *If M satisfies Hypothesis* (1) *then every* $w_{i_1} \ldots w_{i_k}[M]$ *involving* w_3, w_5 or w_9 *is zero also.*

(These particular w_i presumably occur because w_3, w_5, and w_9 map into zero in $H^*(B^{\mathrm{Spin}}; Z_2)$. Compare Thomas [11].)

Proof. Let $(w_a, w_b, \ldots) \subset H^*(B_{\mathrm{O}}; Z_2)$ denote the ideal generated by $w_a, w_b, \ldots$. Let x stand for an arbitrary element of $H^*(B_{\mathrm{O}}; Z_2)$.

The formulae

$$\begin{aligned} Sq^1(w_2 x) &\equiv w_3 x && \mathrm{mod}(w_1, w_2) \\ Sq^2(w_3 x) &\equiv w_5 x && \mathrm{mod}(w_1, w_2, w_3) \\ Sq^4(w_5 x) &\equiv w_9 x && \mathrm{mod}(w_1, w_2, w_3, w_5) \end{aligned}$$

are easily verified. Now if every $w_1 x'[M]$ and every $w_2 x''[M]$ is zero then the Wu relation

$$((Sq^1 + u_1) w_2 x)[M] = 0$$

implies that $w_3 x[M] = 0$. Hence the relation

$$((Sq^2 + u_2) w_3 x)[M] = 0$$

implies that $w_5 x[M] = 0$; and similarly with w_9.

LEMMA (8). *Let M satisfy* (1) *and suppose that the integers* $i_1 \ldots i_k$ *are all either odd, or equal to* 4 *or* 8, *or occur in pairs; and that at least one of the* i_j *is odd. Then* $w_{i_1} \ldots w_{i_k}[M] = 0$.

(For example $w_6^2 w_7 w_8 [M^{27}] = 0$.)

Proof. Suppose that i_1 is odd. Note that

$$Sq^1(w_{i_1 - 1} w_{i_2} \ldots w_{i_k}) \equiv w_{i_1} w_{i_2} \ldots w_{i_k} \,\mathrm{mod}(w_1, w_5, w_9).$$

Thus the Wu relation $(Sq^1(w_{i_1 - 1} w_{i_2} \ldots w_{i_k}))[M] = 0$ completes the proof.

We are now ready to prove Lemma (6). To avoid too much tedium, we will only consider the most difficult dimension, which happens to be 21. Consider then all partitions $i_1 + \ldots + i_k = 21$ which are not excluded by Lemmas (7, 8). There turn out to be seven such partitions, namely:

10, 11; 4, 6, 11; 4, 7, 10; 6, 15; 7, 14; 6, 7, 8; and 4, 4, 6, 7.

To take care of the first, consider the Wu relation

$$(Sq^2(w_8 w_{11}) + u_2 w_8 w_{11})[M] = 0,$$

where

$$Sq^2(w_8 w_{11}) \equiv w_{10} w_{11} + w_8 w_{13}, \quad \text{and} \quad u_2 \equiv 0 \,\mathrm{mod}(w_1, w_2).$$

This proves that $w_{10} w_{11}[M]$ is equal to $w_8 w_{13}[M]$ which is zero by Lemma (8). Next consider the relation

$$(Sq^4(w_6 w_{11}) + u_4 w_6 w_{11})[M] = 0,$$

where

$$Sq^4(w_6 w_{11}) \equiv w_{10} w_{11}, \quad u_4 \equiv w_4 \,\mathrm{mod}(w_1, w_2, w_3)$$

This implies that $w_4w_6w_{11}[M]$ is equal to $w_{10}w_{11}[M]$, which we have just shown is zero. The remaining five partitions are handled similarly, using the Wu relations corresponding to $Sq^1(w_4w_6w_{10})$, $Sq^2(w_4w_{15})$, $Sq^2(w_7w_{12})$, $Sq^2(w_4w_7w_8)$ and $Sq^2(w_4{}^3w_7)$ respectively.

The reader who has enough patience should have no difficulty in carrying out the proof for other dimensions ≤ 23.

REFERENCES

1. P. CONNER and E. FLOYD: *Differentiable Periodic Maps, Ergebn. Math.* **33**, Springer, 1964. (See p. 63.)
2. A. DOLD: Vollständigkeit der Wu-schen Relationen zwischen den Stiefel–Whitneyschen Zahlen differenzierbarer Mannigfaltigkeiten, *Math. Z.* **65** (1956), 200–206.
3. F. HIRZEBRUCH: Uber die quaternionalen projektiven Räume, *S. B. bayer. Akad. Wiss.* (1953), 301–312.
4. F. HIRZEBRUCH: Komplexe Mannigfaltigkeiten, *Proceedings* 1958 *International Congress of Mathematicians*, Cambridge (1960), 119–136.
5. J. MILNOR: A survey of cobordism theory, *Enseign. Math.* **8** (1962), 16–23.
6. J. MILNOR: On the cobordism ring Ω_* and a complex analogue II (in preparation). *
7. J. MILNOR: Remarks concerning spin manifolds (to be published). +
8. J. NASH: Real algebraic manifolds, *Ann. Math., Princeton* **56** (1952), 405–421.
9. N. STEENROD: *The Topology of Fibre Bundles*. Princeton University Press, 1951.
10. R. THOM: Quelques propriétés globales des variétés différentiables, *Comment. Math. Helvet.* **28** (1954), 17–86.
11. E. THOMAS: On the cohomology groups of the classifying space for the stable spinor group, *Bol. Soc Mat. Mex.* (1962), 57–69.
12. C. T. C. WALL, Determination of the cobordism ring, *Ann. Math., Princeton* 72 (1960), 292–311.
13. WU WEN-TSÜN: Les *i*-carrés dans une variété grassmannienne, *C.R. Acad. Sci., Paris* **230** (1950), 918–920.

Princeton, New Jersey, U.S.A.
Princeton, University,

* Never published. (See this volume p. 249.)

\+ This volume pp. 299-306.

A Concluding Amusement: Symmetry Breaking.

This figure below (a doodle generated when my attention wandered during a lecture in the early 60s) shows an immersion of the circle in the plane which extends in two essentially different ways to an immersion of the disk in the plane. (Compare Figure 3. Both extensions can be chosen to be holomorphic by a form of the Riemann mapping theorem.)

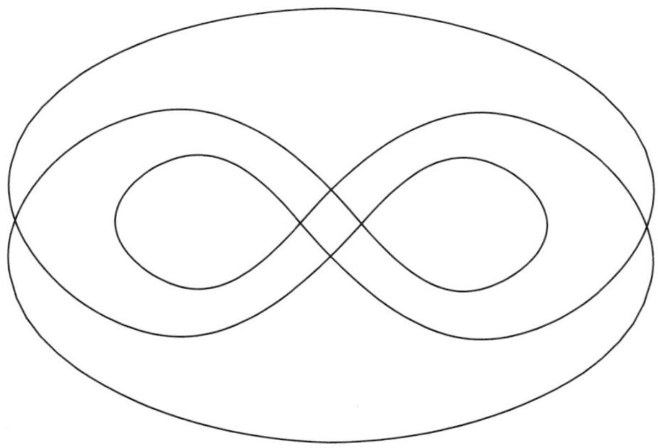

FIGURE 1.

According to SMALE [1959a] and HIRSCH [1959], any smooth immersion $M^n \to N^k$ with $n < k$ is determined, up to regular homotopy, by the homotopy class of the induced vector bundle map $TM^n \to TN^k$. As one particularly famous example of this theory, any two bundle maps $TS^2 \to T\mathbb{R}^3$ are bundle-homotopic, since obstructions to a bundle-homotopy lie in groups $H^k(S^2; \pi_k(\mathrm{SO}))$ which are all zero. It follows that any two immersions of the 2-sphere in 3-space must be regularly homotopic. In other words, we can pass from one to the other through a smooth family of immersions.[1]

[1]Compare the video "Outside In" (Geometry Center, University of Minnesota, 1994; available through AK Peters, Welesley MA), which turns a 2-sphere inside-out through a regular homotopy. For further developments arising from the Smale-Hirsch methods, see PHILLIPS [1967, 1974], as well as GROMOV [1971, 1986].

This theory for immersions $M^n \to N^k$ breaks down completely in the case $n = k$, when the manifold M^n is compact without boundary. For example, there is certainly no immersion $S^1 \to \mathbb{R}^1$ or $S^3 \to \mathbb{R}^3$, although the corresponding bundle maps exist. We are concerned here with immersions of the disk $D^2 \supset S^1$ into the plane. It is not hard to show that a bundle map $TS^1 \to T\mathbb{R}^2$ extends over TD^2 if and only if its winding number is ± 1, and that this extension is unique, up to bundle-homotopy, when it does exist. However, Figure 1 shows an example where this extended bundle map can be realized by two distinct immersions, and Figure 2 shows an immersion of S^1 with winding number ± 1 where no extended immersion is possible.

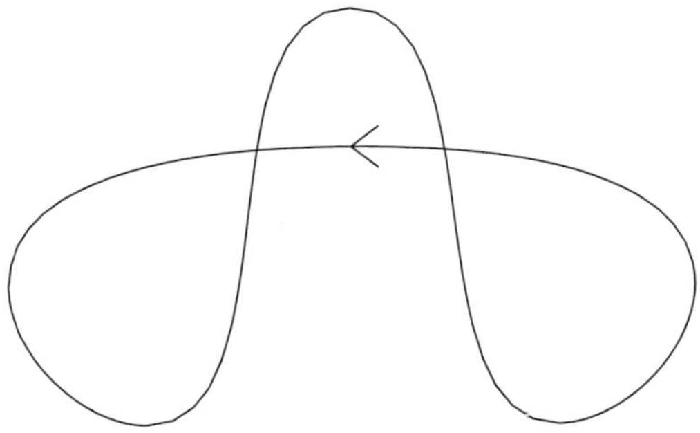

FIGURE 2.

The following figure attempts to illustrate the two ways in which the immersion of Figure 1 can be extended. (See also POENARU [1968, p. 9], GROMOV [1986, p. 28] or ELIASHBERG [1970].)

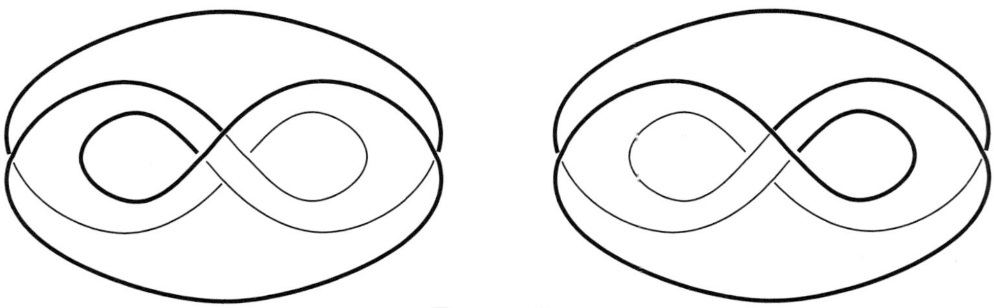

FIGURE 3.

Here is a brief list of references, in chronological order, for the theory of immersions and submersions. (Complete citations may be found in the bibliography which follows.)

1937: WHITNEY, On regular closed curves in the plane.

1956: MILNOR, On the immersion of n-manifolds in $n+1$-space.

1959: SMALE, Classification of immersions of spheres in Euclidean space.

1959: HIRSCH, Immersions of manifolds.

1961: TITUS, The combinatorial topology of analytic functions on the boundary of the disk.

1967: BLANK, Extending immersions of the circle.

1967: PHILLIPS, Submersions of open manifolds.

1968: POENARU, Extension des immersions en codimension 1 (d'apres Samuel Blank).

1971: GROMOV, A topological technique for the construction of solutions of differential equations and inequalities.

1974: PHILLIPS, Smooth maps of constant rank.

1978: WOLIZER, Immersions of surfaces in the plane.

1986: GROMOV, "Partial Differential Relations."

Bibliography

Adachi, M.: [1958], (cited on pp., 303, 306). *On the groups of cobordism* Ω^k, Nagoya Math. J. **13**, 135–156.

Adams, J. F.: [1958], (cited on pp., 224). *On the nonexistence of elements of Hopf invariant one,* Bull. Amer. Math. Soc. **64**, 279–282.

———— : [1958a], (cited on pp., 257, 263, 273). *On the structure and applications of the Steenrod algebra,* Comment. Math. Helv. **32**, 180–214.

———— : [1959], (cited on pp., 257, 263, 264, 273). *Sur la théorie de l'homotopie stable,* Bull. Soc. Math. France **87**, 277–280.

———— : [1965], (cited on pp., 93, 94, 111, 115, 121). *On the groups* $J(X)$, "Differential and Combinatorial Topology. A Symposium in Honor of Marston Morse," Princeton University Press, Princeton, NJ. 121–143.

———— : [1966], (cited on pp., 227, 253). *On the groups* $J(X)$*– IV,* Topology **5**, 21–71.

———— and Walker, G.: [1965], (cited on pp., 93, 94, 121). *On complex Stiefel manifolds,* Proc. Cambridge Philos. Soc. **61**, 81–103.

Adem, J.: [1957], (cited on pp., 258, 273). *The relations on Steenrod powers of cohomology classes,* "Algebraic Geometry and Topology. A Symposium in Honor of S. Lefschetz," 191–238. Princeton University Press, Princeton, NJ.

Ahlfors, L. V. and Bers, L.: [1960], (cited on pp., 144). *Riemann's mapping theorem for variable metrics,* Annals of Math. **72**, 385–404.

Akbulut, S. and McCarthy, J.: [1990], (cited on pp., 10, 20, 84). "Casson's Invariant for Oriented Homology 3-Spheres. An Exposition," Princeton University Press, Princeton, NJ.

Alexander, J. W.: [1930], (cited on pp., 19, 28). *The combinatorial theory of complexes*, Annals of Math. **31**, 292–320.

Alexandroff, P. and Hopf, H.: [1935], (cited on pp., 28). "Topologie," Springer-Verlag, Berlin-New York.

Anderson, D. W., Brown, E. H. and Peterson, F. P.: [1967], (cited on pp., 254). *The structure of the spin cobordism ring,* Annals of Math. **86**, 271–298.

Anderson, P. G.: [1964] (cited on pp., 310). *Cobordism classes of squares of orientable manifolds,* Bull. Amer. Math. Soc. **70**, 818–819.

Antonelli, P. L., Burghelea, D. and Kahn, P. J.: [1972], (cited on pp., 144). *The non-finite homotopy type of some diffeomorphism groups,* Topology **11**, 1–49. (See also Bull. Amer. Math. Soc. **76** [1970], 1246–1250 and *Gromoll groups,* Diff(S^n) *and bilinear constructions of exotic spheres,* Bull. Amer. Math. Soc.

76 [1970], 772–777.)

Arf, C.: [1941], (cited on pp., 120, 121). *Untersuchungen über quadratische Formen in Körpern der Charakteristik* 2, J. Reine Angew. Math. **183**, 148–167.

Atiyah, M.: [1961], (cited on pp., 286, 289). *Bordism and cobordism,* Proc. Cambridge Philos. Soc. **57**, 200–208.

——— Bott, R. and Shapiro, R.: [1964], (cited on pp., 304, 306). *Clifford modules,* Topology, **3**, Suppl. 1, 3–38.

——— and Hirzebruch, F.: [1959], (cited on pp., 297, 298, 304, 306). *Riemann-Roch theorems for differentiable manifolds,* Bull. Amer. Math. Soc. **65**, 276–281.

——— and ——— : [1961], (cited on pp., 224). *Bott periodicity and the parallelizability of the spheres,* Proc. Cambridge Philos. Soc. **57**, 223–226.

Averbuh, B. G.: [1959], (cited on pp., 216, 249, 257, 285, 289). *Algebraic structure of intrinsic homology groups*, Doklady Akad. Nauk. SSSR **125**, 11–14.

Bing, R. H.: [1983], (cited on pp., 10). "The Geometric Topology of 3-manifolds," Amer. Math. Soc. Colloquium Publications, **40** Providence, RI.

Blank, S. J.: [1967], (cited on pp., 317). "Extending Immersions of the Circle," Thesis, Brandeis Univ.

Bochner, S.: [1937], (cited on pp., 22, 28). *Analytic mapping of compact Riemann spaces into Euclidean space*, Duke Math. J. **3**, 339–354.

Borel, A. and Hirzebruch, F.: [1958], (cited on pp., 241, 246, 247, 299, 306). *Characteristic classes and homogeneous spaces,* Amer. J. Math. **80**, 458–538.

——— and ——— : [1959], (cited on pp., 246, 247, 294, 298, 299, 306). *Characteristic classes and homogeneous spaces II,* Amer. J. Math. **81**, 315–382.

——— and ——— : [1960], (cited on pp., 226, 230, 231, 233, 237, 297, 298). *Characteristic classes and homogeneous spaces III,* Amer. J. Math. **82**, 491–504.

——— and Serre, J. P.: [1953], (cited on pp., 43, 45, 267, 269, 271, 273). *Groupes de Lie et puissances réduites de Steenrod,* Amer. J. of Math. **75**, 409–448.

Bott, R.: [1957], (cited on pp., 235, 237, 239, 244, 245, 247). *On the stable homotopy of the classical groups,* Proc. Nat. Acad. Sci. U.S.A. **43**, 933–935.

——— : [1958], (cited on pp., 43, 45, 224, 233, 237). *The space of loops on a Lie group,* Michigan Math. J. **5**, 35–61.

——— : [1959], (cited on pp., 81, 87, 93, 121, 137, 140, 224, 230, 231). *The stable homotopy of the classical groups,* Annals of Math. **70**, 313–337. (See also, Proc. Nat. Acad. Sci. U.S.A. **43** [1957], 933–935.)

——— : [1959a], (cited on pp., 304, 306). *Some remarks on the periodicity theorems,* Bull. Soc. Math. France **89**, 293–310.

——— and Milnor, J.: [1958], (cited on pp., 43, 45, 220, 233, 238, 247). *On the parallelizability of the spheres,* Bull. Amer. Math. Soc. **64**, 87–89.

Boyer, C. P., Galicki, K., Kollár, J. and Thomas, E.: [2005], (cited on pp., 8). *Einstein metrics on exotic spheres in dimensions* 7, 11, *and* 15, Experiment. Math. **14**, 59–64.

Brieskorn, E.: [1966], (cited on pp., 8). *Beispiele zur Differentialtopologie von Singularitäten,* Invent. Math. **2**, 1–14.

Bröcker, T. and Jänich, K.: [1982], (cited on pp., 10). "Introduction to Differential Topology," Cambridge Univ. Press, Cambridge-New York.

Browder, W.: [1972], (cited on pp., 6). "Surgery on Simply-Connected Manifolds," Ergebnisse der Mathematik und ihrer Grenzgebiete, Band 65. Springer-Verlag, New York-Heidelberg.

Brown, M.: [1961], (cited on pp., 126, 140). *The monotone union of open n-cells is an open n-cell.* Proc. Amer. Math. Soc. **12**, 812–814.

———— : [1962], (cited on pp., 24, 288, 289). *Locally flat imbeddings of topological manifolds*, Annals of Math. **75**, 331–341.

Cairns, S. S.: [1940], (cited on pp., 22, 28). *Triangulated manifolds which are not Brouwer manifolds,* Annals of Math. **41**, 809–824.

———— : [1944] (cited on pp., 22, 28). *Introduction of a Riemannian geometry on a triangulable 4-manifold*, Annals of Math. **45**, 218–219.

Cappell, S., Ranicki, A. and Rosenberg, J. (editors): [2000], (cited on pp., 6). "Surveys on Surgery Theory, Vol. 1," Annals of Math. Studies **145**, Princeton U. Press, Princeton, NJ.

———— : [2001], (cited on pp., 6) "Surveys on Surgery Theory, Vol. 2," Annals of Math. Studies **149**, Princeton U. Press, Princeton, NJ.

Cartan, H.: [1954], (cited on pp., 265, 273). *Sur les groupes d'Eilenberg-Mac Lane. II,* Proc. Nat. Acad. Sci. U. S. A. **40**, 704–707.

———— : [1955], (cited on pp., 258, 273). *Sur l'itération des opérations de Steenrod,* Comment. Math. Helv. **29**, 40–58.

Casson, A. J. and Gordon, C. McA.: [1986], (cited on pp., 254). *Cobordism of classical knots. With an appendix by P. M. Gilmer.* Progr. Math., **62**, "À la Recherche de la Topologie Perdue," (editors Guillou and Marin), Birkhäuser Boston, MA, 181–199.

Cerf, J.: [1961], (cited on pp., 90, 121, 126, 140). *Topologie de certains espaces de plongements,* Bull. Soc. Math. France **89**, 227–380.

———— : [1962/63], (cited on pp., 9). *La nullité de* $\pi_0(\mathrm{Diff}\, S^3)$, "Théorèmes de Fibration des Espaces de Plongements. Applications. Séminaire Henri Cartan," **15**, p. 13.

———— : [1968], (cited on pp., 9, 209). "Sur les Difféomorphismes de la Sphère de Dimension Trois ($\Gamma_4 = 0$)," Lect. Notes Math. **53**, Springer-Verlag, New York-Heidelberg.

Chern, S. S.: [1946], (cited on pp., 237). *Characteristic classes of Hermitian manifolds,* Annals of Math. **47**, 85-121.

Chevalley, C.: (cited on pp., 190). "Theory of Lie Groups I," Fifteenth printing. Princeton Mathematical Series, 8. Princeton Landmarks in Mathematics. Princeton University Press, Princeton, NJ.

Chillingworth, D. R. J.: [1976], (cited on pp., 10). "Differential Topology with a View to Applications," Pitman Publishing, London-San Francisco, Calif.-Melbourne.

Conner, P. E. and Floyd, E. E.: [1961], (cited on pp., 288, 290, 307, 308, 314). *Differentiable periodic maps,* Technical Note 12, Dept. of Math., Univ. of Virginia. (See also, *Differentiable periodic maps,* Bull. Amer. Math. Soc. **68** [1962],

76–86, also Ergebnisse der Mathematik und ihrer Grenzgebiete, N. F. Band 33 [1964] Academic Press Inc.)

DeMichelis, S. and Freedman, M. H.: [1992], (cited on pp., 9). *Uncountably many exotic R^4's in standard 4-space,* J. Differential Geom. **35**, 219–254.

Dieudonné, J.: [1989], (cited on pp., 10). "A History of Algebraic and Differential Topology, 1900-1960," Birkhäuser Boston, Inc., Boston, MA.

Dold, A.: [1956], (cited on pp., 284, 289). *Erzeugende der Thomschen Algebra $\mathfrak{N}$,* Math. Z. **65**, 25–35.

———— : [1956], (cited on pp., 308, 314). *Vollständigkeit der Wu-schen Relationen zwischen den Stiefel-Whitneyschen Zahlen differenzierbarer Mannigfaltigkeiten,* Math. Z. **65**, 200–206.

Donaldson, S. K.: [1983], (cited on pp., 9). *An application of gauge theory to four-dimensional topology,* J. Differential Geom. **18**, 279–315.

———— [1987], (cited on pp., 9). *The geometry of 4-manifolds,* Proc. Int. Cong. Math. Berkeley [1986], Amer. Math. Soc., Providence, RI, 43–54.

———— and Kronheimer, P. B.: [1990], (cited on pp., 10). "The Geometry of Four-Manifolds," Oxford Mathematical Monographs. Oxford Science Publications. The Clarendon Press, Oxford University Press, New York

Eells, J. and Kuiper, N.: [1962], (cited on pp., 253, 301, 306). *An invariant for certain smooth manifolds,* Annali di Math. **60**, 93–110.

Ehresmann, C.: [1934], (cited on pp., 268, 273). *Sur la topologie de certains espaces homogènes,* Annals of Math. **35**, 396–443.

Eliashberg, Y.: [1970], (cited on pp., 316). *M. Singularities of folding type,* Izv. Akad. Nauk. SSSR Ser. Mat. **34**, 1110–1126.

———— : [1992], (cited on pp., 9). *Contact 3-manifolds 20 years since Martinet's work,* Ann. Inst. Fourier **42**, 165–192.

Fox, R. H.: [1949], (cited on pp., 179). *A remarkable simple closed curve,* Annals of Math. **50**, 264–265.

———— and Milnor, J.: [1966], (cited on pp., 254, 289). *Singularities of 2-spheres in 4-space and cobordism of knots,* Osaka J. Math. **3**, 257–267. (Also in: "Collected Papers II: The Fundamental Group," Publish or Perish [1995], 59–69. See also the research announcement, Bull. Amer. Math. Soc. **63** [1957], p. 406.)

Freedman, M. H.: [1982], (cited on pp., 9, 209). *The topology of four-dimensional manifolds,* J. Differential Geom. **17**, 357–453.

———— and Quinn, F.: [1990], (cited on pp., 10). "Topology of Four-Manifolds," Princeton Mathematical Series, **39**. Princeton University Press, Princeton, NJ.

Friedman, R. and Morgan, J.: [1994], (cited on pp., 10). "Smooth Four-Manifolds and Complex Surfaces," Ergebnisse der Mathematik und ihrer Grenzgebiete (3) [Results in Mathematics and Related Areas (3)], **27**. Springer-Verlag, Berlin.

———— and ———— : [1998], (cited on pp., 10). "Gauge Theory and the Topology of Four-Manifolds," IAS/Park City Math. Ser. **4**, Amer. Math. Soc., Providence, RI.

Gauld, D.: [1982], (cited on pp., 10). "Differential Topology, an Introduction," Monographs and Textbooks in Pure and Applied Mathematics, **72**. Marcel Dekker, Inc., New York.

Gompf, R. E.: [1993], (cited on pp., 9). *An exotic menagerie,* J. Differential Geom. **37**, 199–223.

——— and Stipsicz, A.: [1999], (cited on pp., 10, 22). "Four-Manifolds and Kirby Calculus," GSM **20**, Amer. Math. Soc., Providence, RI.

Grauert, H.: [1958], (cited on pp., 22). *On Levi's problem and the imbedding of real-analytic manifolds.* Annals of Math. **68**, 460–472.

Griffiths, P. and Harris, J.: [1978], (cited on pp., 204). "Principles of Algebraic Geometry," John Wiley and Sons, New York.

Gromoll, D.: [1966], (cited on pp., 8). *Differenzierbare Strukturen und Metriken positiver Krümmung auf Sphären,* Math. Annalen **164**, 353–371.

——— and Meyer, W.: [1974], (cited on pp., 8). *An exotic sphere with nonnegative sectional curvature,* Annals of Math. **100**, 401–408.

Gromov, M.: [1971], (cited on pp., 315, 317). *A topological technique for the construction of solutions of differential equations and inequalities,* Actes Congr. Int. Math. (Nice 1970), Gauthier-Villars. **2**, 221–223.

——— : [1986], (cited on pp., 315, 316). "Partial Differential Relations," Ergebnisse der Mathematik und ihrer Grenzgebiete (3) [Results in Mathematics and Related Areas (3)], **9**. Springer-Verlag, Berlin.

Grove, K. and Ziller, W.: [2000], (cited on pp., 8). *Curvature and symmetry of Milnor spheres,* Annals of Math. **152**, 331–367.

Guillemin, V. and Pollack, A.: [1974], (cited on pp., 10). "Differential Topology," Prentice-Hall, Inc., Englewood Cliffs, NJ.

Guillou, L. and Marin, A. (editors): [1986], (cited on pp., 9, 10, 227). "À la Recherche de la Topologie Perdue," Progress in Mathematics **62**, Birkhäuser, Boston, MA.

Haefliger, A.: [1961], (cited on pp., 112, 121). *Plongements différentiables de variétés dans variétés,* Comment. Math. Helv. **36**, 47–82.

Hatcher, A.: [1983], (cited on pp., 144). *A proof of the Smale conjecture,* Annals of Math. **117**, 553–607.

Hirsch, M.: [1959], (cited on pp., 315, 317). *Immersions of manifolds,* Trans. Amer. Math. Soc. **93**, 242–276.

——— : [1963], (cited on pp., 8, 204). *Obstruction theories for smoothing manifolds and mappings,* Bull. Amer. Math. Soc. **69**, 352–356.

——— : [1976], (cited on pp., 10). "Differential Topology," Graduate Texts in Mathematics, **33**. Springer-Verlag, New York-Heidelberg.

Hirzebruch, F.: [1953], (cited on pp., 15, 16, 311, 314). *Über die quaternionalen projektiven Räume,* S. B. Math. Nat. Kl. Bayer. Akad. Wiss. 301–312.

——— : [1954], (cited on pp., 226, 235, 237, 280, 281). *Some problems on differentiable and complex manifolds,* Annals of Math. **60**, 213–236.

——— : [1956], (cited on pp., 4, 11, 12, 15, 16, 37, 41, 45, 81, 87, 93, 122, 131, 140, 236, 237, 241, 246, 247, 272, 273, 285, 289, 301, 306). "Neue topologische Methoden in der algebraischen Geometrie," Ergebnisse der Mathematik und ihrer Grenzgebiete (N.F.), Heft 9. Springer-Verlag, Berlin-Gttingen-Heidelberg.

——— : [1959], (cited on pp., 226). "A Riemann-Roch Theorem for Differentiable Manifolds," Séminaire Bourbaki 5 Exp. **177**, 129–149, Soc. Math.

France, Paris.

——— : [1960], (cited on pp., 286, 287, 289, 307, 310, 314). *Komplexe Mannigfaltigkeiten,* Proc. Internat. Congress Math. [1958], 119–136 Cambridge Univ. Press, New York.

——— : [1966], (cited on pp., 4, 71, 210, 212, 220, 226). "Topological Methods in Algebraic Geometry," Third edition. Die Grundlehren der Mathematischen Wissenschaften, Band 131 Springer-Verlag New York, Inc., New York.

——— : [1966/67], (cited on pp., 8). *Singularities and exotic spheres*, Séminaire Bourbaki 10 Soc. Math. France, Paris [1967] **314**, 13–32.

——— : [1971], (cited on pp., 4). *The signature theorem: reminiscences and recreation*, in "Propects in Mathematics," Annals of Math. Studies **70**, Princeton U. Press, 3–31.

——— : [1987], (cited on pp., 7) "Gesammelte Abhandlungen (= Collected Papers)," I, 1951-1962; II, 1963-1987. Springer-Verlag, Berlin.

——— and Mayer, K. H.: [1968], (cited on pp., 8). "O(n)-Mannigfaltigkeiten, exotische Sphären und Singularitäten," Lect. Notes Math. **57**, Springer-Verlag, Berlin-New York.

Hopf, H.: [1935], (cited on pp., 2). *Über die Abbildung von Sphären auf Sphären niedriger Dimension,* Fundam. Math. **25**, 427–440.

Husemoller, D.: [1966], (cited on pp., 165). "Fibre Bundles," McGraw-Hill Book Co., New York-London-Sydney.

Jones, B.: [1950], (cited on pp., 60, 63). "The Arithmetic Theory of Quadratic Forms," Carcus Monograph Series, **10**. The Mathematical Association of America, Buffalo, NY.

Kahn, P. J.: [1965], (cited on pp., 144). *Characteristic numbers and oriented homotopy type,* Topology **3**, 81–95.

Kervaire, M.: [1956], (cited on pp., 82, 87, 243, 244, 247). *Courbure intégrale généralisée et homotopie,* Math. Annalen **131**, 219–252.

——— : [1957], (cited on pp., 70, 87, 102, 110, 122). *Relative characteristic classes,* Amer. J. Math. **79**, 517–558.

——— : [1958], (cited on pp., 81, 87, 243, 247). *On the Pontrjagin classes of certain* SO(n)*-bundles over manifolds,* Amer. J. Math. **80**, 632–638.

——— : [1959], (cited on pp., 67, 83, 87, 97, 245, 247). *An interpretation of G. Whitehead's generalization of Hopf's invariant,* Annals of Math. **69**, 345–365.

——— : [1959a], (cited on pp., 93, 122, 245, 247). *A note on obstructions and characteristic classes,* Amer. J. Math. **81**, 773–784.

——— : [1960], (cited on pp., 4, 116, 120-122). *A manifold which does not admit any differentiable structure,* Comment. Math. Helv. **34**, 257–270.

——— : [1960a], (cited on pp., 81, 87, 94, 122). *Some non stable homotopy groups of Lie groups,* Illinois J. Math. **4**, 161–169.

——— and Milnor, J.: [1960], (cited on pp., 69–71, 81, 82, 87, 93, 94, 114, 115, 122, 132, 137, 140, 239, 296, 297, 298, 305, 306). *Bernoulli numbers, homotopy groups and a theorem of Rohlin,* Proc. Internat. Congress Math., [1958]; Cambridge University Press, 454–458. [This volume, 243–247.]

——— and ——— : [1961], (cited on pp., 78, 112, 122). *On 2-spheres in 4-manifolds,* Proc. Nat. Acad. Sci. U.S.A. **47**, 1651–1657.

——— and ——— : [1963], (cited on pp., 84, 135-137, 140, 301, 305, 306). *Groups of homotopy spheres: I,* Annals of Math. **77**, 504–537. [This volume, 89–122.]

Kirby, R.: [1989], (cited on pp., 9, 10). "The Topology of Four-manifolds," Lecture Notes in Math., **1374**. Springer-Verlag, Berlin.

——— and Siebenmann, L.: [1969], (cited on pp., 9, 10, 20). *On the triangulation of manifolds and the Hauptvermutung,* Bull. Amer. Math. Soc. **75**, 742–749.

———- : [1977], (cited on pp., 24). "Foundational Essays on Topological Manifolds, Smoothings, and Triangulations," Annals of Math. Studies **88**, Princeton Univ. Press, Princeton, NJ.

Kneser, H.: [1929], (cited on pp., 204). *Geschlossene Flächen in dreidimensionalen Mannigfaltigkeiten,* Jahresb. der Deut. Math. Verein. **38**, 248–260.

Kosinski, A.: [1967], (cited on pp., 144). *On the inertia group of π-manifolds,* Amer. J. Math. **89**, 227–248.

——— : [1992], (cited on pp., 10). "Differential Manifolds," Pure and Applied Mathematics, **138**. Academic Press, Inc., Boston, MA.

Lashof, R. K.: [1963], (cited on pp., 252, 291). *Poincaré duality and cobordism,* Trans. Amer. Math. Soc. **109**, 257–277.

Latour, F.: [1977/78], (cited on pp., 20). *Double suspension d'une sphère d'homologie,* [d'après R. Edwards]. Séminaire Bourbaki, 30e anne, Exp. No. **515**, 169–186, Lecture Notes in Math. **710**, Springer-Verlag, Berlin, [1979].

Lawson, B. and Michelsohn, M. L.: [1989], (cited on pp., 253). "Spin Geometry," Princeton University Press, **93**, Princeton, NJ.

Lawson T.: [1973], (cited on pp., 144). *Remarks on the pairings of Bredon, Milnor, and Milnor-Munkres-Novikov,* Indiana Univ. Math. J. **22**, 833–843.

Lefschetz, S.: [1956], (cited on pp., 40, 45, 59, 60, 63). "Topology," (2nd-edition), New York, Chelsea Pub. Co.

Lima, E. L.: [1958], (cited on pp., 263, 273). *Stable Postnikov invariants and their duals,* Amer. Math. Soc. Notices, **5**, abstract number 544-27, p. 215.

Madsen, I. and Milgram, J.: [1979], (cited on pp., 6). "The Classifying Space for Surgery and Cobordism of Manifolds," Annals of Math. Studies **92**, Princeton U. Press, Princeton, NJ.

Mandelbaum, R.: [1980], (cited on pp., 9). *Four-dimensional topology: an introduction,* Bull. Amer. Math. Soc. **2**, 1–159.

Mazur, B.: [1959], (cited on pp., 30, 33, 87). *On embeddings of spheres,* Bull. Amer. Math. Soc. **65**, 59–65. (Extendend version: Acta Math. **105** [1961], 1–17.)

McDuff, D. and Salamon, D. A.: [1995], (cited on pp., 10, 204). "Introduction to Symplectic Topology," Oxford Mathematical Monographs. The Clarendon Press, Oxford University Press, New York. (Second Edition 1998.)

Milnor, J.: [1956], (cited on pp., 234, 317). *On the immersion of n-manifolds in $n+1$-space,* Comment. Math. Helv. **30**, 275–284.

——— : [1956a], (cited on pp., 19, 21, 24, 28, 35, 36, 38, 45, 67, 70, 88, 89, 93, 122, 125, 136, 140). *On manifolds homeomorphic to the 7-sphere*, Annals of

Math. **64**, 399–405. [This volume, 11–17.]

———— : [1956b], (cited on pp., 39, 45, 86, 88). *On the relationship between differentiable manifolds and combinatorial manifolds*, Princeton University [1956], (mimeographed). [This volume, 19–28.]

———— : [1958], (cited on pp., 1, 60, 61, 63, 70, 71, 78, 88, 296, 298). *On simply connected 4-manifolds,* Symposium Internacional de Topología Algebraíca, Univ. Nac. Autónoma de México and UNESCO, Mexico City 122–128 (to be reprinted in Collected Papers, Volume **4**).

———— : [1958a], (cited on pp., 83, 88, 245-247). *On the Whitehead homomorphism,* J. Bull. Amer. Math. Soc., **64**, 79–82. [This volume, 239–242.]

———— : [1958b], (cited on pp., 258-261, 273, 275, 281). *The Steenrod algebra and its dual,* Annals of Math. **67**, 150–171.

———— : [1959], (cited on pp., 32, 33, 70, 88, 217 219, 220). *Differentiable structures on spheres*, Amer. J. Math. **81**, 962–972. [This volume, 35–45.]

———— : [1959a], (cited on pp., 29, 33, 34, 37-40, 44, 45, 89, 90, 98, 102, 122, 218). *Differentiable manifolds which are homotopy spheres,* Princeton University. [This volume, 65-88.]

———— : [1959b], (cited on pp., 52, 63, 89, 90, 122, 125, 136, 140, 289, 290). *Sommes de variétés différentiables et structures différentiables des sphères,* Bull. Soc. Math. France **87**, 439–444. [This volume, 29–34.]

———— : [1960], (cited on pp., 38, 45, 55, 63, 246, 247, 275, 281, 285, 287, 290, 307, 308, 310, 314). *On the cobordism ring* Ω^* *and a complex analogue, I,* Amer. J. Math. **82**, 505-521. [This volume, 257–273.]

———— : [1961], (cited on pp., 20). *Two complexes with are homeomorphic but combinatorially distinct,* Annals of Math. **74**, 575–590. (Also in: "Collected Papers II: The Fundamental Group," Publish or Perish [1995], 123–138).

———— : [1961a], (cited on pp., 89, 93, 94, 98, 99, 102, 104, 111-114, 122, 139, 140, 297, 298, 303, 305, 306). *A procedure for killing homotopy groups of differentiable manifolds,* Proc. Sympos. Pure Math., Vol. III pp. 39–55 American Mathematical Society, Providence, RI. [This volume, 47–63.]

———— : [1962], (cited on pp., 184, 204). *A unique decomposition theorem for 3-manifolds,* Amer. J. Math. **84**, 1–7. (Also in: "Collected Papers II: The Fundamental Group," Publish or Perish [1995], p. 237).

———— : [1962a], (cited on pp., 123, 140, 293, 298, 310, 314). *A survey of cobordism theory,* Enseignement Math. **8**, 16–23. [This volume, 283–290.]

———— : [1963], (cited on pp., 299, 303, 304, 306). *Spin structures on manifolds,* L'Enseignement Math. **9**, 128–203. [This volume, 293-298.]

———— : [1965], (cited on pp., 10). "Topology from the Differentiable Viewpoint," The University Press of Virginia, Charlottesville, Va.

———— : [1965a], (cited on pp., 312). *Remarks concerning spin manifolds,* in "Differential and Combinatorial Topology. A Symposium in Honor of Marston Morse," Princeton University Press, Princeton, NJ., 1965, pp. 55-62. [This volume, 299–306.]

———— : [1966], (cited on pp., 84). *Whitehead torsion,* Bull. Amer. Math. Soc. **72**, 358-426. (Also in: "Collected Papers II: The Fundamental Group," Publish or Perish [1995], 151–219.)

——— : [1968], (cited on pp., 8). "Singular Points of Complex Hypersurfaces," Annals of Math. Studies **61**, Princeton University Press, Princeton, NJ.

——— and Husemoller, D.: [1973], (cited on pp., 78). "Symmetric Bilinear Forms," Ergebnisse der Mathematik und ihrer Grenzgebiete, Band 73. Springer-Verlag, New York-Heidelberg.

——— , Siebenmann, L. and Sondow, J.: [1965], (cited on pp., 9, 10). "Lectures on the h-Cobordism Theorem," Princeton Mathematical Notes, Princeton University Press, Princeton, NJ. (Revised edition in preparation.)

——— , Spivak, M. and Wells, R.: [1963], (cited on pp., 10, 124, 125, 140). "Morse Theory," Annals of Math. Studies **51**, Princeton Univ. Press, Princeton, NJ.

——— and Stasheff, J.: [1974], (cited on pp., 4, 7, 10, 210, 220, 226, 250). "Characteristic Classes," Annals of Math. Studies **76**, Princeton Univ. Press, Princeton, NJ.

Moise, E.: [1952], (cited on pp., 9, 204). *Affine structures in 3-manifolds, V, The triangulation theorem and the Hauptvermutung,* Annals of Math. **56**, 96–114.

Morrey, C. B. Jr.: [1958], (cited on pp., 22). *The analytic embedding of abstract real-analytic manifolds*, Annals of Math. **68**, 159–201.

Morse, A. P.: [1939], (cited on pp., 176). *The behavior of a function on its critical set,* Annals of Math. **40**, 62–70.

Morse, M.: [1925], (cited on pp., 13, 17, 124, 140). *Relations between the critical points of a real function of n independent variables,* Trans. Amer. Math. Soc. **27**, 345–396.

——— : [1931], (cited on pp., 50, 63). *The critical points of a function of n variables,* Trans. Amer. Math. Soc. **33**, 72–91.

——— : [1934], (cited on pp., 50, 63). "The Calculus of Variations in the Large," Amer. Math. Soc. Colloquium Publications, **18**, Providence, RI. (Reprinted 1996.)

Munkres, J.: [1955], (cited on pp., 88). *Some applications of triangulation theorems,* Thesis, University of Michigan.

——— : [1960a], (cited on pp., 8, 32, 34, 92, 122, 204). *Obstructions to the smoothing of piecewise-differentiable homeomorphisms,* Annals of Math. **72**, 521–554. (See also, Bull. Amer. Math. Soc. **65**, [1959] 332–334.)

——— : [1960b], (cited on pp., 9). *Differentiable isotopies on the 2-sphere,* Michigan Math. J. **7**, 193–197.

——— : [1963], (cited on pp., 4, 10, 26, 136, 140). "Elementary Differential Topology," Annals of Math. Studies **54**, Princeton University Press, Princeton, NJ.

——— : [1964], (cited on pp., 8, 204). *Obstructions to imposing differentiable structures*, Ill. J. Math. **8**, 361–376.

Nash, C.: [1991], (cited on pp., 10). "Differential Topology and Quantum Field Theory," Academic Press, Ltd., London.

Nash, J.: [1952], (cited on pp., 308, 314). *Real algebraic manifolds,* Annals of Math. **56**, 405–421.

Newman, M. H. A.: [1926], (cited on pp., 19, 28). *On the foundations of combinatory analysis situs*, Nederl. Akad. Wetensch. Proc. **29**, 611–626.

Nielsen, N.: [1923], (cited on pp., 241). "Traité Élémentaire des Nombres de Bernoulli," Paris, Gauthier-Villars et Cie.

Novikov, S. P.: [1960], (cited on pp., 216, 249, 290). *Some problems in the topology of manifolds connected with the theory of Thom spaces,* Doklady Akad. Nauk. SSSR **132**, 1031–1034 (Sov. Math. Doklady **1**, 717–720).

——— : [1962], (cited on pp., 216, 249). *Homotopy properties of Thom complexes,* Mat. Sbornik **57**, 407–442.

——— : [1965], (cited on pp., 4). *Topological invariance of rational Pontrjagin classes,* Doklady Akad. Nauk. SSSR **163**, 298–300. (Soviet Math. Doklady **6**, 921–923).

Palais, R.: [1960], (cited on pp., 90, 122, 126, 140, 183, 197). *Extending diffeomorphisms,* Proc. Amer. Math. Soc. **11**, 274–277.

——— : [1960a], (cited on pp., 179). *Local triviality of the restriction map for embeddings,* Commentarii **34**, 305–312.

Perelman, G.: [2002], (cited on pp., 9). *The entropy formula for the Ricci flow and its geometric applications,* arXiv: math.DG/0211159v1, November.

——— : [2003a], (cited on pp., 9). *Ricci flow with surgery on three-manifolds,* arXiv: math.DG/0303109, March.

——— : [2003b], (cited on pp., 9). *Finite extinction time for the solutions to the Ricci flow on certain three-manifolds*, arXiv: math.DG/0307245, July.

Peterson, F.: [1956], (cited on pp., 264, 273). *Generalized cohomotopy groups,* Amer. J. Math. **78**, 259–281.

Phillips, A.: [1967], (cited on pp., 315, 317). *Submersions of open manifolds,* Topology **6**, 171–206.

——— : [1974], (cited on pp., 315, 317). *Smooth maps of constant rank,* Bull. Amer. Math. Soc. **80**, 513–517.

Poenaru, V.: [1968], (cited on pp., 316, 317). *Extension des immersions en codimension 1 (d'apres Samuel Blank),* Séminaire Bourbaki **342**, Benjamin 1969.

Pontrjagin, L. S.: [1947], (cited on pp., 283, 290). *Characteristic cycles on differentiable manifolds,* Mat. Sbornik N. S. **21**, 233–284.

——— : [1955] (cited on pp., 95, 122, 138, 140, 287, 290). *Smooth manifolds and their applications in homotopy theory,* Amer. Math. Soc. Transl., Ser. 2, **11**, 1–114. Amer. Math. Soc., Providence, RI.

de Rham, R. G.: [1955], (cited on pp., 24, 28). "Variétés Différentiables. Formes, Courants, Formes Harmoniques," Actualités Sci. Ind., no. 1222 = Publ. Inst. Math. Univ. Nancago III. Hermann et Cie, Paris.

Reeb, G.: [1952], (cited on pp., 13, 17, 124, 140, 208). *Sur certaines propriétés topologiques des variétés feuilletées,* Publ. Inst. Math. Univ. Strasbourg **11**, 5–89, 155–156. Actualités Sci. Ind., no. **1183**, Hermann & Cie., Paris.

Robertello, R. A.: [1965], (cited on pp., 254). *An invariant of knot cobordism,* Comm. Pure Appl. Math. **18**, 543–555.

Rohlin, V. A.: [1951], (cited on pp., 81, 88, 243, 247). *Classification of mappings of S^{n+3} onto S^n,* Doklady Acad. Nauk. SSSR **81**, 19–92.

——— : [1951a], (cited on pp., 283, 290). *A three-dimensional manifold is the boundary of a four-dimensional one,* Doklady Akad. Nauk SSSR (N. S.) **81**, 355–357.

——— : [1952], (cited on pp., 243, 247). *New results in the theory of 4-dimensional manifolds,* Doklady Akad. Nauk. SSSR, **84**, 221–224.

——— and Švarč, S.: [1957], (cited on pp., 289, 290). *The combinatorial invariance of Pontrjagin classes,* Doklady Akad. Nauk SSSR (N. S.) **114**, 490–493.

Rosen, R.: [1960], (cited on pp., 125, 140). *A weak form of the star conjecture for manifolds,* Abstract 570-28. Notices Amer. Math. Soc. **7**, 380.

Sard, A.: [1942], (cited on pp., 151, 176). *The measure of the critical values of differentiable maps,* Bull. Amer. Math. Soc. **48**, 883–890.

Seifert, H.: [1931], (cited on pp., 90, 122). *Konstruktion dreidimensionaler geschlossener Räume,* Ber. Verh. Sächs. Akad. Wiss. Leipzig **83**, 26–66.

——— and Threlfall, W.: [1934], (cited on pp., 108-110, 122). "Lehrbuch der Topologie," Leipzig: B. G. Teubner. Monatsh. Math. Phys. **46** [1937], A18.

Serre, J. P.: [1951], (cited on pp., 43, 45, 97, 122, 138, 141). *Homologie singuliére des espaces fibrés,* Annals of Math. **54**, 425–505.

——— : [1953], (cited on pp., 269, 273). *Groupes d'homotopie et classes de groupes abéliens,* Annals of Math. **58**, 258–294.

Shimada, N.: [1957], (cited on pp., 4, 35, 38, 45, 70, 88). *Differentiable structures on the 15-sphere and Pontrjagin classes of certain manifolds*, Nagoya Math. J. **12**, 59–69.

Siebenmann, L.: [1980], (cited on pp., 9). *Amorces de la chirurgie en dimension quatre: un $S^3 \times R$ exotique [d'après Andrew H. Casson et Michael H. Freedman],* Séminaire Bourbaki [1978/79], **536**, 183–207, Lecture Notes in Math. **770**, Springer, Berlin.

Smale, S.: [1959a], (cited on pp., 315, 317). *Classification of immersions of spheres in Euclidean space,* Annals of Math. **69**, 327–344.

——— : [1959b], (cited on pp., 9, 32, 34, 144, 204). *Diffeomorphisms of S^2,* Proc. Amer. Math. Soc. **10**, 621–626.

——— : [1960], (cited on pp., 209). *The generalized Poincaré conjecture in higher dimensions.* Bull. Amer. Math. Soc. **66**, 373–375.

——— : [1961], (cited on pp., 4, 8, 90, 122, 289, 290). *Generalized Poincaré's conjecture in dimensions greater than four*, Annals of Math. **74**, 391–406.

——— : [1962], (cited on pp., 8, 90, 122, 128, 139, 141). *On the structure of manifolds,* Amer. J. Math. **84**, 387–399.

Spanier, E.: [1959], (cited on pp., 263, 273). *Function spaces and duality,* Annals of Math. **70**, 338–378.

Stallings, J. R.: [1960], (cited on pp., 90, 122, 128, 141, 209). *Polyhedral homotopy-spheres,* Bull. Amer. Math. Soc. **66**, 485–488.

——— : [1962], (cited on pp., 9, 126, 141). *The piecewise-linear structure of Euclidean space,* Proc. Cam. Phil. Soc. **58**, 481–488.

von Staud, K. G. C.: [1840], (cited on pp., 246, 247). *Beweis eines Lehrsatzes die Bernoullischen Zahlen betreffend,* J. Reine Angew. Math. **21**, 372–374.

———— : [1845], (cited on pp., 246, 247). *De numeris Bernoullianis commentatio,* Erlangen.

Steenrod, N.: [1949], (cited on pp., 235, 237). *Cohomology invariants of mappings,* Annals of Math. **50**, 954–988.

———— : [1951], (cited on pp., 13, 14, 17, 37, 45, 56, 59, 63, 66, 88, 176, 187, 209, 217, 234, 237, 239, 283, 286, 290, 308, 314). "The Topology of Fibre Bundles," Princeton University Press, Princeton, NJ.

———— and Whitehead, J. H. C: [1951], (cited on pp., 234, 237). *Vector fields on the n-sphere,* Proc. Nat. Acad. Sci. U.S.A. **37**, 58–63.

Stiefel, E.: [1940], (cited on pp., 234, 237). *Über Richtungsfelder in den projektiven Räumen und einen Satz aus der reelen Algebra,* Comm. Math. Helv. **13**, 201–218.

Stong, R.: [1968], (cited on pp., 10, 254). "Notes on Cobordism Theory," Princeton Math. Notes, Princeton University Press, Princeton, NJ.

Tamura, I.: [1958], (cited on pp., 35, 38, 45). *Homeomorphy classification of total spaces of sphere bundles over spheres,* J. of the Math. Soc. of Japan **10**, 29–43.

Taubes, C.: [1987], (cited on pp., 9). *Gauge theory on asymptotically periodic 4-manifolds,* J. Differential Geom. **25**, 363–430.

Terasaka, H.: [1959], (cited on pp., 254). *On null-equivalent knots,* Osaka Math. J. **11**, 95–113.

Thom, R.: [1952], (cited on pp., 77, 88, 236, 237, 240, 267, 269, 273, 288, 290). *Espaces fibrés en sphères et carrés de Steenrod,* Ann. Sci. Ecole Norm. Sup. **69**, 109–181.

———— : [1954], (cited on pp., 11, 12, 17, 24, 28, 82, 88, 93, 95, 122, 176, 212, 216, 240, 244, 247, 255, 257, 266, 273, 276, 281, 283, 290, 308, 314). *Quelques propriétés globales des variétés différentiables,* Comment. Math. Helvet., **28**, 17–86.

———— : [1957], (cited on pp., 179). *La classification des immersions,* Séminaire Bourbaki; 10e année: 1957/1958. Exposé **157** (mimeographed).

———— : [1958], (cited on pp., 4, 19, 21, 28, 33, 34, 40, 45, 66, 88, 289, 290). *Les classes caractéristiques de Pontrjagin des variétés triangulées,* "Symposium Internacional de Topología Algebraíca (International Symposium on Algebraic Topology)," 54–67, Universidad Nacional Autónoma de México and UNESCO.

———— : [1959], (cited on pp., 5, 32, 34, 208). *Les structures différentiables des boules et des sphères,* Colloque Géom. Diff. Globale (Bruxelles, 1958), 27–35.

Thomas, E.: [1962], (cited on pp., 313, 314). *On the cohomology groups of the classifying space for the stable spinor groups,* Bol. Soc. Mat. Mexicana **7**, 57–69.

Titus, C. J.: [1961], (cited on pp., 317). *The combinatorial topology of analytic functions on the boundary of the disk,* Acta Math. **106**, 45–64.

Toda, H.: [1958a], (cited on pp., 71, 84, 88). *On exact sequences in Steenrod algebra mod 2,* Memoire College of Sci., Kyoto **31**, 33–64.

———— : [1958b], (cited on pp., 71, 83, 88). *p-primary components of homotopy groups III,* Memoirs College of Sci., Kyoto **31**, 191–210.

Veblen, O.: [1931], (cited on pp., 62, 63, 80, 88). "Analysis Situs" (2nd-edition), Amer. Math. Soc. Colloquium Publications, New York.

Wall, C. T. C.: [1959], (cited on pp., 47, 55, 63). *Note on the cobordism ring,* Bull. Amer. Math. Soc. **65**, 329–331.

——— : [1960], (cited on pp., 216, 250, 285, 290, 303, 306, 310-312, 314). *Determination of the cobordism ring,* Annals of Math. **2**, 292–311.

——— : [1962], (cited on pp., 97, 122, 296, 298). *Killing the middle homotopy groups of odd dimensional manifolds,* Trans. Amer. Math. Soc. **103**, 421–433.

——— : [1962a], (cited on pp., 8) *Classification of $(n-1)$-connected $2n$-manifolds,* Ann. of Math. **75**, 163–189.

——— : [1962b], (cited on pp., 8). *The action of Γ_{2n} on $(n-1)$-connected $2n$-manifolds,* Proc. Amer. Math. Soc. **13**, 943–944.

——— : [1964], (cited on pp., 8). *On simply-connected 4-manifolds,* J. Lond. Math. Soc. **39**, 141–149.

——— : [1970], (cited on pp., 6). "Surgery on Compact Manifolds," London Mathematical Society Monographs, No. 1. Academic Press, London-New York.

Wallace, A. H.: [1960], (cited on pp., 6, 47, 63, 98, 122, 297, 298). *Modifications and cobounding manifolds,* Canad. J. Math. **12**, 503–528.

——— : [1968], (cited on pp., 10). "Differential Topology, First Steps," W. A. Benjamin, Inc., New York-Amsterdam.

Weyl, H.: [1923-24], (cited on pp., 130, 141). *Análisis situs combinatorio,* Revista Mat. Hispano-Americana, **5** [1923] and **6** [1924]; Gesammelte Abh. II, 390–432, Springer 1968.

Whitehead, G. W.: [1942], (cited on pp., 235, 238, 239, 243, 247). *On the homotopy groups of spheres and rotation groups,* Annals of Math. **43**, 634–640.

Whitehead, J. H. C.: [1940], (cited on pp., 4, 19, 20, 28, 32, 34, 85, 88, 288, 290). *On C^1-complexes,* Annals Math. **81**, 809–824.

——— : [1940a], (cited on pp., 66, 88, 94, 122, 243, 247). *On the homotopy type of manifolds,* Annals of Math. **41**, 825–832.

——— : [1948], (cited on pp., 68, 88). *On the homotopy type of ANR's,* Bull. Amer. Math. Soc. **54**, 1133–1145.

——— : [1949], (cited on pp., 236, 238). *On simply connected 4-dimensional polyhedra,* Comm. Math. Helv. **22**, 48-92.

——— : [1950], (cited on pp., 85, 88) *Simple homotopy types,* Amer. J. Math. **72**, 1–57.

——— : [1961], (cited on pp., 92, 122). *Manifolds with transverse fields in euclidean space,* Annals of Math. **73**, 154–212.

Whitney, H.: [1935], (cited on pp., 176). *A function not constant on a connected set of critical points,* Duke Math. J. **1**, 514–517.

——— : [1936], (cited on pp., 23, 28, 222). *Differentiable functions defined in arbitrary subsets of Euclidean space*, Trans. Amer. Math. Soc. **40**, 309–317.

——— : [1936a], (cited on pp., 22, 25, 28, 53, 58, 63, 74, 86, 88, 137, 141, 176, 222). *Differentiable Manifolds*, Annals of Math. **37**, 645–680.

——— : [1937], (cited on pp., 317). *On regular closed curves in the plane,* Compositio Math. **4**, 276–284.

——— : [1944a], (cited on pp., 58, 63, 78, 88, 176). *The self intersections of a smooth n-manifold in 2n-space,* Annals of Math. **45**, 220–246.

——— : [1944b], (cited on pp., 176) *The singularities of a smooth n-manifold in (2n − 1) space,* Annals of Math. **45**, 247–293.

——— : [1955], (cited on pp., 50, 63). *On singularities of mappings of euclidean spaces. I. Mappings of the plane into the plane,* Annals of Math. **62**, 374–410.

Wolizer, D. L.: [1978], (cited on pp., 317). "Immersions of Surfaces in the Plane," Thesis, Northeastern Univ.

Wraith, D.: [1997], (cited on pp., 8). *Exotic spheres with positive Ricci curvature,* J. Differential Geom. **45**, 638–649.

Wu W.-T.: [1950], (cited on pp., 70, 88, 235, 238). *Classes caractéristiques et i-carrés d'une variété,* C. R. Acad. Sci. Paris **230**, 508–511.

——— : [1950a], (cited on pp., 233, 238, 314). *Les i-carrés dans une variété grassmannienne,* C. R. Acad. Sci. Paris **230**, 918–920.

——— : [1952], (cited on pp., 11, 15, 17). *Sur les classes caractéristiques des structures fibrées sphériques,* Publ. Inst. Math. Univ. Strasbourg **11**, 5–89, 155–156. Actualités Sci. Ind., no. 1183 Hermann & Cie, Paris.

——— : [1954], (cited on pp., 236, 238). *On Pontrjagin classes. I,* Acta Math. Sinica **3**, 353–367.

——— : [1955], (cited on pp., 236, 238). *On Pontrjagin classes. II,* Acta Math. Sinica **4**, 455–490.

——— : [1954a], (cited on pp., 229, 233, 238). *On Pontrjagin classes. III,* Acta Math. Sinica **4**, 323–346.

Zeeman, C.: [1961], (cited on pp., 90, 122, 128, 141). *The generalized Poincaré Conjecture,* Bull. Amer. Math. Soc. **67**, 270.

Index of Symbols

Main Index

For publications cited, see the bibliography.

ISBN 978-0-8218-4230-0

9 780821 842300

CWORKS/19.3